Haematology and Blood Transfusion

27 ——————————————————————

Hämatologie und Bluttransfusion

Edited by
H. Heimpel, Ulm D. Huhn, München
C. Mueller-Eckhardt, Gießen
G. Ruhenstroth-Bauer, München

Disorders of the Monocyte Macrophage System

Pathophysiological and Clinical Aspects

Edited by
F. Schmalzl, D. Huhn and H.E. Schaefer

With 107 Figures and 57 Tables

Springer-Verlag
Berlin Heidelberg New York 1981

Prof. Dr. med. Franz Schmalzl, Klinik für Innere Medizin, Universität Innsbruck, Anichstraße 35, A-6020 Innsbruck

Prof. Dr. med. Dieter Huhn, Medizinische Klinik III, Klinikum Großhadern der Universität München, Marchioninistraße 15, D-8000 München 70

Prof. Dr. med. Hans Eckart Schaefer, Pathologisches Institut der Universität Köln, Joseph-Stelzmann-Straße 9, D-5000 Köln

Supplement to
BLUT - Journal Experimental and Clinical Hematology
Organ of the *Deutsche Gesellschaft für Hämatologie und Onkologie* der *Deutschen Gesellschaft für Bluttransfusion und Immunohämatologie* and of the *Österreichischen Gesellschaft für Hämatologie und Onkologie*

ISBN-13:978-3-540-10980-8   e-ISBN-13:978-3-642-81696-3
DOI: 10.1007/978-3-642-81696-3

Library of Congress Cataloging in Publication Data
Main entry under title: Disorders of the monocyte macrophage system. (Haematology and blood transfusion ; 27 = Hämatologie und Bluttransfusion ; 27) "Supplement to BLUT - Journal for blood research" -- Bibliography: p. Includes index. 1. Monocytes--Diseases. I. Schmalzl, F. II. Huhn, Dieter. III. Schaefer, H. E. (Hans Eckart), 1936- . IV. BLUT. V. Series: Haematology and blood transfusion ; 27. [DNLM: 1. Macrophages--Physiology--Congresses. 2. Macrophages--Pathology--Congresses. 3. Reticuloendothelial system--Physiopathology--Congresses. Wl HA1655 Bd. 27 / WH 650 D612] RC645.5.D5 616.07'9 81-16595 ISBN-13:978-3-540-10980-8 AACR2

Composition: Fotosatz Service Weihrauch, Würzburg

2127/3321/543210

Contents

List of Participants

Prof. Dr. G. Assmann
Zentrallabor der Universitätsklinik Münster, Westring 3, 4400 Münster, West-Germany

Dr. R.H.J. Beelen
Afdeling Electronenmicroscopie, Faculteit der Geneeskunde, Vakgroep Celbiologie, Vrije Universiteit, Van der Boechorstraat 7, 1007 Amsterdam, The Netherlands

Dr. R. Budde
Patholog. Institut der Universität Köln, Josef-Stelzmann-Straße 9, 5000 Köln 41, West-Germany

Prof. Dr. G. Burg
Dermatolog. Klinik u. Poliklinik der Universität München, Frauenlobstraße 9–11, 8000 München 2, West-Germany

Dr. Daniel Catovsky
MRCPath., Hammersmith Hospital, Royal Postgraduate Medical School Ducane Rd., London W 12 OHS, Great Britain

Prof. Dr. W.Th. Daems
Laboratory for Electron Microscopy, University of Leiden, Rijnsburger-weg 10, 2333 AA Leiden, The Netherlands

Prof. Arthur M. Dannenberg, Jr., M.D.
Dept. of Environmental Health Sciences, The Johns Hopkins University, School of Hygiene & Public Health, 615 N Wolfe Street, Baltimore, Maryland 21205, U.S.A.

Dr. Andreas Feldges
FMH Pädiatrie, spez. Haematologie, Leitender Arzt, Kinderspital, Claudiusstraße 6, 9006 St. Gallen, Switzerland

Dr. R. van Furth, M.D., F.R.C.P.
Professor of Internal Medicine and Infectious Diseases, Academisch Ziekenhuis – Leiden, Rijnsburgerweg 10, 425162 Leiden, The Netherlands

Prof. Dr. R. Haas
Kinderklinik der Universität, Lindwurmstraße, 8000 München 2, West-Germany

Priv. Doz. Dr. W. Heit
Abteilung f. Haematologie, Dept. für Innere Medizin, Steinhövelstraße 9, 7900 Ulm, West-Germany

Doz. Dr. Ch. Huber
Univ. Klinik für Innere Medizin, Anichstraße 35, 6020 Innsbruck, Austria

Prof. Dr. H. Huber
Univ. Klinik für Innere Medizin, Anichstraße 35, 6020 Innsbruck, Austria

Prof. Dr. D. Huhn
Univ. Klinik für Innere Medizin III, Klinikum Großhadern, 8000 München, West-Germany

Dr. G. Janka
Universitäts-Kinderklinik, Lindwurmstraße 4, 8000 München 2, West-Germany

Dr. med. M.L. Lohmann-Matthes
Max-Planck-Institut für Immunbiologie, Stübeweg 51, 7800 Freiburg, West-Germany

Prof. P. Meister
Pathologisches Institut der Univ. München, Thalkirchner Straße, 8000 München 2, West-Germany

Prof. Dr. G. Meuret
St. Elisabeth-Krankenhaus, 7980 Ravensburg, West-Germany

Priv. Doz. Dr. H.K. Müller-Hermelink
Abt. Allgemeine Pathologie und Pathologische Anatomie, Institut für Pathologie, Hospitalstraße 42, 2300 Kiel, West-Germany

Prof. Dr. H.E. Schaefer
Patholog. Institut der Univ. Köln, Lindenthal, Josef-Stelzmann-Straße 9, 5000 Köln, West-Germany

Prof. Dr. F. Schmalzl
Univ. Klinik für Innere Medizin, Anichstraße 35, 6020 Innsbruck, Austria

Dr. J. Schnyder
Wander Ltd., A. Sandoz Research Unit., Res. Inst. Wander, P.O. Box 2747, 3001 Bern, Switzerland

Dr. H.-U. Schorlemmer
Behringwerke AG, Postfach 1140, 3550 Marburg/Lahn, West-Germany

Prof. Dr. J.L. Turk
Dept. of Pathology, Royal College of Surgeons of England, Lincoln's Inn Fields, 35/43, London WC2A 3 PN, Great Britain

Dr. M. Weil
Service d'Oncologie Medicale, Groupe Hopitalies, Pitie Salpetriere, 47 Boulevard de l'Hopital, 75013 Paris, France

Prof. Dr. R.A. Zittoun
Clinique Haematologique Hotel Dieu, 1 Place du Parvis Notre Dame, 75181 Paris, Cedex 04, France

Preface

The origin and function of normal monocytes and macrophages have been clearly defined by extensive investigations in human and in animal models. The central importance of this cell system for the biological defense mechanisms is well established: phagocytosis, inactivation and destruction of organic and inorganic materials, an important role in the initiation of humoral and cell mediated immunological responses, and the secretion of a variety of chemical mediator and effector substances are the most important features of this ontogenetically ancient cell system. However, the data on this cellular system are rather recent, and this may explain why relatively little attention has been payed to the pathology of the monocyte–macrophage system (MMS) until now.

In addition, this monograph should focus attention on the secondary pathophysiological implications of the MMS in disorders not primarily originating from this system. Several techniques are available to identify even abnormal individuals of this cell system and, therefore, can be employed for the study of severely altered or neoplastic monocytic cells.

The monograph is based on the papers presented at the "International Workshop on Disorders of the Monocyte-Macrophage System" held in Innsbruck in October 8th-10th, 1980. We wish to thank to all who helped in organizing this workshop. We are especially indebted to the Government of Tyrol, the City of Innsbruck, and to the Medical Faculty of the University of Innsbruck as well as to the head of the Department of Internal Medicine, Prof. Dr. H. Braunsteiner. We are further indebted to the pharmaceutical industries of the FRG and of Austria for their generous financial support which allowed the organization of this workshop.

Innsbruck, July 1981   F. Schmalzl
München   D. Huhn
Köln   H.E. Schaefer

Acknowledgement

The International Workshop on Disorders of the Monocyte–Macrophage System has been supported by:

Funds "Kampf dem Krebs", Wien – Innsbruck

ASTA-Werke AG, Bielefeld

Bayer-Pharma GesmbH, Wien

Behring Institut GmbH, Wien

Biochemie GesmbH, Wien

Bristol Arzneimittel, Neu-Isenburg

Cyanamid GmbH, Lederle Arzneimittel, Wolfratshausen

Eli Lilly GmbH, Bad Homburg

Gerot Pharmazeutika, Wien

Heinrich Mack Nachf., Chem.-Pharmazeutische Fabrik, Illertissen

Hoechst AG, Frankfurt

Leopold & Co, Graz

Montedison Farmaceutica GmbH, Freiburg

Parke-Davis & Co, München

Sharp & Dohme GmbH, München

Upjohn GmbH, Heppenheim/Bergstraße

Wellcome Pharma GesmbH, Wien

Pharmig – Vereinigung pharmazeutischer Erzeuger, Wien

Pathophysiological Aspects

Haematology and Blood Transfusion Vol 27
Disorders of the Monocyte Macrophage System
Edited by F. Schmalzl, D. Huhn, H.E. Schaefer
© Springer-Verlag Berlin Heidelberg New York 1981

Current View of the Mononuclear Phagocyte System

R. van Furth

During the last 20 years considerable attention has been given to the origin of macrophages. Until recently macrophages localized in various tissues were believed to be a kind of phagocytic connective tissue cell that renews itself by division. It has also been thought by some that fibroblasts transform into macrophages and by others that macrophages derive from lymphocytes.

For modern research on the origin and kinetics of cells adequate definition of the cells under study is obligatory. Formerly, cells were characterized mainly on the basis of morphological characteristics identified by light microscopy. Electron microscopy showed that cells which had seemed to be similar were really different, and still other differences were revealed by (immuno)cytochemical methods. The determination of membrane characteristics (e.g., presence of receptors) or specific (glyco)proteins and functional features of cells has made it possible to define the mononuclear phagocytes more precisely. This subject has been discussed at length elsewhere [1]. Criteria for the characterization of a cell as a mononuclear phagocyte are summarized in Table 1.

1. Origin of Monocytes

The bone marrow origin of the monocytes was firmly established many years ago. The direct precursors of the monocyte, i.e., the promonocyte and the monoblast, were characterized more recently in murine and human bone marrow [1, 2].

Morphological and cytochemical studies have shown that monoblasts and promonocytes have many characteristics in common with monocytes, that these precursor cells have Fc and C3b receptors in their membrane, and that they phagocytose and pinocytose. Their pattern of division has also been determined (Fig. 1). In the mouse the cell-cycle time of the

Table 1. Criteria for the characterization of a cell as mononuclear phagocyte

---

Morphological characteristics
Cytochemical characteristics
  nonspecific esterase
  peroxidase-positive or -negative granules
  lysozyme
Membrane receptors
  Fc receptors
  C receptors
Functional characteristics
  phagocytosis of opsonized bacteria; IgG-coated red cells; latex beads
  no phagocytosis of C-coated red cells
  pinocytosis

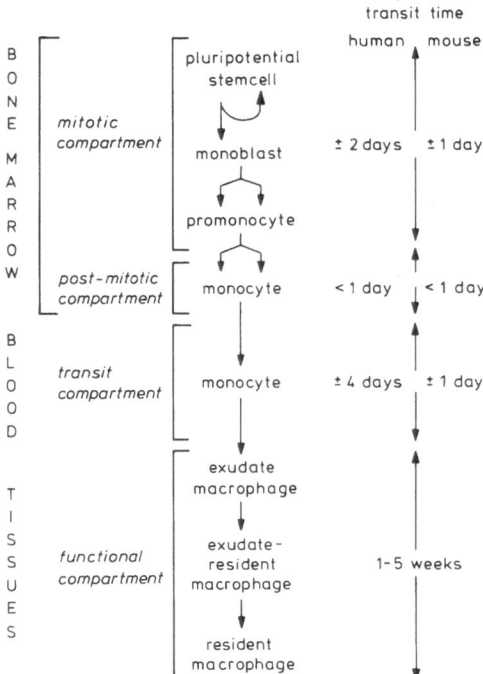

B
O
N
E

M
A
R
R
O
W

mitotic compartment

post-mitotic compartment

B
L
O
O
D

transit compartment

T
I
S
S
U
E
S

functional compartment

pluripotential stemcell

monoblast

promonocyte

monocyte

monocyte

exudate macrophage

exudate-resident macrophage

resident macrophage

human | mouse

± 2 days | ± 1 day

< 1 day | < 1 day

± 4 days | ± 1 day

1 - 5 weeks

Fig. 1. Schematic representation of the production of monocytes in the bone marrow and their transit via the circulation to the tissues

monoblast is 11.9 h, and division of one cell gives rise to two promonocytes; the promonocytes divide only once, after 16.2 h, and give rise to two monocytes. Thus, from monocyte to promonocyte there is a fourfold amplification. Monocytes do not divide further, as demonstrated by the low labeling index found after incubation in vitro with ^{3}H-tymidine or 1 h after a pulse with this DNA precursor [1, 2] and by the absence of colony formation in cultures of normal human peripheral blood monocytes in the presence of colony-stimulating factor (unpublished observations). After monocytes are formed they leave the bone marrow randomly within 24 h. Monocytes remain relatively long in the circulation (mouse: half time 17.4 h; man: half time 71.0 h) compared with granulocytes and leave this compartment randomly [1, 2].

2. Origin of Macrophages

The bone marrow origin of macrophages in the peritoneal cavity, liver, and lung has been shown in a large number of chimera studies (Table 2). Although the results of these studies are convincing, the method has the drawback that chimeras do not represent a true steady-state condition. This problem led the Leiden group to investigate the origin and kinetics of macrophages of mice under normal steady-state conditions. In vitro labeling studies with the DNA precursor ^{3}H-thymidine showed a labeling index of less than 5% for macrophages of the peritoneal cavity, liver, lung, and skin [1] (Table 3). Furthermore, it became clear that the mononuclear phagocytes that label in vitro have very recently (less than 24 to 48 h before harvesting) arrived in the tissue from the bone marrow and do not belong to the resident population of macrophages.

In vivo labeling studies in normal and monocytopenic mice and in irradiated mice with partial bone marrow shielding showed that the monocytes migrate into the tissues, where they become macrophages [1, 3–5].

Table 2. Bone marrow origin of macrophages as demonstrated by chimera studies

	Animal species	Type of chimera	Method used to demonstrate bone marrow origin	Reference number
Peritoneal macrophages	Mouse	Bone marrow	Cytotoxic antibody	25
	Mouse	Bone marrow	Cytotoxic antibody	26
	Mouse	Bone marrow	Chromosome marker	27
	Mouse	Bone marrow	Influenza virus infection	28
Kupffer cells	Human	Liver transplant	Sex chromatin	29
	Mouse	Bone marrow	Chromosome marker	30
	Rat	Bone marrow	Fluorescent antibody	31
	Human	Liver transplant	Sex chromatin	32
	Human	Bone marrow	Sex chromatin	33
	Mouse	Bone marrow	Influenza virus infection	28
Alveolar macrophages	Mouse	Bone marrow	Chromosome marker	34
	Mouse	Bone marrow	Chromosome marker	27
	Mouse	Bone marrow	Esterase marker	35
	Mouse	Bone marrow	Cytotoxic antibody	36
	Mouse	Bone marrow	Giant lysosomes	37
	Dog	Bone marrow	Transaminase	38
	Human	Bone marrow	Sex chromatin	39

The controversy which arose because of different peroxidase-staining patterns of the monocytes and (resident) tissue macrophages [6–8] has been solved by the finding of a transitional cell, a macrophage with peroxidase-positive granules and positive peroxidatic staining of the endoplasmic reticulum and nuclear envelope [9, 10] (Fig. 2).

Recently, the use of another approach confirmed the monocytic origin of macrophages. Studies done in extracts of human monocytes and peritoneal macrophages showed slightly different patterns of nonspecific esterase isoenzymes, but when these monocytes were cultured for 3 days, the cell extracts displayed the macrophage pattern [11, 12].

Our current view on the development of mononuclear phagocytes is summarized in Fig. 3a which gives a schematic representation of the bone marrow origin of tissue macrophages and the occasionally dividing mononuclear phagocytes in the tissues which are immature cells arising from the bone marrow. Quantitative kinetic studies in normal mice have shown that of the monocytes leaving the circulation, 56% become Kupffer cells [4], 15% become pulmonary macrophages [5], and 8% become peritoneal macrophages [13].

Table 3. In vitro and pulse ^{3}H-thymidine labeling of murine mononuclear phagocytes

		In vitro labeling[a]	Pulse labeling[b]
		%	%
Monoblasts		92.0	
Promonocytes		54.0	68.7
Monocytes:	bone marrow	0.3	1.0
	blood	2.3	2.3
Macrophages:	peritoneal	3.6	2.4
	lung	2.8	3.8
	liver	0.8	1.0
	skin	0.7	0.3

[a] In medium with 0.1 µCi/ml ^{3}H-thymidine for 2–24 h.
[b] 1 or 2 h after 1 µCi/g ^{3}H-thymidine given intravenously.

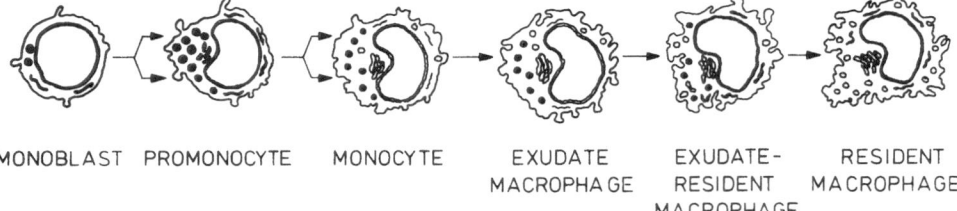

MONOBLAST PROMONOCYTE MONOCYTE   EXUDATE   EXUDATE-   RESIDENT
                                    MACROPHAGE   RESIDENT   MACROPHAGE
                                                   MACROPHAGE

Fig. 2. Peroxidatic activity patterns and the sequence of development of the various types of mononuclear phagocytes of the bone marrow

The alternative concept of the origin of macrophages, i.e., that this population of cells is renewed by the division of mature macrophages (Fig. 3b), must be rejected on the basis of a large body of experimental data collected by our group and others. Furthermore, this alternative concept raises the question as to the fate of the monocytes. In the normal mice weighing 25–30 g the turnover in the circulation amounts to $0.3–0.5 \times 10^5$ monocytes per h; these cells leave the circulation randomly [1]. As mentioned previously, our experiments have shown that these cells migrate to the tissues and transform into macrophages.

In a new mathematical approach to the analysis of the kinetics of macrophages in a tissue compartment, the influx of monocytes, the local division of mononuclear phagocytes, and the efflux from the tissue compartment are taken into account [14]. When this approach was applied, for instance, to the data on pulmonary macrophages from mice in the normal steady state [5] the calculations showed that at least 70% of this macrophage population is maintained by monocyte influx and maximally 30% by local division of immature mononuclear phagocytes originating from the bone marrow. The calculated turnover time of pulmonary macrophages is then about 6 days, which is much shorter than the turnover time previously reported [5] which was based on an incomplete calculation. The same holds for the turnover times of other tissue macrophages (e.g., in the peritoneal cavity and the liver) in the normal steady state. This mathematical approach can also be applied to non-steady-state conditions, e.g., an inflammatory reaction in the lung after an intravenous injection of bacille Calmette-Guérin (BCG) [15].

3. Humoral Control of Monocytopoiesis During an Acute Inflammatory Reaction

During an acute inflammation there is an increased demand for circulating monocytes because these cells must provide the exudate macrophages needed at the site of inflammation. This means an increased production of circulating monocytes, which requires triggering of the dividing cells (monoblasts, promonocytes) in the bone marrow. The humoral regulatory mechanism that controls this extra monocyte production has recently been described [16–20].

Sera from mice and rabbits in which a sterile peritonitis was evoked by latex or silica induced monocytosis in test mice and rabbits, respectively, whereas sera from normal animals or animals injected with saline did not have this effect [17, 20]. The active sera contained a factor that specifically affects the monocytes; the numbers of granulocytes and lymphocytes remained in the normal range [17, 20]. The active sera not only caused a release of monocytes from the bone marrow but also increased the monocyte production in bone marrow. Because this increase is brought about by a decrease in the cell cycle time of the promonocytes and an increase in the number of these cells [17], this serum factor is called factor increasing monocytopoiesis (FIM).

The FIM production and secretion occurs in the peritoneal macrophages at the site of inflammation. Some of the properties of FIM were determined [18, 20]. The effect of FIM is dose dependent, and it is highly thermolabile. The molecular weight of FIM lies between

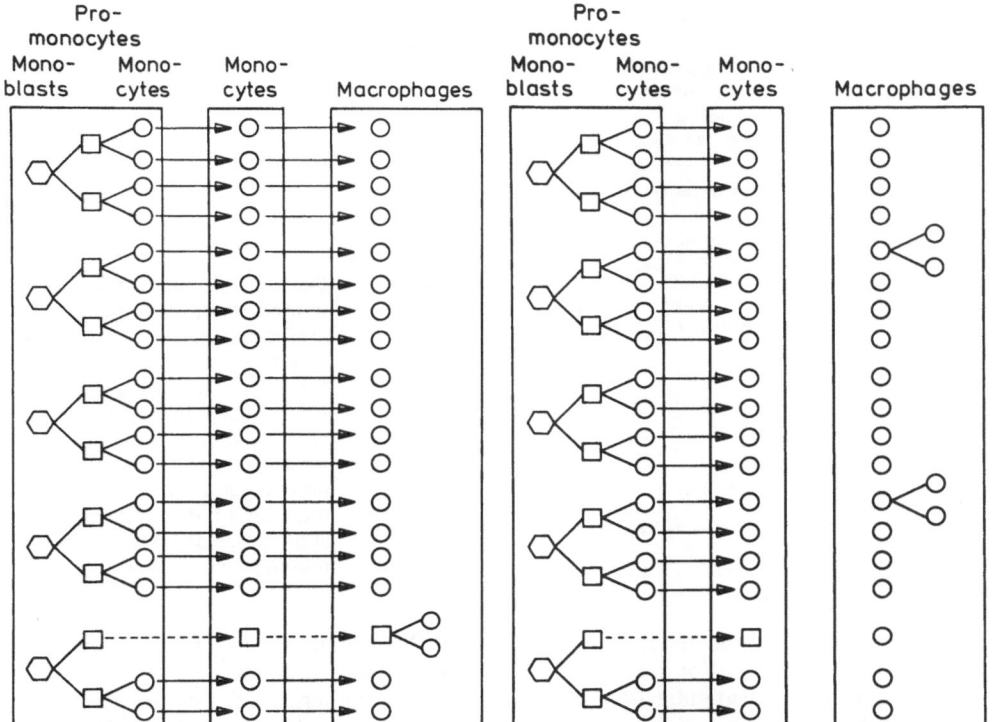

Fig. 3a, b. Concepts on the origin and kinetics of mononuclear phagocytes. **a** Schematic representation of the current concept on the origin and kinetics of mononuclear phagocytes. Monoblasts and promonocytes divide in the bone marrow compartment; the monocytes leave the bone marrow and are transported via the circulation to the tissues, where they become macrophages. Also indicated are the few (immature) mononuclear phagocytes that divide in the tissues, where they have recently arrived from the bone marrow in which they originate. **b** The alternative concept of the origin of the macrophage, now refuted. According to this concept, macrophages renew themselves by division in the tissues; the fate of monocytes formed in the bone marrow is not accounted for in this concept

18 000 and 24 500 daltons. This range of molecular weight makes it unlikely that FIM is one of the mediators of inflammation, for instance bradykinin or a prostaglandin, since these have a lower molecular weight [21, 22]. The possibility that FIM is an intermediate product formed during complement activation (having a similar molecular weight, e.g., C5a [23]) or a clotting product has been excluded. FIM has no colony-stimulating activity for mouse bone marrow cells and is not chemotactic towards macrophages. FIM appeared to be a protein and contains no essential carbohydrate moiety at the active site of the molecule; it is most probably not a glycoprotein. The determination of these properties led to the conclusion that FIM is a new factor and differs from the known mediators of inflammation.

The findings of FIM led to the postulation of a control system for monocytopoiesis (Fig. 4). When an inflammatory reaction is induced in the tissues (e.g., the peritoneal cavity), the macrophages at the site of the lesion phagocytose the inducing substance and then release FIM which is transported via the circulation to the bone marrow where it increases monocyte production. The extra monocytes are then transported by the circulation to the site of inflammation. However, since FIM has no chemotactic effect on macrophages, the influx into the lesion must be governed by another (local) mechanism [24].

When the inflammatory stimulus is eliminated by the mononuclear phagocytes the ex-

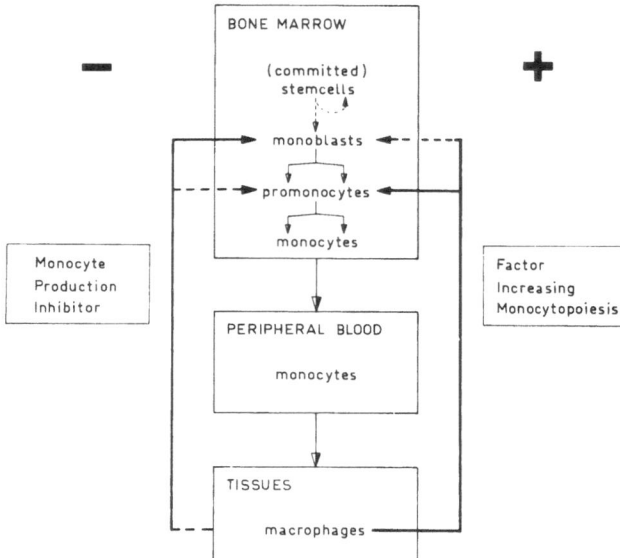

Fig. 4. Schematic representation of the humoral control of monocytopoiesis. After an inflammatory stimulus is applied to a tissue the macrophages release FIM, which is transported via the circulation to the bone marrow, where it increases monocyte production at the level of the monoblast and promonocyte. A second factor (MPI) inhibits monocyte production

tra production of monocytes has to be terminated. This could be brought about by a decreased production and/or release of FIM by macrophages at the site of the lesion. However, there are also indications of the presence of another factor: a monocytopoiesis inhibitor (MPI) [19]. This inhibitory factor has a molecular weight of more than 50 000 daltons, is active in vitro where it reduces monoblast proliferation, and is active in vivo because the number of monocytes in mice decreased after injection of this factor (Fig. 4).

Whether a same regulation by FIM and MPI is also operative under normal steady-state conditions is not yet certain. However, there are indications that serum from normal animals contains humoral factors that regulate the division of mononuclear phagocytes.

4. Summary

The present communication discusses the origin of monocytes and macrophages and gives new mathematical approach to the analysis of results obtained in studies on cell kinetics. It has been shown that the great majority of the macrophages derive from circulating monocytes, whereas the remainder arise by division of immature mononuclear phagocytes in the tissues. However, these dividing cells too originate in the bone marrow and have recently arrived in the tissues. The turnover time of tissue macrophages appears to be of the order of 1 to 5 weeks, which is shorter than had previously been calculated.

During an inflammatory response, monocyte production in the bone marrow is regulated by humoral factors that stimulate or inhibit the mitotic activity of promonocytes and their precursors. Whether such factors are also present during the normal steady state is not yet certain.

References

1. Furth van R (1978) Mononuclear phagocytes in inflammation. In: Vane JR, Fereira SH (eds) Inflammation and anti-inflammatory drugs. Springer, Berlin Heidelberg New York (Handbook of experimental pharmacology, vol 50, p 68)
2. Furth van R, Diesselhoff-den Dulk MMC, Raeburn JA, Zwet van TL, Crofton R, Blussé van Oud Alblas A (1980) Characteristics, origin and kinetics of human and murine mononuclear phago-

cytes. In: Furth van R (ed) Mononuclear phagocytes. Functional aspects. Nijhoff, The Hague Boston London, p 279

3. Furth van R, Cohn ZA (1968) The origin and kinetics of mononuclear phagocytes. J Exp Med 128:415
4. Crofton RA, Diesselhoff-den Dulk MMC, Furth van R (1978) The origin, kinetics and characteristics of the Kupffer cells in the normal steady state. J Exp Med 148:1
5. Blussé van Oud Alblas A, Furth van R (1979) The origin, kinetics, and characteristics of pulmonary macrophages in the normal steady state. J Exp Med 149:1504
6. Daems WT, Brederoo P (1973) Electron microscopical studies on the structure, phagocytic properties and peroxidatic activity of resident and exudate macrophages in the guinea pig. Z Mikrosk Anat Forsch 144:247
7. Daems WT, Wisse E, Brederoo P, Emeis JJ (1975) Peroxidatic activity in monocytes and macrophages. In: Furth van R (ed) Mononuclear phagocytes in immunity, infection and pathology. Blackwell, Oxford London Edinburgh Melbourne, p 57
8. Daems WT, Koerten HI, Soranzo MR (1976) Differences between monocyte-derived and tissue macrophages. In: Reichhard SM, Escobar MR, Friedman H (eds) The reticuloendothelial system in health and disease. Plenum, London New York, p 27
9. Beelen RHJ, Fluitsma DM, Meer van der JWM, Hoefsmit ECH (1980) Development of exudate-resident macrophages, on the basis of the pattern of peroxidatic activity in vivo and in vitro. In: Furth van R (ed) Mononuclear phagocytes. Functional aspects. Nijhoff, The Hague Boston London, p 87
10. Meer van der JWM, Gevel van de JS, Diesselhoff-den Dulk MMC, Beelen RHJ, Furth van R (1980) Long-term cultures of murine bone marrow mononuclear phagocytes. In: Furth van R (ed) Mononuclear phagocytes. Functional aspects. Nijhoff, The Hague Boston London, p 343
11. Radzun HJ, Parwaresch MR (1980) Isoelectric focusing pattern of acid phosphatase and acid esterase in human blood cells, including thymocytes, T lymphocytes, and B lymphocytes. Exp Hematol 8:737
12. Radzun JH, Parwaresch MR, Kulenkampff C, Staudinger M, Stein H (1980) Lysosomal acid esterase: activity and isoenzymes in separated normal human blood cells. Blood 55:891
13. Furth van R, Diesselhoff-den Dulk MMC, Mattie H (1973) Quantitative study on the production and kinetics of mononuclear phagocytes during an acute inflammatory reaction. J Exp Med 138:1314
14. Blussé van Oud Alblas A, Mattie H, Furth van R (to be published) A Quantitative evaluation of pulmonary macrophage kinetics
15. Blussé van Oud Alblas A, Linden van der-Schrever B, Furth van R (1981) Origin and kinetics of pulmonary macrophages during an inflammatory reaction induced by intravenous administration of heat-killed BCG. J Exp Med 154:235
16. Waarde van D, Hulsing-Hesselink E, Furth van R (1978) A serum factor inducing monocytosis during an acute inflammatory reaction caused by newborn calf serum. Cell Tissue Kinet 9:51
17. Waarde van D, Hulsing-Hesselink E, Furth van R (1977) Humoral regulation of monocytopoiesis during an inflammatory reaction caused by particulate substances. Blood 50:141
18. Waarde van D, Hulsing-Hesselink E, Furth van R (1977) Properties of a factor increasing monocytopoiesis (FIM) occurring in the serum during the early phase of an inflammatory reaction. Blood 50:727
19. Waarde van D, Hulsing-Hesselink E, Furth van R (1978) Humoral control of monocytopoiesis by an activator and inhibitor. Agents Actions 84:432
20. Sluiter W, Waarde van D, Hulsing-Hesselink E, Elzenga-Claasen B, Furth van R (1980) Humoral control of monocyte production during inflammation. In: Furth van R (ed) Mononuclear phagocytes. Functional aspects. Nijhoff, The Hague Boston London, p 325
21. Rocha E, Silva M, Garcia Leme J (1972) In: Alexander P, Bacq ZM (eds) Chemical mediators of the acute inflammatory reaction. Pergamon, Oxford, p 101
22. Douglas WW (1975) In: Goodman LS, Gilman A (eds) The pharmacological basis of therapeutics, 5th edn. Macmillan, New York Toronto London, p 589
23. Vollota EH, Müller-Eberhard HJ (1973) Formation of C3a and C5a anaphylatoxins in whole human serum after inhibition of the anaphylatoxin inactivator. J Exp Med 137:1109
24. Wilkinson PC (1974) Chemotaxis and inflammation. Livingstone, Edinburgh London
25. Balner H (1963) Identification of peritoneal macrophages in mouse radiation chimeras. Transplantation 1:217
26. Goodman JW (1964) The origin of peritoneal fluid cells. Blood 23:18

27. Virolainen M (1968) Hematopoietic origin of macrophages as studied by chromosome markers in mice. J Exp Med 127:943
28. Haller O, Arnleiter H, Lindenmann J (1979) Natural, genetically determined resistance toward influenza virus in hemopoietic mouse chimeras. Role of mononuclear phagocytes. J Exp Med 150:117
29. Porter KA (1969) Pathology of the orthotopic homograft and heterograft. In: Starzl TE (ed) Experience in hepatic transplantation. Saunders, Philadelphia London Toronto, p 464
30. Howard JG (1970) The origin and immunological significance of Kupffer cells. In: Furth van R (ed) Mononuclear phagocytes. Blackwell, Oxford London Edinburgh Melbourne, p 178
31. Shand FL, Bell EB (1972) Studies on the distribution of macrophages derived from rat bone marrow cells in exenogeic radiation chimeras. Immunology 22:549
32. Portmann B, Schindler AM, Marray-Lyon IM, Williams R (1976) Histological sexing of a reticulum cell sarcoma arising after liver transplantation. Gastroenterology 70:82
33. Gale RP, Sparkes RS, Gode DW (1978) Bone marrow origin of hepatic macrophages (Kupffer cells) in humans. Science 201:937
34. Pinkett MO, Cowdrey CM, Nowell PC (1966) Mixed hematopoietic and pulmonary origin of "alveolar macrophages" as demonstrated by chromosome markers. Am J Pathol 48:859
35. Brunstetter MA, Hardie JA, Schiff R, Lewis JP, Cross CE (1971) The origin of pulmonary alveolar macrophages. Arch Intern Med 127:1064
36. Godleski JG, Brain JD (1972) The origin of alveolar macrophages in radiation chimeras. J Exp Med 136–630
37. Johnson KJ, Ward PA, Striker G, Kunkel R (1980) A study of the origin of pulmonary macrophages using the Chédiak-Higashi marker. Am J Pathol 101:365
38. Weiden P, Storb R, Tsoi MS (1975) Marrow origin of canine alveolar macrophages. J Reticuloendothel Soc 17:342
39. Thomas ED, Ramberg RE, Sale GE (1976) Direct evidence for a bone marrow origin of the alveolar macrophages in man. Science 192:1016

Haematology and Blood Transfusion Vol 27
Disorders of the Monocyte Macrophage System
Edited by F. Schmalzl, D. Huhn, H.E. Schaefer
© Springer-Verlag Berlin Heidelberg New York 1981

The Kinetics of Mononuclear Phagocytes in Man

G. Meuret

1. Abstract

This paper summarizes the essential steps of a systematic cell kinetic analysis of monocytopoiesis in bone marrow and blood monocytes in man [14]. The results are interpreted on the basis of current concepts hypothesizing that blood monocytes have only one source, i.e., monocytopoiesis in bone marrow [8, 9], that cells move undirectionally from bone marrow to blood and tissue, and that monocytopoiesis is governed by several control mechanisms. Macrophage kinetics are not discussed because comprehensive studies in man are lacking and due to the present controversy between the notion that all tissue macrophages arise from blood monocytes and the other one which assumes that macrophage precursors exist in tissue [6].

2. Normal Conditions

2.1 Mononuclear Phagocytes in Bone Marrow

In order to detect mononuclear phagocytes among other bone marrow cells we applied the simultaneous cytochemical demonstration of NaF-sensitive and NaF-resistant naphthol-AS-D-acetate esterase [26]. With this method it is not possible to discriminate between monoblasts, promonocytes, and monocytes as in studies which were carried out by van Furth et al. [7–11]. Therefore, the whole pool of enzyme-positive cells was designated as "mononuclear phagocytes (MNP) of bone marrow." It must be considered, however, that early monocytopoietic cells which might possess very low activities of the marker enzyme may escape detection by the method used in our studies.

In healthy individuals the fraction of MNP in bone marrow averaged about 3% corresponding to a total medullar pool of 6×10^8 cells per kg body weight. The mean ^{3}H-thymidine labeling index (^{3}H-TDR LI) of these cells was 12% (Table 1) [19].

Table 1. MNP in bone marrow of healthy individuals [19]

MNP in bone marrow	No. examined	Mean ± SD
Relative number	20	$2.9 \pm 0.6\%$
Total pool	9	5.8 ± 1.2 cells $\times 10^8$/kg
^{3}H-TDR LI in vitro	7	$12.0 \pm 1.8\%$
DNA synthesis time	23	9.5 ± 1.9 h
Cell birth rate	7	6.8 ± 1.7 cells $\times 10^6$/kg $\times$ hour

	Nucleus Morphology			
	round or oval small	round or oval large	slightly folded	distinctly folded
Mononuclear Phagocytes in Bone Marrow				
Frequency distribution (%)	5	31	51	13
^{3}H-TDR labeling index (%)	7	10	10	25
Blood Monocytes				
Frequency distribution (%)	–	8	34	58
^{3}H-TDR labeling index (%)	–	3.28	0.54	0.05
Naphthol-AS-D-chloro-acetate esterase (activity index)	–	128	53	26

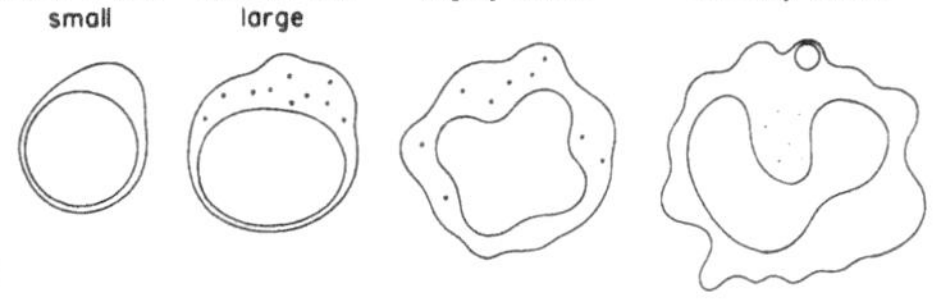

Fig. 1. Different types of MNP in bone marrow and blood grouped according to nuclear morphology [15, 19]

The MNP of bone marrow were divided into four types of cells according to nuclear morphology (Fig. 1) [19]. Type I cells showing lymphocyte-like nuclei comprised the smallest fraction and had the lowest ^{3}H-TDR LI (about 7%). The largest fractions consisted of Type II cells with large round nuclei and Type III cells with large and slightly folded nuclei. The morphologic features of Type IV cells with distinctly folded nuclei were similar to those of the most frequently occurring blood monocytes. In Type IV cells ^{3}H-TDR LI reached the maximum of about 25%.

Phase contrast time-lapse microcinematography of living cells in culture showed that motility of nuclei and cytoplasma increases with maturation [16]. From these observations which are similar to those being reported for granulocytopoietic cells it is concluded that early monocytopoietic precursors are fixed with the highest probability on smears with round nuclei and more mature cells with folded nuclei. Therefore, the different morphologic types of MNP express different degrees of maturation.

2.2 Stem Cell to Blood Transit Time of Monocytopoiesis

Using ^{3}H-Diisoprophylfluorophosphate (^{3}H-DFP) it is possible to label human blood

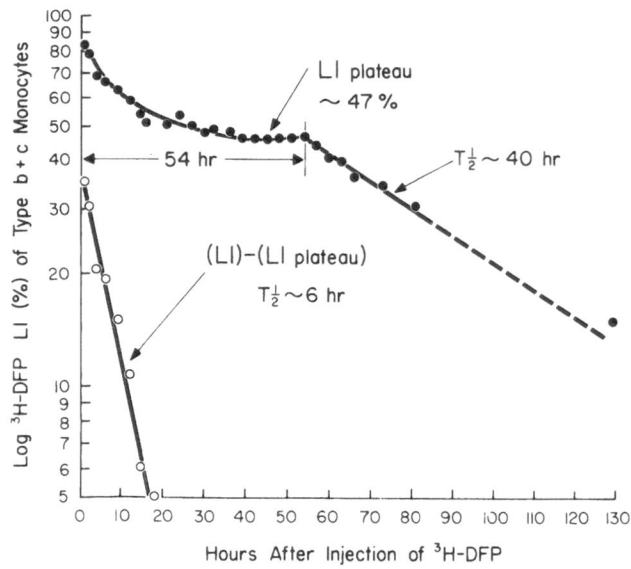

Fig. 2. Time course of ^{3}H-DFP LI in blood monocytes after intravenous injection of ^{3}H-DFP in a hematologically normal individual [17]

monocytes and the vast majority of MNP in bone marrow [12]. Thus, upon intravenous injection of this marker, a phase of high influx rates of labeled monocytes into the blood follows, reflecting the minimal passage time of monocytopoietic cells through the medullar precursor compartment (Fig. 2) [13, 17]. In a hematologically normal individual this phase lasted about 54 h. The egress of monocytes from the marrow into the tissue can be derived from the initial, steeply declining curve component [2]. It proved to be an exponential process with a half time of 6 h.

2.3 DNA Synthesis Time and Cell Cycle Time of MNP in Marrow

DNA synthesis time of MNP in bone marrow was measured in hematologically normal individuals by a ^{3}H-TDR and ^{14}C-TDR double labeling technique [19]. The average of 23 determinations was 9.5 h, and they ranged from 6.6 to 13.3 h.

Further information on precursor kinetics was obtained by observing the influx chronology of labeled monocytes into the blood following ^{3}H-TDR pulse labeling (Fig. 3) [17]. The phases of steeply increasing influx rates of labeled monocytes (t_2) approximate DNA synthesis time of monocyte precursors. In 11 determinations t_2 averaged 11 h and ranged from 10 to 13 h. This corresponds quite well with the results of the double labeling technique.

The time intervals between onsets and peaks of two consecutive influx waves (t_3, t_4) give a rough estimate of cell cycle time. The average of eight determinations of t_3 und t_4 was 29 h; they ranged from 22 to 35 h.

2.4 Cell Egress from Precursor Compartments into the Blood

Information on this process was derived from different approaches. Two important facts emerged from the analysis of the influx chronology of labeled monocytes following ^{3}H-TDR pulse labeling (Fig. 3) [14, 17].

1. Labeled monocyte influx into the blood began after a lag period (t_1) of 6 h (range 5–7 h; three determinations) after ^{3}H-TDR injection. This short period of time allows the cells to complete the cell cycle and to enter the blood. However, there is no time left for mature cells to rest in bone marrow and to accumulate in a storage pool.

2. Influx waves of labeled monocytes overlapped and showed different patterns for the

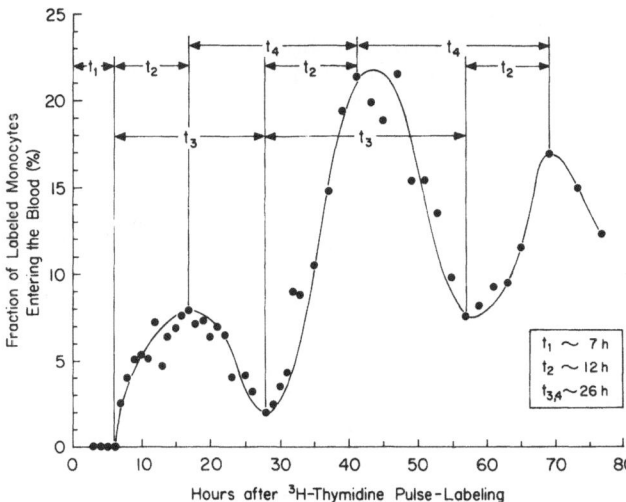

Fig. 3. Chronology of labeled monocyte influx into the blood after ^{3}H-TDR pulse labeling in a hematologically normal patient [17]

13

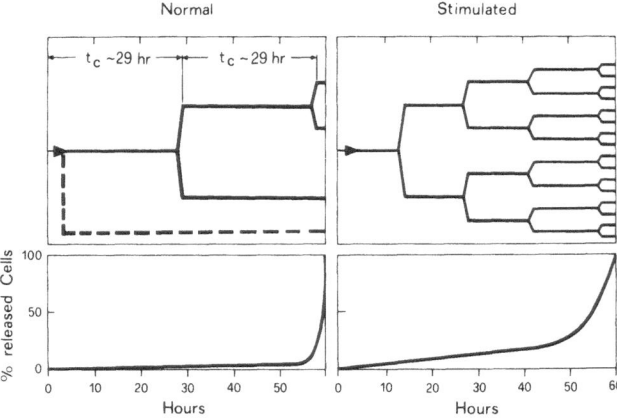

Fig. 4. Model of human monocytopoiesis [13, 17]

different types of blood monocytes [14, 17]. These phenomena could be interpreted by assuming that cell transit from the precursor compartments into the blood basically is possible at any time during precursor maturation. However, egress probability is low for immature cells. It increases slowly during the first 50 h of maturation to reach its maximum value at 50–60 h when the majority of the cells is released into the blood (Fig. 4) [14, 17].

The characteristics of marrow cell egress are reflected in different degrees of blood monocytes maturation (Fig. 1). Using morphologic criteria for cell differentiation, blood monocytes can be grouped into a number of cell categories having similar appearances as their counterparts in bone marrow. There is one exception, namely, lymphocyte-like Type I cells which are but rarely found in blood of normal individuals [15]. Cells of identical morphology in bone marrow possess a lower grade of differentiation than their counterparts in peripheral blood. A similar finding was reported for band and segmented granulocytes of both of these cell compartments [1, 21]. As judged by ^{3}H-TDR LI, morphologically similar types of cells exhibit much lower proliferation activity in blood as compared to bone marrow (Fig. 1). Monocytes with large round nuclei comprise the most immature cell population occurring in the blood. They are characterized by the following pattern [15]: high activity of naphthol-AS-D-chloroacetate esterase and peroxidase and relatively high ^{3}H-TDR LI, combined with low activity of naphthol-AS-D-acetate esterase. The most differentiated blood monocytes exhibit distinctly folded nuclei, a reverse enzyme pattern, and extremely low ^{3}H-TDR LI.

Cytophotometric measurements carried out on Feulgen-stained blood monocytes demonstrated that 98% of the cells possess diploid DNA content (Tab. 2) [14]. This finding as well as low ^{3}H-TDR LI of blood monocytes (Fig. 1) indicates that cell release from the precursor compartment into the blood is rather selectively restricted to nonproliferating cells.

2.5 Model of Normal Human Monocytopoiesis

Marrow sojourn of the majority of MNP lies between 50 and 60 h. During this period

Table 2. DNA content of blood monocytes determined by cytophotometric measurements of 5017 Feulgen-stained cells in 5 healthy individuals [14]

Relative DNA content	Percent of measured blood monocytes
Diploid	98.42
Intereuploid	1.40
Tetraploid	0.18

14

proliferating monocyte precursors whose mean cell cycle time amounts to about 30 h are able to complete a maximum of two consecutive mitotic divisions. These divisions amplify the cell flow rate entering the system by stem cell differentiation. Cell birth rate (BR) within the proliferating precursor pool, calculated by the following formula [5], averaged 7 million cells per kg body weight per hour:

$$BR = \frac{\text{medullar pool of MNP} \times {}^3\text{H-TDR LI}}{\text{DNA synthesis time}}$$

$$= \frac{5.8 \times 10^8 \text{ cells/kg} \times 0{,}12}{9.5 \text{ h}} = 7.3 \times 10^6 \text{ cells/kg} \times \text{h}$$

The growth fraction (GF) of MNP in marrow was lower than in other hematopoietic cell systems:

$$GF = \frac{\text{cell cycle time} \times {}^3\text{H-TDR LI}}{\text{DNA synthesis time}}$$

$$= \frac{29 \text{ h} \times 12\%}{9.5 \text{ h}} = 37\%$$

This result supports the assumption that under normal conditions only a fraction of the MNP in marrow utilizes their proliferation potential.

2.6 Blood Monocyte Kinetics

Detailed studies of blood monocyte kinetics were carried out using autotransfusion of blood monocytes which were labeled in vitro by ^{3}H-DFP. The intravascular fate of labeled monocytes was followed by autoradiographic determinations of monocyte ^{3}H-DFP LI (Fig. 5) [12]. The behavior of ^{3}H-DFP LI indicated that blood monocytes interchange between a circulating and a marginal pool. The marginal monocyte pool was three and a half times larger than the circulating monocyte pool (Table 3). Labeled monocytes disappeared from circulation in an exponential fashion (Fig. 5), the mean half time being 8.4 h. This is in accord with the results obtained by in vivo labeling with ^{3}H-DFP (Fig. 2). Monocyte turnover rate (which equals total blood monocyte pool $\times 2/T\frac{1}{2}$) averaged 7.5×10^6 cells per kg body weight per hour.

2.7 Macrophages

Very little work has been dedicated to the analysis of kinetics of blood monocytes after their emigration into the tissue. Present knowledge is based almost exclusively on animal

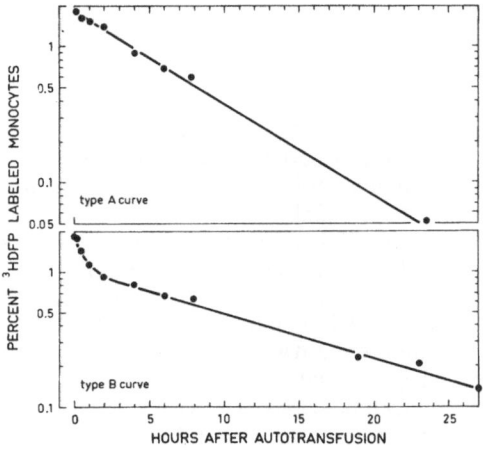

Fig. 5. Disappearance of ^{3}H-DFP-labeled monocytes from the circulating blood following autotransfusion [12]

15

Table 3. Blood monocyte kinetics in eight healthy individuals [12]

Blood monocytes	Mean ± SD
Circulating pool	18 ± 6 cells $\times 10^6$/kg
Marginal pool	63 ± 28 cells $\times 10^6$/kg
Total blood pool	81 ± 31 cells $\times 10^6$/kg
Half disappearance time	8.4 ± 1.9 h
Turnover rate	7.5 ± 4.0 cells $\times 10^6$/kg $\times$ hour

Table 4. DNA synthesis activity and bacteriostatic potential of blood monocytes, macrophages in pleural effusions, ascites, and alveolar macrophages [25]

	^{3}H-TDR LI (%)	Bacteriostatic index (%)
Blood monocytes	0.4 ± 0.2	28 ± 4
(healthy individuals)	(No. = 9)	(No. = 16)
Macrophages in effusions	1.5 ± 2.1	86 ± 3
(benign and malignant diseases)	(No. = 27)	(No. = 6)
Alveolar macrophages	0.05 ± 0.1	87 ± 9
(normal and diseases states)	(No. = 40)	(No. = 33)

studies. Van Furth et al [4, 7, 8, 11] demonstrated that 56% of monocytes leaving the blood enter the liver and give rise to Kupffer cells, 8% appear in the peritoneal cavity, and 15% are transformed into alveolar macrophages. The remaining fraction is supposed to enter other tissues.

In man only macrophages in effusion fluids and alveolar macrophages are easily accessible for investigation. Both cell types possess significantly higher bacteriostatic indexes than blood monocytes. This reflects a higher degree of maturation (Table 4) [25]. In some cases ^{3}H-TDR LI of macrophages in effusions or ascites were definitively higher than in blood monocytes. This was especially pronounced with malignant diseases where these macrophages reached ^{3}H-TDR LI of 4.0%–9.6%. In contrast, alveolar macrophages seemed to have lost their proliferative potential almost completely.

3. Increased Monocyte Production

3.1 Monocytopoiesis in Bone Marrow

Monocytopoiesis is not equipped with a medullar storage pool of mature cells. We therefore analyzed the reaction of the cell system in a situation of abruptly increasing monocyte demand. This occurs in acute inflammation when all macrophages colonizing the inflammatory reaction site are recruited from blood monocytes [28].

Monocytopoiesis was studied just before and about 15 h after partial gastrectomies in patients suffering from chronic ulcers [20]. Within this short time interval the number as well as the ^{3}H-TDR LI of MNP in bone marrow increased considerably (Table 4). Both changes contributed to doubling the monocyte birth rate within the medullar pool of MNP. It is interesting to note that maximum increase in proliferation activity was associated with the small lymphocyte-like Type I cells, whereas Type IV cells did not change.

These observations demonstrated that monocytopoiesis has a considerable proliferation reserve which is quickly recruited when monocyte demand increases. This

Table 5. Mononuclear phagocytes in bone marrow of four patients with gastric or duodenal ulcers before and 13-17 h after beginning of partial gastrectomy (mean values) [20]

Mononuclear phagocytes in bone marrow	Medullar pool (cells $\times 10^8$/kg)	^{3}H-TDR LI (%)
Pooled Type I–IV cells:		
Preoperative	6.6	23
Postoperative	8.9	32
Type I cells:		
Preoperative	0.4	11
Postoperative	1.3	24
Type IV cells:		
Preoperative	1.4	39
Postoperative	1.7	37

reserve resides mainly in the compartment of Type I MNP which under normal conditions demonstrated the lowest proliferation activity.

The mechanism of cell kinetics underlying the rise in proliferation activity proved to be similar to the ones described for promonocytes in mice [7] or other hematopoietic cells such as myelocytes [5]. In acute inflammation, cell cycle time shortens (Table 6), thus allowing the precursor cells to undergo additional cell cycles during their sojourn in bone marrow (Fig. 4). In addition, the growth fraction increases, indicating that nonproliferating precursor cells are triggered into cell cycle.

Table 6. Shortening of cell cycle time and rise in growth fraction of MNP in bone marrow during septicaemia [19]

MNP in bone marrow	Normal	Septicaemia
Cell cycle time	29 h	13 h
Growth fraction	37%	50%

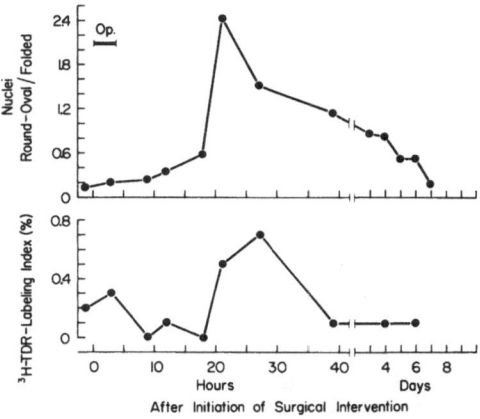

Fig. 6. Postoperative "shift" of blood monocytes in favor of immature cells being characterized by round or oval nuclei and relatively high ^{3}H-TDR LI [20]

17

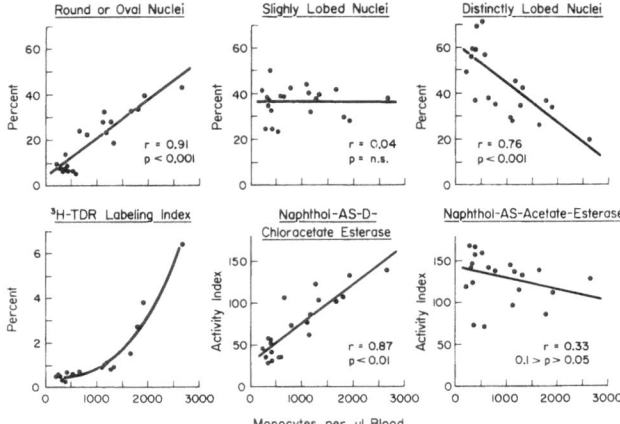

Fig. 7. Premature cell release from precursor compartments increases with monocyte blood counts [15] or monocyte turnover rate (Fig. 8)

3.2 Cell Transit from Precursor Compartments into Blood

Concomittant with the increasing monocytopoietic proliferation activity cell release from precursor compartments into blood shifts in favor of immature cells. This shift manifests itself by a rise in the fraction of immature blood monocytes with round nuclei and relatively high ³H-TDR LI (Fig. 6) [15, 20].

The examination of blood monocytes in patients with chronic infectious diseases demonstrated that the "shift to the left" is positively correlated with monocyte blood counts (Fig. 7) [15]. As monocyte counts rise in parallel to monocyte turnover rate (Fig. 7) [12], it was concluded that increases in monocyte production rate are always quantitatively coupled with premature cell release from precursor compartments.

3.3 Kinetics of Blood Monocytes

Increase in monocyte production or in monocyte turnover rate are paralleled by a rise in monocyte blood count (Fig. 8) [12]. In some patients increased monocyte turnover was associated with a slight prolongation of the intravascular half time (Fig. 9). However, changes in turnover did not influence cell distribution between the circulating and the marginal monocyte pool [12].

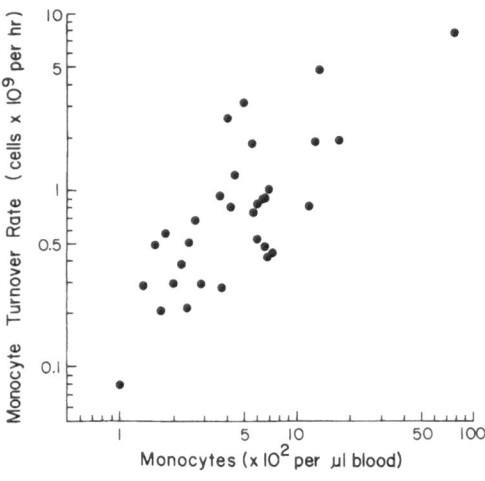

Fig. 8. Monocyte blood counts are positively correlated with monocyte turnover rates [12]

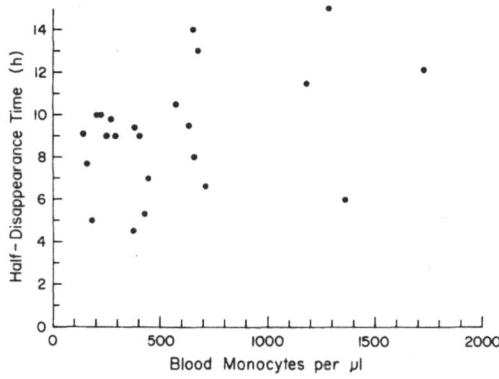

Fig. 9. Half disappearance time of blood monocytes in relation to blood monocyte counts [12]

4. Regulation of Monocytopoiesis

The observations in acute and chronic states of inflammation indicated the presence of regulatory mechanisms which adapt monocyte production to monocyte demand (Table 6). Our studies indicated that two processes are controlled, i.e. both proliferation activity of precursors in bone marrow and cell release from precursor compartments (Fig. 10).

De Waarde et al. [29, 30] isolated a factor appearing in the serum of mice after induction of acute inflammation. This factor is produced by inflammatory macrophages. It stimulates proliferation activity of promonocytes in marrow. Thus, the factor increasing monocytopoiesis (FIM) seems to mediate a control loop which plays a role in tuning monocyte production to monocyte demand.

The observation of low amplitude oscillations of monocyte blood counts in healthy individuals (Fig. 10) were considered indicative of a second control process [18]. The most frequently occurring period of the oscillation was 5 days. It could be simulated by a computer model of a feedback control system linking monocyte-macrophage pools with stem cell differentiation (Figs. 11 and 12) [18]. The results substantiated the hypothesis that stem cell differentiation is modulated by soluble factors being secreted by monocytes and/or macrophages.

5. Monocytopoiesis in Disease States

Increased monocyte-macrophage demand is reflected in bone marrow by a corresponding rise in MNP proliferation activity. Therefore, it is possible to estimate monocyte consumption by determining the fraction and ^{3}H-TDR LI of MNP in bone marrow. The product of both parameters correlates with monocyte birth rate within the medullar pool of MNP.

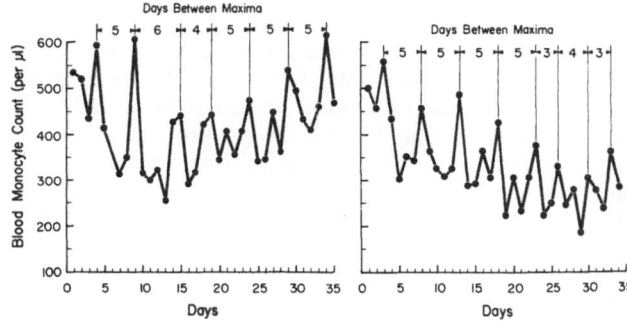

Fig. 10. Low amplitude oscillations of monocyte blood counts in healthy individuals. The most frequently occurring period was 5 days [18]

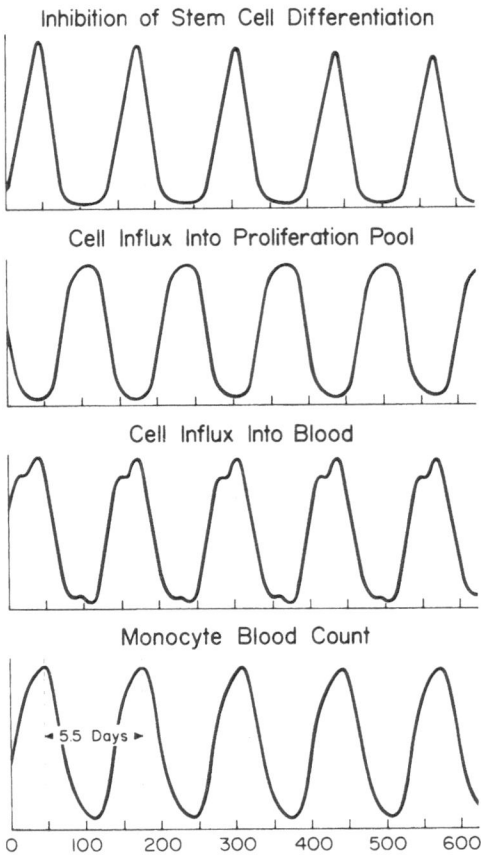

Inhibition of Stem Cell Differentiation

Cell Influx Into Proliferation Pool

Cell Influx Into Blood

Monocyte Blood Count

◄ 5.5 Days ►

0   100   200   300   400   500   600
Hours

Fig. 11. Computer simulation of a mathematical model describing a control loop linking the monocyte-macrophage pools with stem cell differentiation [18]

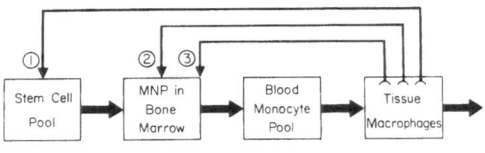

Controlled Processes
① Stem Cell Differentiation
② Proliferation Activity of MNP in Bone Marrow
③ Marrow Cell Release

Fig. 12. Models of several control mechanisms regulating monocytopoiesis

Patients with untreated tuberculosis showed a marked monocytopoietic hyper-proliferation (Fig. 13) [27] which was assumed to compensate a high macrophage loss due to macrophage toxicity of tubercle bacilli. Sarcoidosis patients did not demonstrate significant deviations from normal monocytopoiesis, indicating that this disease produces low turn-over granuloma. Monocytopoietic proliferation activity was definitively increased in all patients with widely spread eczematous diseases [21], whereas hyperproliferation was found only in some of the patients with Crohn's disease, ulcerative colitis [24], and disseminated psoriasis vulgaris [22].

Among the malignant diseases investigated so far (Fig. 14) pronounced hyperactivity of monocytopoiesis was associated with untreated mycosis fungoides [22] and Hodgkin's

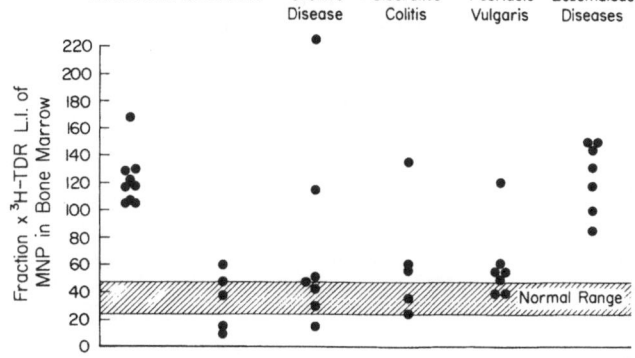

Fig. 13. Monocytopoietic proliferation activity in inflammatory disease [22, 24, 27]

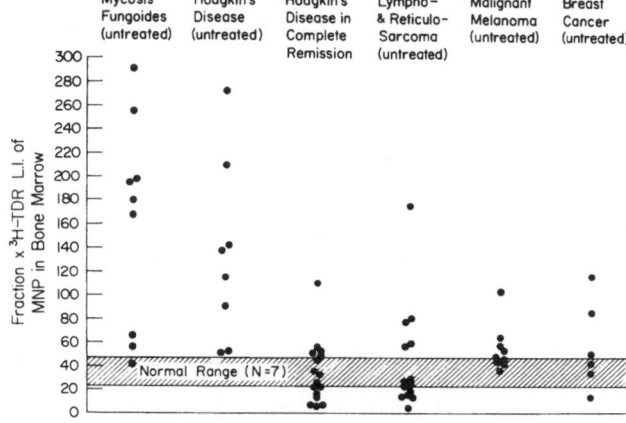

Fig. 14. Monocytopoietic proliferation activity in malignant diseases [22–24]

disease [23]. Monocyte production normalized in Hodgkin's disease during long-term complete remission. In contrast, increase in monocytopoiesis was observed only in some of the patients with untreated non-Hodgkin's lymphoma. Moderate hyperproliferation occurred in part of the patients with untreated, superficially spreading melanoma and untreated breast cancer [23].

References

1. Altman AJ, Stossel TP (1974) Functional immaturity of bone marrow band polymorphonuclear leukocytes. Br J Haematol 27:241
2. Athens JW, Mauer AM, Ashenbrucker H, Cartwright GE, Winthrobe MM (1959) Leukocyte kinetic studies with diisoprophylfluorophosphate (DFP 32). Blood 14:303
3. Boll I, Kühn A (1965) Granulocytopoiesis in human marrow cultures studies by means of kinematography. Blood 26:449
4. Crofton RW, Diesselhoff den Dulk MMC, van Furth R (1978) The origin kinetics and characteristics of the Kupffer cells in the normal steady state. J Exp Med 148:1
5. Cronkite EP, Vincent PC (1969) Granulocytopoiesis. Ser Haematol 2:3
6. Daems WT, van der Rhee HJ (1980) Peroxidase and catalase in monocytes, macrophages, epithelioid cells and giant cells of the rat. In: van Furth R (ed) Mononuclear phagocytes. Functional aspects, part I. Nijhoff, The Hague Boston London, p 43

7. Furth R van (1978) Mononuclear phagocytes in inflammation. In: Vane JR, Ferreira SH (eds) Inflammation. Springer, Berlin Heidelberg New York (Handbook of experimental pharmacology, vol 50/I, p 68)
8. Furth R van (1980) Cells of the mononuclear phagocytes system nomenclature in terms of sites and conditions. In: van Furth R (ed) Mononuclear phagocytes. Functional aspects, part I. Mijhoff, The Hague Boston London, p 1
9. Furth R van, Diesselhoff den Dulk MMC (1970) The kinetics of promonocytes and monocytes in the bone marrow. J Exp Med 132:813
10. Furth R van, Diesselhoff den Dulk MMC, Mattie H (1973) Quantitative study on the production and kinetics of mononuclear phagocytes during an acute inflammatory reaction. J Exp Med 138:1314
11. Furth R van, Diesselhoff den Dulk MMC, Raeburn JA, van Zwet TL, Crofton R, Blusse A, Alblas O (1980) Characteristics, origin and kinetics of human and murine mononuclear phagocytes. In: van Furth R (ed) Mononuclear phagocytes. Functional aspects, part I. Nijhoff, The Hague Boston London, p 279
12. Meuret G (1973) Monocyte kinetic studies in normal and disease states. Br J Haematol 24:275
13. Meuret G (1973) Human monocytopoiesis. Exp Haematol 2:238
14. Meuret G (1974) Die Monocytopoese beim Menschen. Blut [Suppl 13]
15. Meuret G, Djawari D, Berlet R, Hoffmann G (1971) Kinetics, cytochemistry and DNA-synthesis of blood monocytes in man. In: Di Luzio (ed) The reticuloendothelial system and immune phenomena. Plenum, New York, p 33
16. Meuret G, Bundschu-Lay A, Senn HJ, Huhn D (1974) Functional characteristics of chronic monocytic leukemia. Acta Haematol (Basel) 52:95
17. Meuret G, Bammert J, Hoffmann G (1974). Kinetics of human monocytopoiesis. Blood 44:801
18. Meuret G, Bremer C, Bammert J, Ewen J (1974) Oscillation of blood monocyte counts in healthy individuals. Cell Tissue Kinet 7:223
19. Meuret G, Batara E, Fürste HO (1975) Monocytopoiesis in normal man: pool size, proliferation activity and DNA-synthesis time of promonocytes. Acta Haematol (Basel) 54:261
20. Meuret G, Detel U, Kilz HP, Senn HJ, van Lessen H (1975) Human monocytopoiesis in acute and chronic inflammation. Acta Haematol (Basel) 54:328
21. Meuret G, Bammert J, Gessner U (1976) Neutrophil marrow release. A system analysis. Blut 33:389
22. Meuret G, Schmitt E, Hagedorn M (1976) Monocytopoiesis in chronic eczematous diseases, psoriasis vulgaris, and mycosis fungoides. J Invest Dermatol 66:22
23. Meuret G, Schmitt E, Tseleni S, Widmer M (1978) Monocyte production in Hodgkin's disease and Non-Hodgkin's lymphoma. Blut 37:193
24. Meuret G, Bitzi A, Hammer B (1978) Macrophage turnover in Crohn's disease and ulcerative colitis. Gastroenterology 74:501
25. Meuret G, Schildknecht O, Joder P, Senn HJ (1980) Proliferation activity and bacteriostatic potential of human blood monocytes, macrophages and pleural effusions, ascites and of alveolar macrophages. Blut 40:17
26. Schmalzl F, Braunsteiner H (1968) On the origin of monocytes. Acta Haematol (Basel) 39:177
27. Schmitt E, Meuret G, Stix L (1977) Monocyte recruitment in tuberculosis and sarcoidosis. Br J Haematol 35:11
28. Spector WG, Walters MNJ, Willoughby DA (1965) The origin of the mononuclear cell in inflammatory exsudates induces by fibrinogen. J Pathol Bacteriol 90:181
29. Waarde D van, Hulsing-Hesselink E, van Furth R (1977) Properties of a factor increasing monocytopoiesis (FIM) occurring in the serum during the early phase of an inflammatory reaction. Blood 5:727
30. Waarde D van, Hulsing-Hesselink E, van Furth R (1978) Humoral control of monocytopoiesis by an activator and an inhibitor. Agents Actions 8:432

Haematology and Blood Transfusion Vol 27
Disorders of the Monocyte Macrophage System
Edited by F. Schmalzl, D. Huhn, H.E. Schaefer
© Springer-Verlag Berlin Heidelberg New York 1981

Biochemical Properties of Human and Murine Mononuclear Phagocytes and Their Changes on Activation

J. Schnyder

"Mononuclear phagocytes" is the term generally used to refer to a very large family of phagocytic cells which are all derived from the bone marrow or blood monocytes. These cells are long lived and capable of differentiating into a variety of macrophages or macrophage-like cells. The macrophages themselves appear to exist at different levels of activation. Several eliciting agents have been shown in experimental animals to recruit populations of activated macrophages, and in vitro culture experiments have demonstrated that quiescent macrophages can be induced to differentiate in response to particulate and soluble stimuli. The latter process is usually called macrophage activation, although everybody knows that activation is very hard to define. Morphological, biochemical, and functional parameters are often used to characterize the state of activation of macrophages. A review of such criteria and a discussion of their possible meanings has been presented recently [4, 26, 34].

In our laboratory we have studied the properties of quiescent and inflammatory macrophages. We have paid particular attention to the mechanisms involved in the differentiation of a normal, resting macrophage into an inflammatory macrophage, and we have monitored these changes by testing for a number of enzymic activities. If at all, enzymic parameters are only indirect reflections of cellular functions; in contrast to the latter, however, they are easy to determine and provide exact quantitative data. All experiments from our laboratory were performed on mouse peritoneal macrophages, the type of mononuclear phagocyte which has been studied most widely and in most detail. These cells are obtained by peritoneal lavage from untreated mice or from mice which were treated intraperitoneally with phlogogenic agents. The macrophages from untreated animals are called resident, nonelicited, or quiescent, those from treated animals are called elicited. Since eliciting induces inflammation, the term inflammatory macrophages is also used.

In this overview the mouse peritoneal macrophage will serve as a model on which to outline the major biochemical changes occurring during macrophage activation. This will be the framework on which we shall later discuss biochemical information now available on human mononuclear phagocytes and some clinical disorders which are associated with anomalies of the mononuclear phagocyte enzymes.

1. Enzymes of Mouse Peritoneal Macrophages

Depending on their actual catalytic properties, their intracellular localization, and their fate under culture conditions, different groups of macrophage enzymes may be distinguished. *Lysozyme* may be considered a typical phagocyte enzyme. It is present in mononuclear and polymorphonuclear phagocytes and is not found in cells of fibroblastic, lymphoid, and epithelioid lineage. In neutrophil leukocytes, lysozyme is stored both in the azurophil and

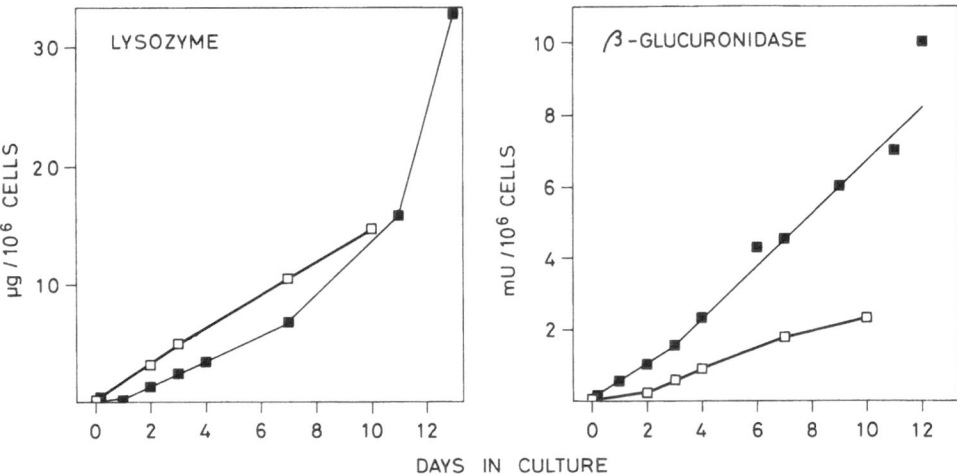

Fig. 1. Lysozyme and β-glucuronidase secretion by resident (*white symbols*) and thioglycollate-elicited macrophages (*black symbols*). Cumulative mean values from triplicate cultures

the specific granules [1]. Like the other granule components, it is synthesized during neutrophil maturation and it is not produced in the mature cells found in circulation and in the tissues. The situation is quite different in mononuclear phagocytes which have a very low lysozyme content. Nevertheless, these cells continuously secrete large amounts of lysozyme by a mechanism which is apparently directly coupled to de novo enzyme synthesis [16]. The rate of lysozyme secretion is similar in resident and elicited macrophages (Fig. 1). Under standard culture conditions, this rate remains constant for at least two weeks. For this reason, lysozyme secretion is a useful parameter to assess the viability of macrophages in culture.

The *lysosomal hydrolases* are a large group of enzymes which are found in all nucleated animal cells. In the monocytes, they are localized within the primary granules, which resemble both biochemically and functionally the azurophil granules of neutrophils [33]. In the macrophages, which are free of granules, those hydrolases are confined within the lysosomal system. It was established long ago that the lysosomal hydrolases are released into endocytic vacuoles where they serve the purpose of intracellular digestion [5, 6]. Only recently, it was found that macrophages secrete large amounts of lysosomal hydrolases [41] and that the rate of secretion appears to reflect the degree of activation of the cell. This is illustrated in Fig. 1. The amounts of lysosomal hydrolases which are released into the culture media clearly exceed the amounts of these enzymes which are found intracellularly. This suggests that lysosomal enzyme secretion depends on active enzyme synthesis, and indeed, when protein synthesis is blocked with cycloheximide, secretion ceases almost completely [41].

The *neutral proteinases* are proteolytic enzymes with optimum activity around neutral pH. In contrast to the acid proteinases (e.g., cathepsin B and cathepsin D) these enzymes are not localized within lysosomes but appear to be associated with small secretory vesicles. As is the case of lysozyme, their intracellular levels are very low. The best known neutral proteinase of macrophages is plasminogen activator which converts plasminogen into plasmin [17, 45]. Other enzymes of this type found in mouse macrophages are elastase, a metalloproteinase [49], and specific collagenase [48]. Neutral proteinases are secreted by activated macrophages only. For this reason, plasminogen activator secretion is often used as a parameter to distinguish between quiescent and activated cells. All these proteinases are likely to act in the pericellular space. Plasminogen activator (via plasmin) and elastase have a fairly broad substrate specificity and are thought to be involved in tissue degradation and tissue damage.

Ectoenzymes are constituents of the cellular membrane with their catalytic site exposed to the extracellular space. Two of these, alkaline phosphodiesterase I and 5'-nucleotidase, are of particular interest in the macrophage as cell activation is accompanied by a rise in alkaline phosphodiesterase I and a decrease in 5'-nucleotidase activity [11, 12].

2. Enzyme Changes During Macrophage Activation

The differences which we and others observed in normal and inflammatory or activated macrophages made it worth-while to attempt to study changes in enzyme activities as a consequence of macrophage stimulation in culture. As the stimulus we adopted phagocytosis. We followed a lead by Davies et al. [8] who had found that macrophages which were cultured in the presence of phagocytosable material acquired the morphological aspect of activated cells and secreted considerable amounts of lysosomal glycosidases. In our experiments, phagocytosis was limited to 1 h at the onset of culture and was followed by a culture period of 10 to 14 days during which the cells were exposed to standard medium. The effect of the phagocytic stimulus is shown in Fig. 2. As already shown in Fig. 1, during the first days of culture normal peritoneal macrophages secrete only lysozyme. Their secretion of lysosomal hydrolases begins between the 2nd and the 3rd day. However, when such cells are subjected to phagocytosis of zymosan, they immediately start to secrete lysosomal hydrolases. Secretion continues for several days, despite the fact that the phagocytic uptake was stopped very early on. An even more dramatic consequence of the phagocytic stimulus is observed some days later. After 3 to 4 days, the cells which have phagocytosed start to produce and secrete plasminogen activator. In fact, phagocytosis induces a whole differentiation program in the macrophage involving major changes in enzyme activities. Some of the biochemical properties of nonelicited macrophages which were stimulated by zymosan phagocytosis are compared in Fig. 3 with those of nonphagocytosing controls and of inflammatory macrophages elicited with thioglycollate medium. Clearly, quiescent cells which engage in phagocytosis acquire many features of their inflammatory counterparts. In addition to the changes in secretory activity which were already pointed out, the phagocytic stimulus induces a marked elevation of the levels of enzymes associated with different subcellular compartments like lactate dehydrogenase, β-glucuronidase, and alkaline phosphodiesterase I.

Figure 3 may be regarded as a useful synopsis of biochemical parameters which are easily assayed and which provide reliable and quantitative data for distinguishing between quiescent and activated macrophages. Indeed, biochemical changes similar to those which

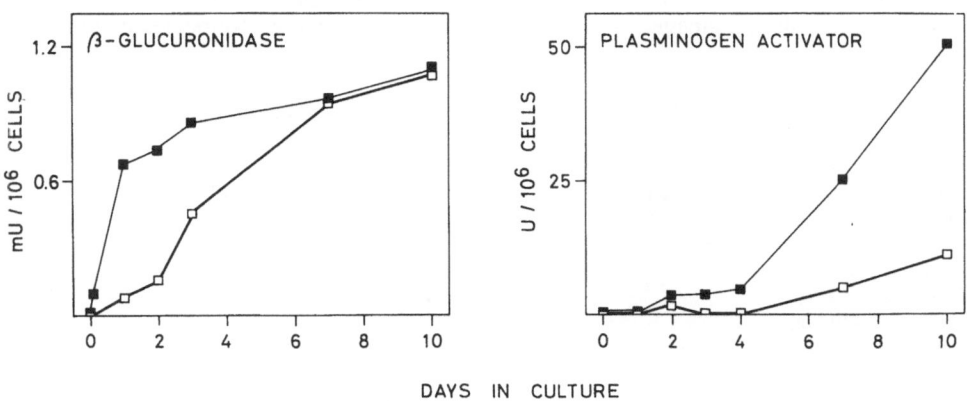

Fig. 2. Beta-glucuronidase and plasminogen activator secretion by resident (*white symbols*) and zymosan-stimulated macrophages (*black symbols*). Cumulative mean values from triplicate cultures

25

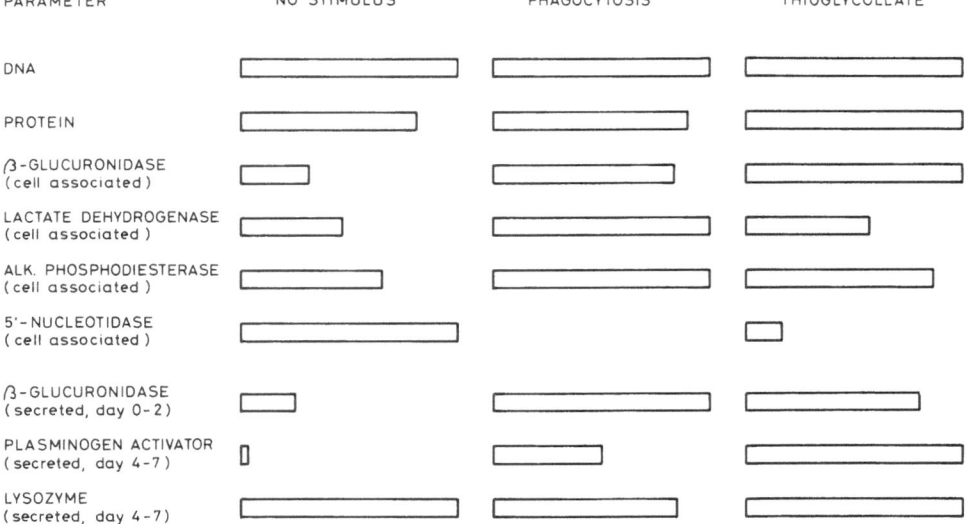

Fig. 3. Biochemical properties of quiescent and zymosan-stimulated nonelicited, and thioglycollate-elicited macrophages. Cell-associated enzyme activities and content of DNA and protein were determined on day 10 in the nonelicited macrophages and on day 0 in thioglycollate-elicited macrophages. The enzymes secreted are shown for the culture periods indicated

are shown here to occur as a consequence of zymosan phagocytosis under culture conditions or of eliciting with thioglycollate medium are induced by a great variety of agents and/or conditions, as indicated in Table 1. A complete list of the conditions known to induce an increase in the cellular levels of lysosomal hydrolases in mouse peritoneal macrophages is given by Baggiolini [2].

3. Biochemical Markers of Human Mononuclear Phagocytes and Changes Induced by Activating Conditions

Most of the work on human mononuclear phagocytes has been performed on blood monocytes and alveolar macrophages. Peritoneal macrophages have also been obtained by peritoneal dialysis or drainage, but biochemical data are still scanty. Monocytes are obtained

Table 1. Conditions which induce macrophage activation in vivo and in vitro

In vivo Intraperitoneal treatment of the mice (i. e., eliciting) with the following agents	In vitro Exposure of the cells to the following agents added to the culture media
Thioglycollate medium	Erythrocytes (± opsonins)
Proteose peptone	Zymosan
Sodium caseinate	Bacteria
Serum	Complement components
Endotoxin	Phorbol myristate acetate
Strepto. A cell walls,	Concanavalin A
BCG and other microorganisms	Redox compounds
or fractions therefrom	Lymphokines

Table 2. Enzyme activities described in human mononuclear phagocytes and determined by biochemical methods

Enzyme	References	
	Monocytes	Alveolar macrophages
Myeloperoxidase	23	
β-Glucuronidase	14, 31, 50	22, 25, 30, 40
N-Ac-glucosaminidase	14, 31, 50	40
Acid phosphatase	31	22, 30, 40
Cathepsin D		19, 30
Plasminogen activator	15, 17	
Elastase	15	9, 19, 22, 38
Angiotensin converting enzyme	13	20
5'-Nucleotidase	23, 31	
Lysozyme	16, 23, 31	19, 30

from the blood by differential fractionation of the white cell layer. Although they are their recognized precursors, these cells differ significantly from the macrophages since they have two types of granules, one of which is characteristically peroxidase positive [33]. During phagocytosis the granules are discharged and are not regenerated. This explains why the macrophages are granule free. Alveolar macrophages have the appearance of highly activated cells, since they are large and contain a great number of digestive vacuoles [32].

Enzymes which have been reported to occur in human mononuclear phagocytes are listed in Table 2. With the exception of the angiotensin-converting enzyme, which hydrolyzes angiotensin I to the vasoactive angiotensin II, all enzyme activities listed have also been described to occur in mouse peritoneal macrophages [2, 7].

Limited information is available on biochemical changes induced in human mononuclear cells by compounds and treatments which stimulate mouse cells. As shown in Table 3, however, human mononuclear phagocytes appear to respond to stimulation in similar fashion as those from the mouse. Indeed, zymosan, immune complexes, products of mitogen-stimulated lymphocytes, and phorbol myristate acetate (PMA) induce an increased production and release of lysosomal glycosidases and plasminogen activator as well as an enhanced hexose monophosphate shunt (HMPS) activity and superoxide production both in human monocytes and alveolar macrophages.

Table 3. Biochemical changes induced by in vitro stimulation in human and murine mononuclear phagocytes

Parameter assayed	Stimulus	References		
		Human monocytes	Human alveolar macrophages	Mouse peritoneal macrophages
Lysosomal enzyme release	None	14		41
	Zymosan	14, 21		8, 42, 44
	Immune complexes		25	3
Plasminogen activator	Lymphokine	18		27, 46
HMPS activity	Zymosan	35, 39	35	42
	Lymphokine	37		29
	PMA		21	43
Superoxide production	Zymosan	24, 36		10, 24
	PMA	24	21	24

27

Table 4. Biochemical changes in human mononuclear phagocytes in inflammatory diseases and smoking

Enzymes or function	Changes [References]
Superoxide production	Increased in monocytes of psoriasis patients [28]
	Increased in alveolar macrophages of smokers [19, 25]
Acid hydrolases formation and/or release	Increased in monocytes in autoimmune diseases [14]
	Increased in alveolar macrophages of smokers [25, 30]
Neutral proteinases	Increased formation and release of elastase in smokers [19, 38]
	Increased levels of angiotensin-converting enzyme in smokers and sarcoidosis patients [20]

The data obtained in the mouse system show clearly that similar degrees of activation can be induced by stimulation in vitro (e.g., by phagocytosis) or by eliciting in vivo. Examples of in vivo activation in man are encountered in disease or in cases of over-exposure to irritants. As presented in Table 4, superoxide production and the synthesis and release of acid hydrolases and neutral proteinases are enhanced in inflammatory conditions, e.g., psoriasis, autoimmune diseases, and sarcoidosis. Smoking comes very close to the eliciting procedures used in animals, and actually macrophages obtained from smokers by bronchial lavage are probably the best example of human macrophages activated in vivo (Table 4).

4. Conclusions

Most of our knowledge about the biochemical aspects of macrophage activation stems from experiments which were performed either in vivo or in vitro with mouse peritoneal macrophages. A comparison of the large amount of data available on the mouse cell with a much more limited information on human mononuclear cells strongly suggests that the biochemical properties of the cells of the two systems are similar and that the concepts about the differentiation of quiescent into inflammatory macrophages which have been put forward in the mouse system are likely to be valid for the human cells as well.

Acknowledgment

I thank Prof. Marco Baggiolini for his help in preparing the manuscript.

References

1. Baggiolini M (1980) The neutrophil. In: Weissmann G (ed) The cell biology of inflammation. Elsevier/North-Holland Biomedical Press, Amsterdam (Handbook of inflammation, vol II, p 163)
2. Baggiolini M (to be published) Lysosomal hydrolases. In: Escobar M, Reichard S, Friedman H (eds) The reticuloendothelial system: a comprehensive treatise. Plenum, New York
3. Cardella CJ, Davies P, Allison AC (1974) Immune complexes induce selective release of lysosomal hydrolases from macrophages. Nature 247:46
4. Cohn ZA (1978) The activation of mononuclear phagocytes: fact, fancy and future. J Immunol 121:813
5. Cohn ZA, Wiener E (1963) The particulate hydrolases of macrophages. I. Comparative enzymology, isolation, and properties. J Exp Med 118:991

6. Cohn ZA, Wiener E (1963) The particulate hydrolases of macrophages. II. Biochemical and morphological response to particle ingestion. J Exp Med 118:1009
7. Davies P, Bonney RJ (1980) The secretion of hydrolytic enzymes of mononuclear phagocytes. In: Weissmann G (ed) The cell biology of inflammation. Elsevier/North-Holland Biomedical Press, Amsterdam (Handbook of inflammation, vol II, p 497)
8. Davies P, Page RC, Allison AC (1974) Changes in cellular enzyme levels and extracellular release of lysosomal acid hydrolases in macrophages exposed to Group A streptococcal cell wall substance. J Exp Med 139:1262
9. De Cremoux M, Hornebeck W, Jaurand MC, Bignon J, Robert L (1978) Partial characterization of an elastase-like enzyme secreted by human and monkey alveolar macrophages. J Pathol 125:171
10. Drath DB, Karnovsky ML (1975) Superoxide production by phagocytic leukocytes. J Exp Med 141:257
11. Edelson PJ, Cohn ZA (1976) 5'-Nucleotidase activity of mouse peritoneal macrophages. I. Synthesis and degradation in resident and inflammatory populations. J Exp Med 144:1581
12. Edelson PJ, Erbs C (1978) Plasma membrane localisation and metabolism of alkaline phosphodiesterase I in mouse peritoneal macrophages. J Exp Med 147:77
13. Friedland J, Setton C, Silverstein E (1978) Induction of angiotensin converting enzyme in human monocytes in culture. Biochem Biophys Res Commun 83:843
14. Ganguly NK, Kingham JGC, Lloyd B, Lloyd RS, Price CP, Triger DR, Wright R (1978) Acid hydrolases in monocytes from patients with inflammatory bowel disease, chronic liver disease, and rheumatoid arthritis. Lancet 2:1073
15. Godoshian Ragsdale C, Arend WP (1979) Neutral protease secretion by human monocytes. Effect of surface-bound immune complexes. J Exp Med 149:954
16. Gordon S, Todd J, Cohn ZA (1974) In vitro synthesis and secretion of lysozyme by mononuclear phagocytes. J Exp Med 139:1228
17. Gordon S, Unkeless JC, Cohn ZA (1974) Induction of macrophage plasminogen activator by endotoxin stimulation and phagocytosis. Evidence for a two-stage process. J Exp Med 140:995
18. Greineder DK, Connorton KJ, David JR (1979) Plasminogen activator production by human monocytes. I. Enhancement by activated lymphocytes and lymphocyte products. J Immunol 123:2808
19. Harris JO, Olsen GW, Castle JR, Maloney AS (1975) Comparison of proteolytic enzyme activity in pulmonary alveolar macrophages and blood leukocytes in smokers and nonsmokers. Am Rev Respir Dis 111:579
20. Hinman LM, Stevens C, Matthay RA, Gee JBL (1979) Angiotensin convertase activities in human alveolar macrophages: Effects of cigarette smoking and sarcoidosis. Science 205:202
21. Hoidal JR, Fox RB, Repine JE (1979) Defective oxidative metabolic responses in vitro of alveolar macrophages in chronic granulomatous disease. Am Rev Respir Dis 120:613
22. Janoff A, Rosenberg R, Gladston M (1971) Elastase-like, esteroprotease activity in human and rabbit alveolar macrophage granules. Proc Soc Exp Biol Med 136:1054
23. Johnson WD Jr, Mei B, Cohn ZA (1977) The separation, long-term cultivation, and maturation of the human monocyte. J Exp Med 146:1613
24. Johnston RB Jr, Godzik CA, Cohn ZA (1978) Increased superoxide anion production by immunologically activated and chemically elicited macrophages. J Exp Med 148:115
25. Joseph M, Tonnel AB, Capron A, Voisin C (1980) Enzyme release and superoxide anion production by human alveolar macrophages stimulated with immunoglobulin E. Clin Exp Immunol 40:416
26. Karnovsky ML, Lazdins JK (1978) Biochemical criteria for activated macrophages. J Immunol 131:809
27. Klimetzek V, Sorg C (1977) Lymphokine-induced secretion of plasminogen activator by murine macrophages. Eur J Immunol 7:185
28. Krueger GG, Jederberg WW, Ogden BE, Reese DL (1978) Inflammatory and immune cell function in psoriasis. II. Monocyte function, lymphokine production. J Invest Dermatol 71:195
29. Lazdins JK, Kühner AL, David JR, Karnovsky ML (1978) Alteration of some functional and metabolic characteristics of resident mouse peritoneal macrophages by lymphocyte mediators. J Exp Med 148:746
30. Martin RR (1973) Altered morphology and increased acid hydrolase content of pulmonary macrophages from cigarette smokers. Am Rev Respir Dis 107:596
31. Musson RA, Shafran H, Henson PM (1980) Intracellular levels and stimulated release of lysosomal enzymes from human peripheral blood monocytes and monocyte-derived macrophages. J Reticuloendothel Soc 28:249

29

32. Nichols BA (1976) Normal rabbit alveolar macrophages. I. The phagocytosis of tubular myelin. J. Exp Med 144:906
33. Nichols BA, Bainton DF (1973) Differentiation of human monocytes in bone marrow and blood: Sequential formation of two granule populations. Lab Invest 29:27
34. North RJ (1978) The concept of the activated macrophage. J Immunol 121:806
35. Papermaster-Bender G, Whitcomb ME, Sagone AL Jr (1980) Characterization of the metabolic responses of the human pulmonary alveolar macrophage. J Reticuloendothel Soc 28:129
36. Reiss M, Roos D (1978) Differences in oxygen metabolism of phagocytosing monocytes and neutrophils. J Clin Invest 61:480
37. Rocklin RE, Winston CT, David JR (1974) Activation of human blood monocytes by products of sensitized lymphocytes. J Clin Invest 53:559
38. Rodriguez RJ, White RR, Senior RM, Levine EA (1977) Elastase release from human alveolar macrophages: Comparison between smokers and nonsmokers. Science 198:313
39. Sagone AL Jr, King GW, Metz EN (1976) A comparison of the metabolic response to phagocytosis in human granulocytes and monocytes. J Clin Invest 57:1352
40. Scharfman A, Houdret N, Roussel P, Biserte G (1978) Action du corynebacterium parvum sur les activités de glycosidases et de protéases de macrophages péritonéaux de souris. C R Acad Sci [D] (Paris) 287:317
41. Schnyder J, Baggiolini M (1978) Secretion of lysosomal hydrolases by stimulated and nonstimulated macrophages. J Exp Med 148:435
42. Schnyder J, Baggiolini M (1978) Role of phagocytosis in the activation of macrophages. J Exp Med 148:1449
43. Schnyder J, Baggiolini M (1980) Induction of plasminogen activator secretion in macrophages by electrochemical stimulation of the hexose monophosphate shunt with methylene blue. Proc Natl Acad Sci USA 77:414
44. Schorlemmer HU, Edwards JH, Davies P, Allison AC (1977) Macrophage responses to mouldy hay dust, Micropolyspora faeni and zymosan, activators of complement by the alternative pathway. Clin Exp Immunol 27:198
45. Unkeless JC, Gordon S, Reich E (1974) Secretion of plasminogen activator by stimulated macrophages. J Exp Med 139:834
46. Vassalli JD, Reich E (1977) Macrophage plasminogen activator: Induction by products of activated lymphoid cells. J Exp Med 145:429
47. Weissmann G, Dukor P, Zurier RB (1971) Effects of cyclic AMP on release of lysosomal enzymes from phagocytes. Nature [New Biol] 231:131
48. Werb Z, Gordon S (1975) Secretion of a specific collagenase by stimulated macrophages. J Exp Med 142:346
49. Werb Z, Gordon S (1975) Elastase secretion by stimulated macrophages. Characterization and regulation. J Exp Med 142:361
50. Yatziv S, Epstein LB, Epstein CJ (1978) Monocyte-derived macrophages: an in vitro system for studying hereditary lysosomal storage diseases. Pediat Res 12:939

Haematology and Blood Transfusion Vol 27
Disorders of the Monocyte Macrophage System
Edited by F. Schmalzl, D. Huhn, H.E. Schaefer
© Springer-Verlag Berlin Heidelberg New York 1981

Macrophages in the Regulation of Immunity

C. Huber and G. Stingl

1. Abstract

According to the classical view macrophages are prone to ingest and to destruct antigens and intracellular pathogens. More recently it became apparent that macrophages, besides their phagocytic capacities, are involved in the activation and regulation of lymphocyte functions. This article aims to summarize some recent findings on the role of macrophages during the afferent phase of immune responses.

2. Introduction

Macrophages (MØ) represent the first population of mononuclear blood cells to which well-defined functional capacities were ascribed. According to the classical view originally proposed by Metschnikoff MØ are prone to ingest and to destruct antigens or intracellular pathogens. More recently it has been shown that MØ – besides their phagocytic capacities – are crucially involved in the activation and regulation of lymphocyte functions. The aim of this article is to briefly summarize recent findings on the role of MØ during the afferent phase of immune responses (IR). The points on which we would specifically like to focus concern [1] the role of MØ in the presentation of antigens to immunocompetent lymphocytes and the genetic control of this cooperation, [2] the role of MØ in determining the balance between different populations of regulatory T cells, and [3] the heterogeneity of the MØ system with particular reference to the Langerhans cells of the skin. Because of the topic of this meeting we shall largely restrict the presentation of data to experiments performed with human cells.

3. The Role of MØ as an Antigen Presenting Cell in Antibody-Mediated IR

Findings in experimental animal systems demonstrated that most antigens, if recognized by purified B-lymphocytes, are not capable of inducing a strong production of IgG antibodies. In order to elicit optimal IgG antibody production the respective antigens have to be recognized by a specialized subset of T lymphocytes. These helper T cells (T_H) will then deliver an antigen-specific signal which is required by the B cell for induction of IgG antibody production. Details on the fine specificity of this T–B cell cooperation and its genetic control are given in reference [3]. More recently, however, it appeared that the interaction of T-dependent antigens with B and T_H cells alone does not lead to an optimal IgG antibody

[1] This work was financially supported by the austrian funds "Zur Förderung der wissenschaftlichen Forschung", projects nos. 3417, 3859 and 3791.

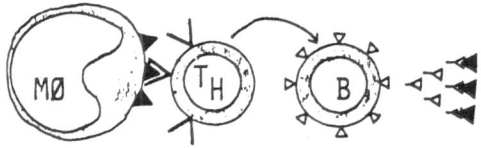

▲ antigen

∧ receptor

△ ᵃ antibody

Fig. 1. Schematic presentation of cooperation of MØ and helper T cells (T_H) in response to T cell dependent antigens

production. MØ are required to present the antigen in a suitable form to T_H cells. Details on the specificity and the genetic control of this MØ-T cell cooperation will be discussed in Sect. 5 (see also Sect. 9). A schematic presentation illustrating the role of the MØ during the recognitive phase of an IgG antibody mediated immune response is given in Fig. 1.

In experimental animal systems co-operation between MØ and T_H cells has been shown in response to a variety of soluble protein antigens. Such antigens include synthetic amino acid polymers and polypeptide hormones as well as xenogeneic antibodies. In any case cooperation can be demonstrated by measurement of specific antibody titers or by evaluation of antigen-specific proliferation of T_H cells. An example of the induction of proliferation of antigen-specific T_H cell proliferation with antigen-pulsed human MØ is shown in Fig. 2. In this experiment purified T cells were sensitized to rabbit IgG (RIgG) pulsed MØ in a primary culture. Secondary proliferative responses were subsequently induced by restimulation of sensitized T cells with antigen-pulsed or native MØ. As demonstrated, secondary proliferative responses with peaks on day 2 can only be induced with autologous antigen pulsed MØ. Moreover, autologous antigen pulsed MØ which have been depleted of HLA-DR⁺ cells were no longer capable of presenting this antigen to T_H cells in an

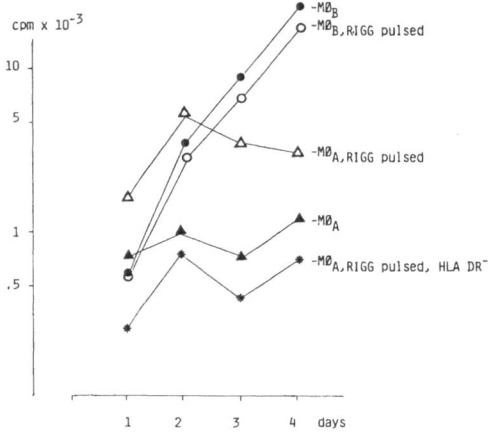

Fig. 2. Secondary proliferative responses of human T_H cells with specificity for rabbit gammaglobulin (RIgG) is restricted for the expression of self-HLA DR on the level of the antigen-presenting MØ. Unsensitized T cells were isolated from human peripheral blood by E rosetting and were incubated in primary cultures with 10% mitomycin-C treated and rabbit gammaglobulin pulsed MØ. MØ were obtained by surface adherence. Subsequently they were pulsed by a 60 min incubation at 37° with RIgG. After 12 days of culture memory T cells were restimulated with autologous *(A)* or allogeneic *(B)* MØ which have been exposed or not exposed to RIgG. Individual A and B are HLA nonidentical. Furthermore, MØ were depleted from HLA-DR⁺ cells by means of complement-dependent lysis with a xenoantibody with specificity for common determinants expresses on this molecule. Proliferative responses of 2.5×10^3 responder T cells after a 12-h ³H-thymidine pulse (2 µCi/cell, specific activity mCi/mmol) are shown

appropriate manner. Allogeneic MØ with and without RIgG, in contrast, did not induce secondary proliferative responses with an early peak on day 2 but primary responses with delayed peaks directed against MØ associated alloantigens.

In conclusion, B lymphocytes require the help of T cells for production of IgG antibodies. These T_H cells are primarily triggered by antigen presented on the surface of MØ in association with certain autologous surface structures coded for by the major histocompatibility complex (MHC). The nature of these structures and their genetic control will be discussed in Sect. 5.

4. The Role of MØ as an Antigen Presenting Cell in Delayed Hypersensitivity Reactions (DLH)

In analogy to humoral IR, MØ were shown to play a similar role in the induction of DLH reactions. Certain antigens such as simple chemical haptens or PPD do not induce strong humoral antibody production but induce DLH reactions. These antigens are primarily recognized by T cells if presented on the surface of MØ. Details of the experimental animal data are found in reference [10]. The antigens presenting to T cells are exclusively recognized when associated on the surface of the antigen presenting cell with certain autologous surface structures coded for by the MHC. An example of the genetically restricted response of human T cells to trinitrophenyl (TNP) haptenized MØ is shown in Fig. 3. Here human T cells were sensitized to autologous TNP-modified MØ in a primary culture. Sensitized T cells were subsequently restimulated with autologous or allogeneic TNP treated MØ which differed in the degree of histocompatibility. As shown, antigen specific secondary proliferative responses can only be induced by autologous or histocompatible MØ, whereas HLA nonidentical MØ failed to do so. In conclusion, cells responding in DLH reactions preferentially recognize antigens which are presented in association with certain autologous surface structures coded for by the MHC. The genetic control of this interaction will be discussed in Sect. 5.

5. The Role of MØ as an Accessory Cell in Lectin-Induced T Cell Proliferations

A broad panel of lectins are known to represent polyclonal lymphocyte activators. Among these certain lectins such as phytohaemagglutinine (PHA) or Concanavaline A (Con A) were found to almost selectively induce T cell proliferation. When, however, T cells were highly depleted of MØ by passage through columns of nylon wool or Sephadex G 10, these purified T cells were no longer capable of proliferating in the presence of these lectins [4]. A

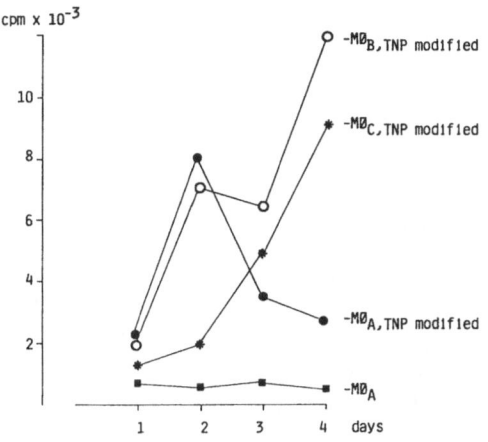

cpm x 10^{-3}

Fig. 3. Secondary proliferative responses of human T cells to hapten-modified MØ are restricted for the expression of self HLA on the level of the haptenized MØ. Unsensitized T cells were isolated as described in Fig. 2. They were co-cultivated with identical numbers of TNP-modified autologous MØ. For details of the modification procedure see reference [5]. After 10 days of culture sensitized cells were restimulated with autologous *A* or allogeneic *B, C* TNP modified or native MØ. Individuals *A* and *B* are haplotype identical sibling children, individual *C* is HLA nonidentical. Proliferative responses were evaluated as described in Fig. 2

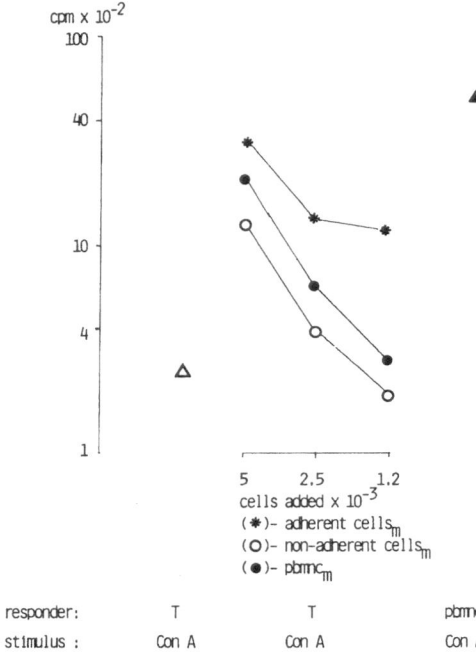

cpm x 10^{-2}

100

40

10

4

1

5   2,5   1.2
cells added x 10^{-3}
(*)- adherent cells$_m$
(O)- non-adherent cells$_m$
(●)- pbmc$_m$

responder: T   T   pbmc
stimulus :   Con A   Con A   Con A

Fig. 4. MØ are accessory cells for Con A induced T cell proliferation. T cells were purified as described in Fig. 2 and were further freed of MØ by passage through a Sephadex G 10 column. Peripheral blood mononuclear cells or MØ depleted T cells were then assessed for their proliferative capacity in the presence of 12 µg/ml Con A in the absence or presence of mitomycin-C treated autologous peripheral blood mononuclear cells, non-adherent or adherent cells. Proliferation was assessed as described in Fig. 2. Results on day 4 are shown

representative experiment with human T cells is depicted in Fig. 4. As shown, proliferative capacity of MØ-depleted T cells could be restored by mitomycin-C treated autologous MØ. It should be mentioned that different lectins differ from each other with respect to their MØ dependency. Whereas Con A responses, particularly when induced by suboptimal amounts of lectin, are highly MØ dependent, PHA responses are much more independent. In addition this cooperation of MØ and T cells in response to lectin stimuli is not restricted by the origin of the MØ in mice. It is, however, dependent on the presence of MØ expressing Ia molecules [1]. A similar requirement for the HLA-DR$^+$ cells is also seen in man and is demonstrated in the experiment shown in Fig. 5. Whereas the response to Con A is markedly diminished after removal of HLA-DR bearing accessory cells, PHA responses and alloantigen-induced mixed leukocyte cultures (MCL) were only marginally affected.

responder	pretreatment	stimulator	cpm (day 5)	deviation (%)
A	PS	0	257	
		A$_m$	311	
		U 255$_m$	10 225	
		PHA	92 201	
		Con A	10 506	
A	PS plus RC˙	0	836	
		A$_m$	1 267	
		U 255$_m$	12 837	plus 25
		PHA	114 440	plus 24
		Con A	10 176	minus 4
A	RAHLA-DR plus RC˙	0	642	
		A$_m$	1 137	
		U 255$_m$	12 837	plus 38
		PHA	78 008	minus 15
		Con A	3 148	minus 71
A	0	0	287	
0	0	A$_m$	105	

Fig. 5. Accessory cells for Con A induced T cell proliferation express HLA-DR molecules. Peripheral blood mononuclear cells were freed of viable HLA-DR bearing cells by pretreatment with anti-HLA-DR serum and rabbit complement. For details see legend to Fig. 2. Cells were subsequently analyzed for their capacity to proliferate in the presence of autologous (A$_m$) or allogeneic (U 255$_m$) stimulator cells or in the presence of optimal amounts of the lectins PHA and Con A. Proliferation was evaluated as described in Fig. 2. Results on day 5 are demonstrated

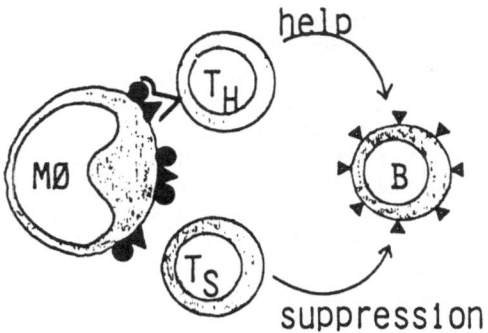

 receptor

▲ antigen

● HLA-DR

Fig. 6. Schematic presentation of the associative recognition of antigen and self surface structures on the level of the antigen presenting MØ

6. The Genetic Control of the Co-operation Between Antigen-Presenting MØ and Responding T Cells

Recent findings of immunogenetic and biochemical analyses have unravelled the functional role of a large group of surface glycoproteins (GP) coded for by the H_2 region of the mouse or the HLA region in man. For review see references [2, 6]. Polymorphic genes located within this region code for two different groups of structurally and functionally different surface glycoproteins. One type of GP exhibits an apparent molecular weight of 44 000 daltons and is expressed in association with b2 microglobulin on almost every cell.

These structures are the main target antigens for allogeneic killer cells and for killer cells with specificity for virally or chemically modified self antigens. Serologically they correspond to the HLA A and B antigens in man and the K and D antigens in mouse. The other type of GP represents a two chain molecule with apparent molecular weight of 34 000 and 28 000 daltons. Serologically they correspond to the HLA-DR antigens in man and the I region associated antigens (Ia antigens) in the mouse. These antigens are selectively expressed on certain MØ and B cells. Evidence in congenic mouse strains and recent findings in man suggest that HLA-DR or Ia antigens are identical to those structures on the surface of MØ which control co-operation with T cells. Results already discussed in Fig. 2, 3, and 5 illustrate these conclusions.

It thus appears that T cell dependent antigens or antigens inducing DLH reactions are preferentially recognized by T cells in association with HLA-DR or Ia molecules on the surface of the antigen presenting MØ. A schematic presentation of such an associative recognition of antigens with self structures expressed on the surface on the antigen-presenting MØ is given in Fig. 6. The practical implication of this experimentally well supported concept is that genetic control of immune responses takes place at the level of the MØ. This should be kept in mind when discussing disorders of the monocyte–macrophage system.

7. The Role of MØ in Determining the Balance Between Different Populations of Regulatory T Cells

Recent results in experimental animal systems obtained by Barjuch, Benaceraff, and co-workers suggested another important aspect of the role of MØ. These authors found that IgG antibody responses to certain synthetic polymers of L amino acids are under genetic control of a single dominant gene (for review see Sect. 3). Upon immunizations with such polypeptides which present the immune system with antigenic challenges of limited heterogeneity, random bred animals as well as inbred strains of experimental animals distributed themselves into easily separable "responder" and "nonresponder" animals. Further studies revealed that the failure of nonresponder strains to produce IgG antibodies is not due to the lack of responsive immunocompetent cells but due to an excessive induction of suppressor (T_S) cells. Moreover, in responder strains both T_H and T_S cells were induced, but the extent of T_H cell activation was strong enough to overcome suppression mediated by the simultaneously activated T_S system [3]. Further studies have concentrated on the role of the antigen-presenting MØ in determining the balance between T_H and T_S cells in such "responder" animals. When soluble amino acid polymers were presented to MØ–depleted spleen cell suspensions of responder animals strong induction of antigen-specific T_S cells was observed. These findings, which are schematically depicted in Fig. 6, indicated that antigen presentation by MØ is critical to the suppressor-helper T cell balance [3].

8. Characterization of the Antigen-presenting Cell

After having discussed the crucial role of macrophages in the afferent limb of the immune response, it appears germane to ask whether the capacity to initiate T cell dependent immune reactions is a functional property of all macrophages or resides only in a particular subset. Macrophages belong to the mononuclear phagocyte system. This system encompasses several cell types whose common denominator is the capacity to firmly adhere to glass surfaces and avid phagocytosis [11]. Mainly from experiments in the guinea pig and mouse system we learned that not all cells which fulfil these criteria necessarily participate in the initiation of the immune response.

Yasmashita and Shevach [13] provided evidence that virtually all guinea pig peritoneal exudate macrophages avidly ingest latex particles and firmly attach to glass surfaces but that only a rather small subpopulation acts as potent stimulators of antigen-specific and allogeneic T cell activation. In contrast to other peritoneal exudate macrophages, this subpopulation bears Ia antigens. The critical role of Ia antigen expression at the level of the antigen-presenting cell has since then been convincingly demonstrated for several other species, including man, and from the data listed above it seems evident that the Ia type of the antigen-presenting cell determines the genetic restriction of the immune response. It is of interest that there exist specific cell types which are particularly potent accessory cells in several antigen-dependent T cell responses in vitro. These include murine dendritic cells which constitute a small but distinct subpopulation of mononuclear cells in several lymphoid organs of mice [7] and epidermal Langerhans cells which represent a small subset of mammalian epidermal cells [12]. Inspite of certain differences between these two cell populations they have several important features in common. Both cell types derive from the bone marrow, are highly dendritic in shape, synthesize Ia antigens, and act as powerful stimulators of antigen-specific and allogeneic T cell activation [8, 9]. Interestingly, however, their endocytotic capacities are rather restricted, and Langerhans cells, in particular, only poorly adhere to glass surfaces. Inspite of certain similarities to cells from the mononuclear phagocyte system, murine dendritic cells and Langerhans cells do therefore not qualify as typical "macrophages."

Although most investigators would agree that macrophages (or mononuclear phagocytes) and antigen-presenting cells do represent overlapping cell populations, we feel that those cells which are critical in the induction or sensitization phase of the immune response should not be simply referred to as a "macrophage" until their lineage has been clearly defined.

References

1. Ahmann GB, Sachs DH, Hodes RS (1978) Requirement for an Ia bearing accessory cell in Con A induced T cell proliferation. J Immunol 121:1981

2. Barnstable CS, Jones EA (1978) Isolation structure and genetics of HLA-A, -B, -C and -DRw (Ia) antigens. Br Med Bull 34:241
3. Benacerraf B, Germain RN (1978) The immune response genes of the major histocompatibility complex. Immunol Rev 38:70
4. Rosenstreich DL, Farrar JJ, Dougherty S (1976) Absolute macrophage dependency of T lymphocyte activation by mitogens. J Immunol 116:131
5. Shearer GM, Rehn TG, Garbarino CA (1975) Cell mediated lympholysis of trinitrophenyl-modified autologous lymphocytes. Effector cell specificity of modified cell surface components controlled by the H2K and H2D serological regions of the murine major histocompatibility complex. J Exp Med 141:1427
6. Snell GD (1978) T cells, T cell recognition structures and the major histocompatibility complex. Immunol Rev 38:3
7. Steinman RM, Cohn ZA (1973) Identification of a novel cell type in peripheral lymphoid organs of mice. I. Morphology, quantitation, tissue distribution. J Exp Med 137:1142
8. Steinman RM, Witmer MD, Nussenzweig MC, Chen LL, Schlesinger S, Cohn ZA (1980) dendritic cells of the mouse: identification and characterization. J Invest Dermatol 75:14
9. Stingl G (1980) New aspects of Langerhans cell function. Int J Dermatol 19:189
10. Thomas DW, Yamashita U, Shevach EM (1977) The role of Ia antigens in T cell activation. Immunol Rev 35:97
11. Van Furth R, Cohn ZA, Hirsch JG, Humphrey JH, Spector WG, Langevoort HL (1972) The mononuclear phagocyte system: a new classification of macrophages, monocytes and their precursor cells. Bull WHO 46:845
12. Wolff K (1972) The Langerhans cell. Curr Probl Dermatol 4:79
13. Yamashita U, Shevach EM (1977) The expression of Ia antigens on immunocompetent cells in the guinea pig. II. Ia antigens on macrophages. J Immunol 119:1584

Haematology and Blood Transfusion Vol 27
Disorders of the Monocyte Macrophage System
Edited by F. Schmalzl, D. Huhn, H.E. Schaefer
© Springer-Verlag Berlin Heidelberg New York 1981

The Role of Macrophages as Effector Cells

H. Huber, M. Ledochowski and G. Michlmayr

Mononuclear phagocytes (MP) are commonly found in chronic inflammatory reactions and often represent the most important effector cell in these areas. This review will discuss the following topics:
1. Factors that are capable of activating mononuclear phagocytes.
2. Some effector mechanisms of mononuclear phagocytes.
3. The pathophysiologic functions of MP in inflammatory reactions.

1. Activation of Mononuclear Phagocytes

Many functional activities of MP are fully expressed only after prior activation. Different stimuli are able to activate these cells and they may affect various effector cell functions to a different degree [2, 11]. Mouse peritoneal macrophages, activated by thioglycolate injections, have in particular been studied extensively. Some of these data, comparing activated macrophages with resting MP [11], are summarized in Table 1.

„Inflammatory" MP are characterized, e.g., by their increase in size, spreading, phagocytic potential, superoxide production, and particularly by their pronounced secretory activity (Table 1). MP activated by complement (C)-derived factors differ in some of their effector functions (e.g., after activation of MP by C3b, MP liberate lysosomal enzymes in large quantities, but a stimulation of plasminogen activator secretion was not observed under these conditions).

Immune complexes, C components, or mediators produced by sensitized lymphocytes are considered as activating stimuli of particular clinical importance [2]. Some features of these activating factors are described and their significance discussed.

Table 1. Properties of resident and inflammatory macrophages obtained from the peritoneal cavity [11]

	Resident	Inflammatory	Ratio I/R
Cell protein ($\mu g/10^6$ cells)	80	130	1.6
Spreading (% cells/h)	4	95	24
Fluid phase pinocytosis ($nl/10^6/h$)	46	247	5.3
Phagocytosis			
E-IgG	600	1600	2.7
E-IgM-C	40	1000	25
Superoxide anion (nmole/mg protein/90 min)	45	520	12
Plasminogen activator (U/mg protein)	1	800	800
Collagenase $U/10^7$ cells	1	15	15
Elastase $U/10^7$ cells	1.8	68	38

Table 2. Fc receptor on mononuclear phagocytes

Specific binding of immune complexes containing IgG (particularly IgG$_1$ and IgG$_3$) [28]

Membrane binding of immune complexes followed by:
1. phagocytosis ("zipper mechanism" [23])
2. increased formation of oxygen metabolites; intracellular killing requires interaction of the Fc-receptor with fluid phase IgG [36]
3. enzyme secretion (e. g., lysosomal enzymes [9])

1.1 Activation of MP by Immune Complexes

Immune complexes have the capacity to activate MP. Under certain in vitro conditions this activation was observed without the addition of C components [9, 36]. One of the prerequisites for the interaction of immune complexes with MP was the presence of Fc receptors on the surface membrane of MP [1, 25, 37].

A few hundred IgG molecules bound to an antigen can already cause the attachment of immune complexes to human monocytes [25]. On mouse macrophages the binding capacity increases when they are activated [6]. Fc receptors mediating the binding of antigen-IgG-antibody complexes are distributed randomly over the whole surface of MP. They are mobile in the membranes' liquid phase and are continuously resynthesized [22]. The IgG molecules in humans that are preferentially bound are Ig-G1 and Ig-G3 [1, 16, 25, 28]. Fc receptors and C3 receptors can evoke a synergistic effect (Fig. 1).

Effector cell function of MP initiated by binding of immune complexes to the Fc receptor include phagocytosis (see Sect. 2.1) and the secretion of various lysosomal enzymes. Under certain conditions the bactericidal activity of MP required a membrane activating stage mediated by the Fc and/or C3 receptor [36]. A summary of these effector cell functions in relation to the Fc receptor are listed in Table 2.

Using immune (ferritin-antiferritin) complexes a dose-dependent activation of MP in respect to the secretion of lysosomal enzymes has been documented [9]. The role of the Fc receptor in respect to phagocytosis of antigen-IgG-antibody complexes will be discussed more extensively in Sect. 2.1.

1.2 The Activation of MP by Complement Components

The role of C components as stimuli initiating the activation of MP has been widely discussed recently [2, 45, Schorlemmer, this volume p. 59]. MP have a surface membrane receptor for the third component of complement [27, 35]. This was first shown with erythrocytes-antibody-complement complexes. There were only about 100 molecules of C3b necessary to bind these complexes to human monocytes [25]. C3b formed by the alternative pathway also elicited an efficient binding to macrophages. These experiments indicate that the membrane of MP functions as a very effective receptor for activated C3 (Fig. 1).

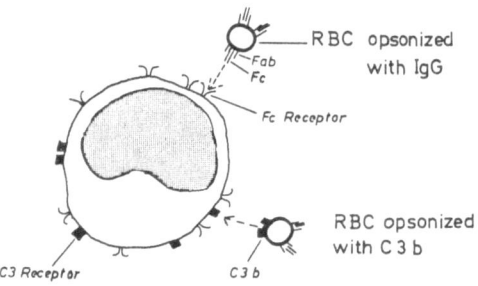

R BC opsonized with IgG

Fab
Fc

Fc Receptor

RBC opsonized with C 3 b

C3 Receptor

C3 b

Fig. 1. Activation for MP by immune complexes and by complement components

1.3 The Activation of MP by Lymphocyte-derived Products

After contact with the corresponding antigen, sensitized lymphocytes may liberate several products which affect the functions of MP. These substances immobilize these cells at the site of inflammation and activate them for their participation in cellular immune reactions [5, 13, 42, this volume p. 31]. At present it would seem that these lymphocyte-derived products (migration inhibition factor, macrophage activating factor and others) are the most important stimuli for the activation of macrophages. These "lymphokines" are liberated by sensitized lymphocytes in the presence of specific antigens, by unsensitized lymphocytes after they have been incubated with different mitogens, or in mixed lymphocyte cultures [39]. T lymphocytes and, at least under certain conditions, B-lymphocytes are the source of migration inhibition factor and perhaps of other lymphokines [10].

Migration inhibition factor (MIF) is a dialyzable macromolecule with a molecular weight of 23 000 daltons. In electrophoresis MIF migrated with the albumin fraction and is inactivated by trypsin and chymotrypsin [12, 42, 43].

The macrophage activating factor (MAF) is similar, though not identical with MIF. This factor is of particular interest, as it may play an important role in macrophage activation leading to cytotoxic effects against tumor-cell targets (Lohmann-Matthes, this volume p. 49).

2. Effector Mechanisms of MP

The main effector cell functions of MP are their phagocytic potential, their capacity for secretion of a variety of enzymes and other substances, and their microbicidal (and tumoricidal) activities.

2.1 Phagocytosis

Phagocytosis is a two-stage process (Fig. 2). In the first phase, the particle is attached to the surface membrane of the phagocyte ("attachment"). In the second phase, the particle is engulfed by the cell and digested by intracellular enzymes ("engulfment").

2.2 Opsonins and Phagocytosis

Opsonins facilitate the attachment phase of phagocytosis and in this way enhance particle engulfment. Opsonins have a thermostabile and a thermolabile component [32]. Most authors agree [32] that the thermostabile component is primarily IgG, which interacts with the phagocytes by the Fc receptor. The thermolabile component is considered to be equivalent to complement cleavage products (particularly C3b), interacting with the C3b receptor. The Fc receptors present on MP facilitate binding and phagocytosis of immune complexes containing IgG. After attachment to the macrophage, the surface membrane moves over the immune complex until it is covered (zipper mechanism 23). The C3b receptor also enhances the attachment of the immune complexes containing C3 but is not necessarily followed by actual engulfment [15, 27].

The Fc- and C3 receptors show a possible in vivo significance in respect to the clearance of immune complexes [17]. *Red cell-IgG* antibody complexes are primarily removed by the spleen, particularly if not C-binding. The splenic reservoir provides favorable conditions for Fc receptor formation (*Fc receptor mediated clearance*).

Red cell-IgM antibody complexes required the participation of C components for accelerated destruction and the pattern of clearance was distinctly different from that observed with red cell-IgG antibody complexes. Binding throughout the macrophage system (particularly to Kupffer cells) with marked return back of sensitized RBC into the circulation was observed. This finding indicated the transient attachment, particularly at lower

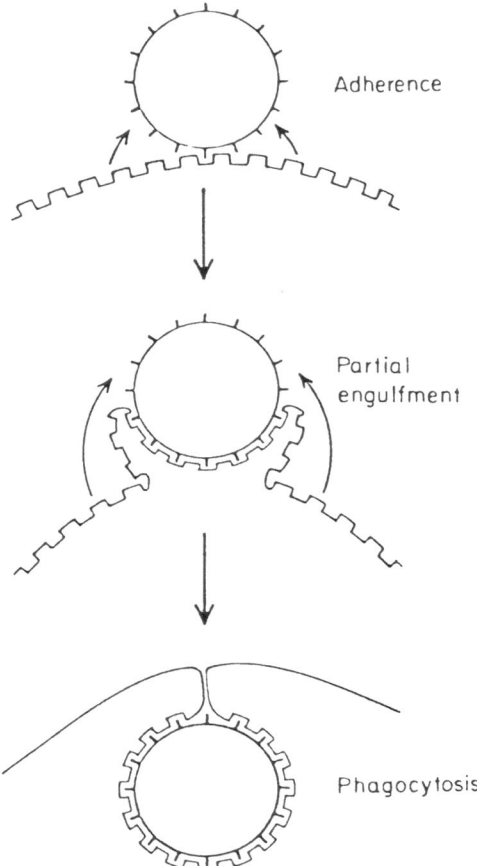

Adherence

Partial
engulfment

Phagocytosis

Fig. 2. The zipper model of phagocytosis [23]

antibody concentrations. IgM antibodies were only required for C activation *(C3 receptor mediated clearance)*. Red cell-IgG antibody complexes are removed more effectively in case of C-binding antibodies. Accelerated clearance required much lower antibody concentrations *(Fc + C3 receptor mediated clearance)*.

Under our experimental conditions an approximately tenfold higher C3b concentration on the immune complex was required to induce ingestion and not only attachment to human monocytes in vitro [27]. It was shown that complement-coated erythrocytes were ingested almost exclusively by activated macrophages, whereas resting macrophages were quite ineffective [11]. If both Fc and C3b receptors take part in phagocytosis, as is the case for antigen-IgG-C complexes, a highly efficient phagocytosis can be observed requiring low concentrations of IgG antibodies and activated C3 [15, 27]. Under these circumstances the two receptors were considered to have a co-operative effect [26, 27].

Table 3. Oxygen dependent antimicrobial systems [32]

---

1. Myeloperoxidase-mediated
2. Myeloperoxidase-independent
   - H_2O_2
   - Superoxide anion (O_2^-)
   - Hydroxyl radical (OH.)
   - Singlet oxygen ($^1O_2$)

42

2.3 Microbicidal Systems in Mononuclear Phagocytes

Mononuclear phagocytes bind and effectively destroy a variety of microorganisms. Some of them (e.g., mycobacteria, listeria, salmonella, brucella, and some viruses and protozoa) might require a prior activation stage of these cells. The killing of these and other microorganisms are mediated primarily by oxygen-dependent toxic mechanisms (see Table 3).

2.3.1 Oxygen-Dependent Systems

Phagocytosis is followed by a metabolic burst with an increase of oxygen utilization. The oxygen is used for the production of superoxide ($.O^-_2$), in order to produce hydrogen superoxide (H_2O_2), hydroxyl radicals (.OH), and other oxygen radicals (like $^1O_2$). The bactericidal effects of these metabolites have been examined on isolated granulocytes [32] and monocytes [29] and on macrophages [33].

Patients with chronic granulomatous disease not only showed bactericidal defects in granulocytes but also in monocytes [14]. In these patients the bactericidal dysfunction is due to a defect of the "metabolic burst" [32], and they lack the pronounced increase in oxygen consumption normally observed after phagocytosis.

The microbicidal properties of H_2O_2 are highly enhanced by the presence of myeloperoxidase [32]. The myeloperoxidase in human monocytes resembles in many ways the myeloperoxidase in human neutrophils. During maturation of MP, however, the content of peroxidase decreases markedly [van Furth, this volume p. 5]. As shown under several test conditions oxygen-dependent antimicrobial systems are enhanced by products of sensitized lymphocytes ("lymphokines"). This effect was particularly well documented in respect to toxoplasma gondii [40], and trypanosoma cruzi [41].

2.3.2 Oxygen-Independent Antimicrobial Mechanisms

Under certain conditions oxygen-independent cytotoxical reactions have been observed, but the mechanisms causing this cytotoxic effect require further investigations. Whereas oxygen-dependent killing is often a postphagocytic event, extracellular cytotoxic reactions are not always dependent upon the generation of O_2 metabolites [30, 47]. Very likely the cytotoxic effect of MP against some tumor cells is particularly mediated by such oxygen-independent mechanisms [Lohmann-Matthes, this volume p. 49].

2.4 Secretion of Enzymes

Mononuclear phagocytes in an activated stage function as secretory cells and liberate a variety of enzymes in large quantities. The secretory potential of activated macrophages is probably comparable to the secretory activity of hepatocytes [41]. Particularly neutral proteases play a major role in various inflammatory processes and may effectively destroy constituents of the connective tissues. The best documented enzymes secreted by activated macrophages in fairly large quantities are plasminogen activator, elastase, and collagenase [Schnyder, this volume p. 25].

2.4.1 Plasminogen Activator

Plasminogen activators transform the inactive precursor plasminogen into the fibrinolytically active plasmin. Besides the fibrinolytical activity, plasmin has the capacity to activate the complement (e.g., by cleaving C3) and the kinin system [32]. Plasmin therefore seems to be one of the major mediators in inflammatory reactions. This neutral protease is secreted in large amounts by activated macrophages [21], whereas the intracellular content of this enzyme is rather low [21]. An up to 800-fold increase of plasminogen activator secretion by activated macrophages has been reported [11]. The secretory capacity of MP, evaluated in terms of plasminogen activator liberation, therefore may serve as a marker for activated macrophages. Monocytes in culture medium also have the capacity to secret this enzyme [20].

2.4.2 Elastase

This neutral protease hydrolizes and solubilizes native elastinfibrils. Elastin is the major protein of elastic structures, e.g., large blood vessels, skin, and lung [2]. The amount of neutral proteases which is secreted by mononuclear phagocytes depends on the stage of their activation. Human monocytes in culture medium also secret elastase [49, 50]. Alpha-2-macroglobulin and alpha-1-antitrypsin are very effective inhibitors of this enzyme. In congenital deficiencies of alpha-1-antitrypsin the uncontrolled activity of elastase causes the well-known severe alterations of pulmonary tissues. The activity of elastase that is produced by mononuclear phagocytes seems to be more specific than the activity of "elastase-like proteases" produced by neutrophils [49, 50].

2.4.3 Collagenase

Collagenase is a neutral protease that catalyzes the hydrolysis of collagen. This enzyme seems to be effective in hydrolyzing type I and III collagen [24]. Some other neutral proteases secreted by mononuclear phagocytes contribute to the further cleavage of collagen fragments by cleaving them to lowmolecular peptides [49, 50]. The collagenase produced by macrophages seems to play an important role in the remodeling of tissues (e.g., in wound repair), chronic inflammatory reactions, and arthropathies. It is mainly secreted by activated macrophages.

3. The Pathophysiology of Mononuclear Phagocytes in Inflammatory Sites

3.1 Chronic Inflammatory Reactions: Collaboration of MP with Other Cells and Humoral Factors

In acute inflammatory reactions cells of the granulocyte series predominate [32]. In the chronic stage of inflammation macrophages in various forms of their functional activity are present together with other cell types [2]. Some features of MP relevant to inflammatory responses are schematically summarized in Fig. 3.

In some inflammatory diseases of clinical importance which are autoimmune in nature (e.g., rheumathoid arthritis) lymphocytes predominate and MP are present in lower numbers. In some infectious

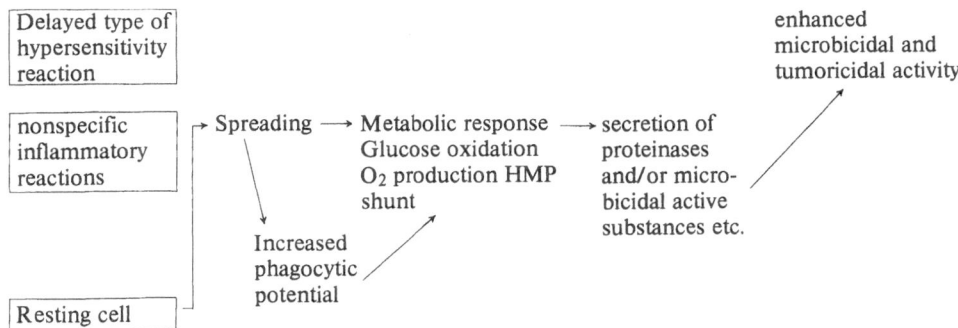

Fig. 3. Stages in the activation of mononuclear phagocytes [11]

diseases that are caused by intracellular localized microorganisms (e.g., mycobacterium tuberculosis, toxoplasma gondii, listeria monocytogenes) numerous lymphocytes together with many MP are consistently found. Chronic inflammatory reactions caused by substances resistant to digestion (e.g., cell walls of group A streptococci, zymosan particles, and asbestos) show predominantly cells of the macrophages series [7]. Very likely, products of lymphocytes and/or C-associated components together with other substances induce the activation of macrophages in the course of these various inflammatory reactions.

3.2 Delayed Hypersensitivity Reactions and Effector Mechanisms of Mononuclear Phagocytes

The development of an antimicrobial resistance is often associated with the manifestation of delayed hypersensitivity reactions [38]. Under such conditions products derived from sensitized lymphocytes play a major role in macrophage activation. The stepwise increase of functional activities of MP by nonspecific and lymphocyte-derived factors is shown in Fig. 3.

Obligatory intracellular organisms, e.g., mycobacteria [18], toxoplasma gondii [40], and trypanosoma cruzi [41], were effectively destroyed only after the development of a specific immunity. Sensitized lymphocytes were required to induce effective killing of these organisms by the macrophages, which under the influence of lymphocyte-derived products develope into an activated state and liberate H_2O_2 and other toxic oxygen products [40, 41].

In response to various microorganisms and other stimuli, subgroups of MP show distinct response patterns. Cells recently migrated into inflammatory sites exhibit a high sensitivity for lymphokines [44]. Peroxidase-positive mononuclear cells migrate quickly into inflammatory areas and replace resident peroxidase-negative macrophages. Some MP apparently lose their phagocytic potential (e.g., the Langerhans cell), but effectively present antigens to T lymphocytes [Huber and Stingl, this volume p. 31].

3.3 Complement-Associated Systems and Effector Mechanisms of Mononuclear Phagocytes

C-associated factors may activate MP. Moreover, MP have been shown to synthesize several C-associated factors [51]. The intimate relationship between these cells and components of the alternative pathway of C activation are considered as an important amplification system in the course of inflammatory responses [Schorlemmer, this volume p. 59].

Many substances that cause inflammatory reactions like endotoxins, polysaccharides on cell surface membranes, immune complexes, and others may induce the activation of the C system. This leads to the stimulation of macrophages and enhances their capacity of secretion [Schorlemmer, this volume p. 59]. This activated factor B of the alternative pathway is a very effective spreading factor for MP [11]. Neutral proteases secreted by activated macrophages have the capacity of cleaving C components, especially C3. Mononuclear phagocytes are considered to be the most important amplification system for C associated factors [2]. Monocytes themselves have the capacity to produce various complement components. Especially the components of the alternative pathway and their inhibitors (like C3b inactivator and beta-1-H) are produced by MP [51].

Besides C-associated components there are factors of the kinin and the coagulation system that interact with MP under certain conditions. Some prostaglandins (E_2 and F_2) are synthesized by macrophages, particularly in their activated state [2]. The inhibiting effect of prostaglandin E_1 and E_2 on the proliferation of colony forming cells as precursors of granulocytes and monocytes has been documented [34].

4. Concluding Remarks

A brief comparison of effector functions of MP and polymorphonuclear granulocytes may conclude this review. The neutrophil as a quickly mobilized effector cell shows only a very short life span at the inflammatory sites. On the other hand MP arrive later but stay in these areas for a prolonged period of time. Neutrophils contain effective antimicrobial systems. After phagocytosis they liberate, mainly oxygen-dependent, cytotoxic substances. They show poor secretory activity. By contrast MP secrete a variety of biologically important substances, particularly after activation. These products affect invading micro-organisms, connective tissues, and a variety of cells. Membrane receptors for the Fc fragment of IgG and for activated C3 are present on both cell types. However the MP recognize immune complexes more effectively and increase this membrane function quite remarkably after activation. Interacting in many ways with sensitized lymphoid cells and components of the C system MP show very effective amplification mechanisms. Polymorphonuclear granulocytes fulfill rather primitive effector functions with fast and high killing potentials, whereas the many functional activities of effector macrophages with the capacity for adjustment to different inflammatory stimuli only recently became fully estimated.

References

1. Abramson N, Gelfand EW, Jandl JH, Rosen FS (1970) The interaction between human monocytes and red cells. Specificity for IgG subclasses and IgG fragments. J Exp Med 132:1207
2. Allison AC (1979) Macrophage activation in relation to the pathogenesis of chronic inflammation. Behring Inst Mitt 63:38–51
3. Atkinson JP, Frank MM (1974) Complement independent clearance of IgG sensitized erythrocytes: inhibition by cortisone. Blood 44:629-637
4. Atkinson JA, Frank MM (1974) Studies on in vivo effects of antibody: interaction of IgM antibody and complement in the immune clearance and destruction of erythrocytes in man. J Clin Invest 54:339–348
5. Bendtzen K (1978) Biological properties of lymphokines. Allergy 33:105
6. Bianco C, Griffin FM, Silverstein SC (1975) Studies of the macrophage complement receptor: alteration of receptor function on macrophage activation. J Exp Med 141:1278–1290
7. Bonney RJ, Gery I, Tsau-Yen Lin, Meyenhofer MF, Acevedo N, Davies P (1978) Mononuclear phagocytes from Carrageenan-induced granulomas. J Exp Med 701:261
8. Brown DL, Lachmann PJ, Dacie JV (1970) The in vivo behavior of complement-coated red cells: studies in C6 deficient, C3 depleted, and normal rabbits. Clin Exp Immunol 7:401–421
9. Cardella CJ, Davies P, Allison AC (1974) Immune complexes induce selective release of lysosomal hydrolases from macrophages. Nature 247:46
10. Chess L, Schlossmann SF (1977) Human lymphocyte subpopulations. Adv Immunol 25:213–241
11. Cohn ZA (1978) The activation of mononuclear phagocytes: Facts, fancy and future. J Immunol 121:813
12. David J (1975) A brief review of macrophage activation by lymphocytic mediators. In: The phagocytic cell in host resistance. Raven, New York, p 143
13. David J, David R (1972) Cellular hypersensitivity and immunity. Prog Allergy 16:300–449
14. Davies WC, Huber H, Douglas SD, Fudenberg HH (1968) A defect in circulating mononuclear phagocytes in chronic granulomatous disease of childhood. J Immunol 101:1093–1095
15. Ehlenberger AG, Nussenzweig V (1977) The role of membrane receptors for C3b and C3d in phagocytosis. J Exp Med 145:357
16. Engelfriet CP, Pondman KW, Wolters G, von dem Borne AEG, Beckers D, Misset-Groenveld G, Loghem JJ (1970) Autoimmune haemolytic anemias: III. Preparation and examination of specific antisera against complement components and products, and their use in serological studies. Clin Exp Immunol 6:721–732
17. Frank MM et al. (1977) Pathophysiology of immune hemolytic anemia. Ann Intern Med 87:210–222
18. Godal T, Rees RJW, Lamvik JO (1971) Lymphocyte-mediated modification of blood-derived macrophage function in vitro; Inhibition of growth of intracellular mycobacteria with lymphokines. Clin Exp Immunol 8:625–637
19. Goetze O, Medicus RG, Müller-Eberhard HJ (1977) Alternative pathway of complement: non-enzymic reversible transition of precursor to active properdin. J Immunol 118:525–528
20. Gordon S (1975) The secretion of lysozyme and a plasminogen activator by mononuclear phagocy-

tes. In: Mononuclear phagocytes in immunity, infection and pathology. Blackwell Scientific Publications, Oxford London Edinburgh, Melbourne, p 463

21. Gordon S, Unkeless JC, Cohn ZA (1974) Induction of macrophage plasminogen activator by endotoxin stimulation and phagocytosis. J Exp Med 140:995

22. Griffin FM, Bianco CJ, Silverstein SC (1975) Characterization of the macrophage receptor for complement and demonstration of its functional independence from the receptor for the Fc-portion of immunoglobulin G. J Exp Med 141:1269

23. Griffin FM, Griffin JA, Silverstein SC (1976) Studies of the mechanism of phagocytosis. II. The interaction of macrophages with anti-immunoglobulin IgG-coated bone marrow-derived lymphocytes. J Exp Med 144:788

24. Horwitz AL, Hance AJ, Crystal RG (1977) Granulocyte collagenase: Selective digestion type I relative to type III collagen. Proc Natl Acad Sci USA 74:897–901

25. Huber H, Fudenberg HH (1968) Receptor sites of human monocytes for IgG. Int Arch Allergy 34:18–31

26. Huber H, Holm G (1975) Surface receptors of mononuclear phagocytes: Effect of immune complexes on in vitro function in human monocytes. In: Mononuclear phagocytes in immunity, infection and pathology. Blackwell, Oxford London Edinburgh Melbourne, p 291

27. Huber H, Polley MJ, Linscott WD, et al. (1968) Human monocytes: distinct receptor sites for the third component of complement and for immunoglobulin G. Science 162:1281–1283

28. Huber H, Douglas SD, Nussbacher J, Kochwa S, Rosenfield RE (1971) IgG subclass specificity of human monocyte receptor sites. Nature 229:419–420

29. Johnston RB, Keele BB, Misra HP, Lehmeyer JE, Webb LS, Baehner RL, Rajagopalan KV (1975) The role of superoxide anion generation in phagocytic bactericidal activity. J Clin Invest 55:1357–1372

30. Katz P, Simone CHB, Henkart PA, Fauci AS (1980) Mechanisms of antibody-dependent cellular cytotoxicity. J Clin Invest 65:55–63

31. Klebanoff SJ (1978) Chemotaxis. In: The neutrophil: Klebanoff SJ, Clark RA (eds) Function and clinical disorders. North-Holland, Amsterdam New York Oxford, p 104

32. Klebanoff SJ, Clark RA (1978) The neutrophil: Function and clinical disorders. North-Holland, Amsterdam New York Oxford

33. Klebanoff SJ, Hamon CB (1975) Antimicrobial systems of mononuclear phagocytes. In: Mononuclear phagocytes in immunity, infection and pathology. Blackwell, Oxford London Edinburgh Melbourne, p 507

34. Kurland JI, Bockmann R (1978) Prostaglandin E production by human blood monocytes and mouse peritoneal macrophages. J Exp Med 147:952–957

35. Lay WH, Nussenzweig V (1968) Receptors for complement on leukocytes. J Exp Med 128:991

36. Leijh PC, Van den Barselaar MT, Van Zwet TL, Daha MR, Van Furth R (1978) Requirement of extra cellular complement and immunoglobulin for intracellular killing of micro-organisms by human monocytes. J Clin Invest 63:772–784

37. Lobuglio AF, Cotran RS, Jandl JH (1967) Red cells coated with immunoglobulin G: binding and sphering by mononuclear cells in man. Science 158:1582–1585

38. Mackaness GB (1964) The immunological basis of acquired cellular resistence. J Exp Med 120:105

39. Morley J, Wolstengroft RA, Dumonde DC (1978) Measurement of lymphokines. In: Handbook of experimental immunology. Blackwell, Oxford London Edinburgh Melbourne, p 27.1

40. Murray HW, Juangbhanich CW, Nathan CF, Cohn ZA (1979) Macrophage oxygen-dependent antimicrobial activity. J Exp Med 150:938–964

41. Nathan CF, Brukner LH, Silverstein SC, Cohn ZA (1979) Extracellular cytolysis by activated macrophages and granulocytes. J Exp Med 149:84–99

42. Rocklin RE (1976) Mediators of cellular immunity, their mature and assay. J Invest Dermatol 67:372

43. Rocklin RE, McDermott RP, Chess L, Schlossmann SF, David JR (1974) Studies on mediator production by highly purified human T- and B-lymphocytes. J Exp Med 140:1303

44. Ruco LP, Meltzer MS (1977) Macrophage activation for tumor cytotoxicity: increased lymphokine responsiveness of peritoneal macrophages during acute inflammation. J Immunol 120:1054–1062

45. Schorlemmer HU, Davies P, Hylton W, Gugig M, Allison AC (1977) The selective release of lysosomal acid hydrolases from mouse peritoneal macrophages by stimuli of chronic inflammation. Br J Exp Pathol 58:315

46. Schreiber AD (1977) An experimental model of immune hemolytic anemia. Ann Intern Med 87:210–222

47. Sorrel TC, Lehrer RI, Cline MJ (1978) Mechanism of nonspecific macrophage-mediated cytoto-xicity: evidence for lack of dependence upon oxygen. J Immunol 120:347–352
48. Waksman BH, Namba Y (1976) On soluble mediators of immunologic regulation. Cell Immunol 21:161
49. Werb Z, Gordon S (1975) Elastase secretion by stimulated macrophages. Characterization and regulation. J Exp Med 142:361–377
50. Werb Z, Gordon S (1975) Secretion of a specific collagenase by stimulated macrophages. J Exp Med 142:346–360
51. Whaley K (1980) Biosynthesis of the complement components and the regulatory proteins of the alternative complement pathway by human peripheral blood monocytes. J Exp Med 151:501–516

Haematology and Blood Transfusion Vol 27
Disorders of the Monocyte Macrophage System
Edited by F. Schmalzl, D. Huhn, H.E. Schaefer
© Springer-Verlag Berlin Heidelberg New York 1981

The Macrophage as a Cytotoxic Effector Cell

M.-L. Lohmann-Matthes

1. Introduction

Strong evidence has been accumulating during the last years that cells of the mononuclear phagocyte system play a crucial role as effector cells in the body's defense against foreign cells and microorganisms. In three different systems the generation of cytotoxic macrophages has been reported. We reported in 1972 that during the course of an intraperitoneal stimulation with allogeneic cells macrophages were rendered cytotoxic at the site of rejection [23]. Similar data were reported for a syngeneic mouse tumor system by the group of Evans using a cytostasis assay [5]. In a third situation, animals which were infected by an intraperitoneal application of BCG or Toxoplasma were demonstrated to have strongly cytotoxic peritoneal macrophages [11]. Later experiments revealed that in all three situations macrophages were activated by a lymphokine secreted by sensitized T lymphocytes, when they meet antigen [6, 14, 24]. We called this lymphokine macrophage cytotoxicity factor (MCF) [24].

Several other groups also reported on nonspecific macrophage-mediated cytotoxicity induced by a lymphokine [3, 7, 29], which was called macrophage activating factor (MAF). Macrophage cytotoxicity factor (MCF) and MAF are used in an identical way for the same mediator. Macrophages activated by MCF have aquired the capacity to destroy tumor targets by direct cell to cell contact. The percentage of cytotoxicity was calculated according to the formula:

$$\% \text{ release} = \frac{\text{cpm supernatant}}{\text{cpm supernatant} + \text{cpm sediment}} \times 100$$

The calculation of specific cytotoxicity was based on the difference between total release of an experimental culture and spontaneous release of a control culture in the presence of nonactivated macrophages.

2. Characteristics of Macrophage Activation by MCF

The MCF activity is absorbed onto the surface of macrophages within 3 h of incubation in the cold. The presence of MCF on the surface of macrophages after 3 h of adsorption in the cold can be shown using specific antisera against MCF in a radioimmunoassay (Sun and Lohmann-Matthes, in preparation). MCF activity can be completely absorbed out of a given lymphokine preparation by subsequent absorption procedures on macrophage monolayers. After the 3-h absorption period, the unbound MCF can be washed off, and the cells then need a further incubation period of 12–18 h in order to develop full cytotoxicity [30, 33]. Lymphokine-induced macrophage cytotoxicity is a rather slow lytic reaction, since 18–24 h of cytotoxicity assay are needed to completely lyse target cells like P815X, 5178Y, and YAC-1. For tough targets like MethA fibrosarcoma cells 48 h of cytotoxicity assay are necessary to obtain complete lysis.

Macrophage cytotoxicity factor has been tested on many target cells for direct cytotoxi-

Table 1. ^{51}Cr release from various tumor target cells and their corresponding normal counterpart[a]

Target cell	% spec. ^{51}Cr release in the presence of activated macrophages
3 T3 SV40	30 ± 6
3 T3 fibroblasts	0
BALBc MethA fibrosarcoma cells	15 ± 4
BALBc fibroblasts	0
DBA/2 5178Y	42 ± 5
DBA/2 5-day-old bone marrow blasts	18 ± 4

[a] 2×10^5 bone-marrow macrophages from an 8-day-old culture were seeded into flat-bottomed glass tubes. They were activated with 100 µl of an titrated batch of MCF. After 24 h the cells were washed and ^{51}Cr-labeled targets were added. Assay time: 18 h. Effector/target ratio: 8 : 1

city. None of these targets showed any lysis except for the YAC-1 tumor. However, in the case of the YAC-1, the direct toxic effect turned out to be due to Con A present in the crude lymphokine preparation, since Con A alone is toxic for YAC-1, whereas the purified MCF showed no more direct cytotoxicity on YAC-1 target cells.

Thus, MCF seems to act exclusively by activating macrophages. These activated macrophages in our experiments need direct contact with the target in order to exert their cytotoxicity. We have never been able to find any supernatant factor secreted from activated macrophages which would kill our targets.

We spent a considerable degree of effort on the question whether activated macrophages would lyse only malignant targets, as has been postulated by several groups [11, 13, 27, 29], or whether normal cells may also be affected. For all these experiments bone marrow derived macrophages, which are available in large quantities, and the same batch of lymphokine were used [25]. All experiments were performed under identical conditions. Table 1 summarizes our results.

All pairs of targets were carefully compared with regard to their lysability using allogeneically sensitized T lymphocytes as effector cells [25]. Whereas 3 T3 fibroblasts, BALBc fibroblasts, and DBA/2 mature macrophages were tough targets compared to their corresponding counterparts, bone marrow blast cells and 5178Y lymphoma cells showed an identical pattern of lysability with T lymphocytes as effectors [25]. Thus, in most situations where macrophages pretend to have a selective cytotoxicity for tumor targets, the explanation lies in the different lysability of the targets compared. However, from the data with bone marrow blasts and 5178Y lymphoma cells one can deduce that macrophages have indeed a preference for some tumor lines but that quickly proliferating normal blasts are also affected to a certain degree. These data are in line with results published by Keller [15] who also showed that activated macrophages act on both normal and malignant cells.

3. Antibody-dependent Macrophage-mediated Cytotoxicity

Young macrophages and macrophage precursor cells obtained from bone marrow cultures and also young macrophages from the induced peritoneal cavity are capable of killing antibody-coated tumor cells [4, 21]. The killing efficiency depends on the dosage of the antibody used (Fig. 1). Twelve-day old bone marrow macrophages of C57BL/10 origin were incubated with ^{125}I-labeled P815 cells in the presence of varying concentrations of antitarget cell antibody. Fig. 1 shows that macrophages exert good cytotoxicity against antibody-coated P815 cells. Also for this type of killing direct contact between effector and target cell is necessary. The spontaneous release of target cells incubated in the presence or absence of antibody was identical.

Thus, there are two completely unrelated mechanisms inducing macrophages to kill tar-

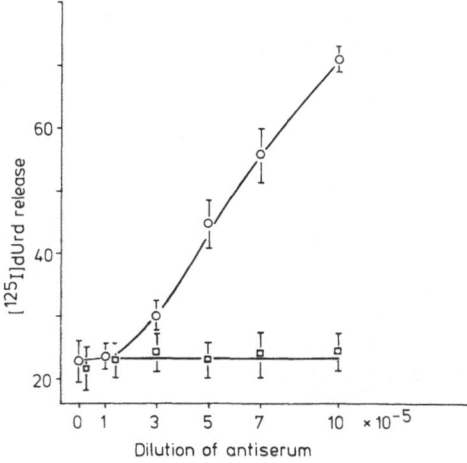

Fig. 1. Titration of C57Bl/10 anti-DBA/2 antiserum on C57BL/10 bone marrow macrophages. 2×10^5 adherent cells of an 8-day-old culture were incubated with ^{125}IdUrd-labeled DBA/2 P815X cells in the presence (*circles*) or absence (*squares*) of antiserum at dilutions as indicated

get cells. The next obvious question was whether these two mechanisms function in a co-operative way. To test this, both the activating lymphokine and the antibody coating the target cells were diluted to subthreshold concentrations at which no target cell lysis occurred.

Fig. 2 demonstrated that, indeed, the combination of the two subthreshold stimuli produced a clear-cut synergistic effect resulting in target cell lysis [18]. A similar synergistic effect has been reported when lymphokine activation is combined with LPS treatment of the macrophages [12]. Ruco et al. [31] have shown that the responsiveness of macrophages to MCF (MAF) and to LPS is controlled by the same gene. In contrast, in the situation described here the synergistic effects are performed by two completely different mechanisms. Ruco et al. [31] and Meltzer et al. [27] have presented evidence that the activation of macrophages to cytotoxicity by lymphokines requires a sequence of signals. From our data on the synergism between antibody-dependent and lymphokineinduced macrophage cytotoxicity, we assume the following sequence of events. During a first step the macrophage is activated to develop a higher lytic potential. During a second step, nonspecific surface receptors for target cells are induced which are responsible for the contact between the activated macrophages and the target cell. LPS or high dosages of lym-

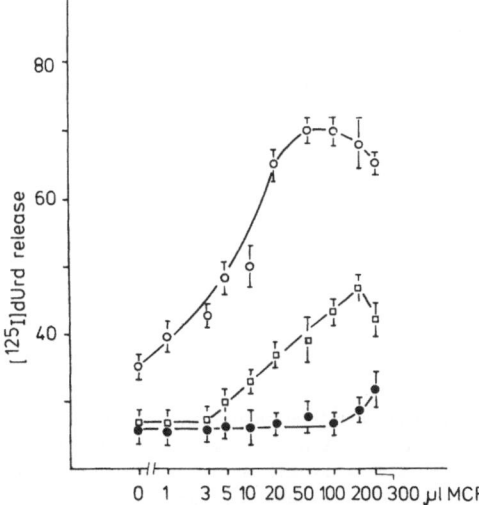

Fig. 2. Co-operation of lymphokine-induced and antibody-dependent cytotoxicity tested on bone marrow macrophages of the lymphokine low responder strain C3H/HeJ. 2×10^5 macrophages were activated for 24 h with various amounts of a titrated MCF preparation. After washing the macrophages were incubated for 18 h with antiserum and ^{125}IdUrd labeled P815X cells. Antiserum dilution $1 : 10^5$, *squares*, 1×10^4, *open circles*. *Closed circles* are lymphokine-preactivated macrophages without antiserum

51

phokine can induce this second step. This second step can be replaced by antibody which coates the target cells and establishes contact with the preactivated macrophage. Since the lytic potential of the macrophage is elevated by the preactivation with lymphokine, far fewer antibody molecules are required than in the case of antibody-dependent cytotoxicity with normal macrophages as effector cells. This assumption is supported by the fact that macrophages from C3HHeJ mice show a co-operative behavior with similar sensitivity to the activating lymphokine as macrophages from C57Bl/10 mice, although the macrophages from C3HHeJ mice are known to be low responders for the activation with lymphokines [18, 31] (Fig. 2). Thus, the elevated lytic potential of C3HHeJ mice is obtained by activation with similar dosages as in a high responder strain. However, the second step of lymphokine activation which is responsible for the development of receptors for target cells may be defective in C3HHeJ mice. In the synergistic system, this defect is fully compensated since contact is established by antibody on the surface of the target cell. The specificity of the synergistic effect is characterized by the specificity of the antiserum and only at high lymphokine concentrations does the nonspecific effect of the lymphokine dominate [20].

4. In Vivo Effects with MCF (MAF)

C57BL/10 mice were injected intraperitoneally with three different MCF preparations: 1. crude Con A induced spleen cell supernatants containing MCF (MAF) activity, 2. MCF (MAF) purified on a Phenylalanin-Sepharose column, a procedure which removes about 90% of the contaminating protein [16], and 3. MCF (MAF) purified on Phenylsepharose and subsequently analyzed by isoelectric focusing in the presence of 6-M urea, a procedure which separates MCF from MIF (MCF focuses at pH 8.5, whereas MIF focusses at pH 5.5 [17]). All three preparations were injected into the peritoneal cavity of mice and induced strongly cytotoxic peritoneal macrophages. The cytotoxicity of the peritoneal macrophages was not due to the Concanavalin A present in preparation (1), since when methyl α-D-mannoside was added similar results were obtained. Also preparation (2) and (3), which are devoid of Con A, are similarly effective in inducing cytotoxic macrophages in vivo (Table 2).

When preparation (1) or (2) were injected a three- to fourfold increase in the number of peritoneal macrophages was obtained which was likely to be due to the MIF activity present

Table 2. Cytotoxicity of peritoneal macrophages of mice injected three times with various MCF-containing or control preparations[a]

Material injected intraperitoneally	% ^{51}Cr release from 5178Y cells
3×2 ml crude Con A supernatant	60 ± 4
3×2 ml MCF purified on phenylalanin-Sepharose	63 ± 5
3×2 ml MCF purified by isoelectric focusing in the presence of 6-M urea	65 ± 4
Inactive controls	25 ± 2

[a] C57BL/10 mice were injected three times with MCF-containing preparations. Twelve h after the last injection peritoneal cells were collected and tested against 5178Y DBA/2 lymphoma cells at a ratio of 8 : 1. Assay time: 18 h. ConA supernatants were prepared by stimulating 10×10^6 spleen cells with 1 ug Con A for 24 h in the absence of serum.
Purification of MCF on phenylalanin-Sepharose and analysis by isoelectric focusing was done as described by Kniep et al. [17]. The different MCF preparations were titrated to identical activity.
For the collection of peritoneal cells the peritoneal cavity was rinsed with 5 ml of culture medium supplemented with 200 E. heparin. After 1 h cells were washed and ^{51}Cr-labeled 5178Y cells were added

Table 3. Number and cytotoxicity of peritoneal macrophages of mice injected with different fractions of a lymphokine preparation analyzed by isoelectric focusing[a]

Fraction from isoelectric focusing analysis injected i. p.	Number of peritoneal macrophages	% spec. cytotoxicity P815
Fraction pH 8.5 containing MCF	$9 \times 10^6 \pm 1$	31 ± 4
Fraction pH 5.5 containing MIF	$22 \times 10^6 \pm 2$	5 ± 1
Control fraction pH 11	$6 \times 10^6 \pm 1$	0 ± 4

[a] The lymphokine preparation (ConA induced spleen cell supernatants) was first passed through an Phenyl-Sepharose column and subsequently applied to an isoelectric focusing analysis in the presence of 6-M urea [13]. The eluted fractions were dialyzed overnight against phosphate buffered saline. C57BL/10 mice were injected three times i.p. within 36 h. Twelve h after the last injection peritoneal cells were collected and tested for cytotoxicity at an effector to target ratio of 8 : 1 for 18 h

in these preparations. When the MCF-containing fraction of preparation (3) was injected intraperitoneally the number of peritoneal macrophages was only slightly increased above the controls, but the macrophages were highly cytotoxic. In contrast, when the MIF activity containing fraction of preparation (3) was injected, the mice had high numbers of peritoneal macrophages which, however, showed hardly any cytotoxicity (Table 3) [19].

So far all mice survived multiple injections of MCF in perfect health. Nevertheless it must be checked most carefully whether there are negative side effects induced by multiple MCF injections. In further experiments we applied the described system in a tumor and in a bacterial system. We obtained complete in vivo rejection of the MethA Balbc fibrosarcoma and a fourfold prolonged mean survival time of salmonella typhimurium infected mice when these animals were injected with MCF 2 days prior to tumor or bacterial inoculum (Lang and Lohmann-Matthes, in preparation).

Also other groups have reported successful in vivo application of lymphokines. Thus Gezey et al. [9] showed in 1978 that a local application of MIF resulted in a skin infiltration resembling a delayed-type hypersensitivity reaction. Fidler [8] reported recently that he reduced the number of metastases in the lung significantly by injecting the tumor-bearing animals with crude MAF preparation embedded in liposomes.

These data taken together indicate that there is increasing evidence that in vitro prepared lymphokines can induce strong in vivo effects. In the experiments reported here the lymphokine MCF (MAF) was most effective in rendering peritoneal macrophages in vivo cytotoxic. In addition the expriments show a clear-cut differential effect of preparations containing MCF as compared to MIF. In consequence the results suggest that the application of purified MCF (MAF) preparations may in the future become of important therapeutic value in tumor and infectious disease.

5. Natural Killing Activity of Macrophage Precursor Cells

We have reported previously that the promonocyte, a nonadherent and nonphagocytic precursor cell of the macrophage lineage, has two strongly cytotoxic properties: it kills antibody-coated targets in a K cell like manner (4) like the young monocyte and in addition it kills certain susceptible tumor cells like the YAC-1 in an NK-like manner [21].

We first obtained such cells, which are an intermediate maturation stage between the monoblast and the monocyte, from liquid bone marrow cultures of mice. This intermediate maturation stage, which may be called promonocyte or macrophage precursor cell, has not yet developed the typical characteristics of a mature monocyte or macrophage such as adherence, phagocytosis, and positive staining for nonspecific esterase. This cell, however, has strong natural killer activity and kills spontaneously susceptible targets such as the

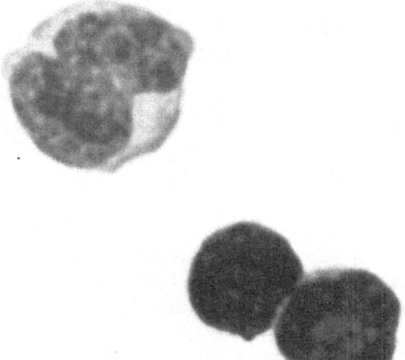

Fig. 3. Peritoneal exsudate cells of a mouse stimulated 3 days previously with corynebacterium parvum were depleted of all mature monocytes and macrophages by passage through nylon wool columns. Shown are two lymphocytes and one medium sized cell, presumably belonging to the monocytic lineage

YAC-1 tumor cell in the mouse [21]. Such macrophage precursor cells are not an in vitro artifact but they can also be isolated under in vivo conditions. They are present preferentially at the site of an immune reaction or inflammation and in the peripheral blood [22].

Two in vivo situations where particularly high amounts of cells with NK activity are present are the peritoneal cavity of an corynebacterium parvum stimulated mouse [28] and the spleen of a nu/nu mouse. Figure 3 shows cells of an peritoneal exudate of a corynebacterium parvum stimulated mouse from which all mature macrophages were removed by adherence columns. Figure 4 shows the same cell population after 3 days of culture. The macrophage precursor cells have now matured to typical adherent macrophages. Table 4 shows an experiment where these cells were treated with an allogeneic antimacrophage serum + C' and also with an recently developed rat anti-mouse monoclonal antibody. Both antimacrophage antibodies show a strong reduction of the NK activity, demonstrating that cells of the macrophage lineage are responsible for natural killing. Similar results are obtained with nu/nu spleen cells [22, 32].

We have also prepared monocyte-free cell populations from human peripheral blood. Such a population has high NK activity and consists of typical small lymphocytes and cells which have a morphology identical to the promonocyte of the mouse system (Fig. 5). When

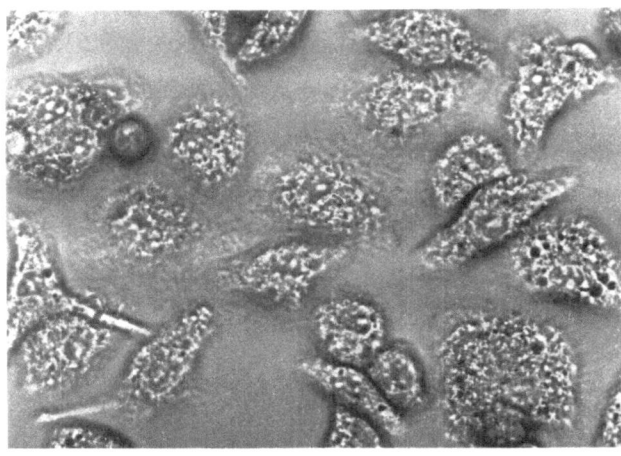

Fig. 4. Same population as Fig. 3 after 3 days in vitro culture. A monolayer of mature macrophages has developed

Table 4. Effect of pretreatment with antimacrophage antibodies + C′ on the NK activity of corynebacterium parvum induced peritoneal cells[a]

Pretreatment of CP-induced peritoneal cells passed through adherence columns	% ^{51}Cr release from YAC-1
–	59 ± 3
Medium + C′	61 ± 3
Alloantimacrophage-serum Mph 1.2. + C′	27 ± 2
Rat anti-mouse monoclonal antibody	
M 102 + C′	25 ± 3

[a] Spontaneous release from YAC-1 was 19 ± 2. Effector/target ratio was 10 : 1. Assay time 8 h. Peritoneal cells were collected from mice which 3 days previously had received 100 μg corynebacterium parvum suspension (Deutsche Wellcome). Cells were passed extensively through adherence columns. Passed cells were tested for NK activity with or without treatment with anti-macrophage serum + C′

this population matures, within a period of 5 days a monolayer of adherent esterase-positive cells has developed [22].

6. Conclusions

Cells of the macrophage lineage possess three different mechanisms to act as cytotoxic effector cells. The lymphokine-induced macrophage cytotoxicity, which is nonspecific in nature, is directed against all proliferating target cells but shows in addition a clear preference for tumor targets. This mechanism co-operates and acts synergistically with the second mechanism, the antibody-dependent macrophage cytotoxicity. This co-operation appears very useful, since under most in vivo conditions during the course of an immune reaction both lymphokine and antibody are produced and secreted. Chemical purification of MCF, which has already partially been achieved [17], will in the future become much easier with the use of the cell fusion technique both for the production of MCF and also of monoclonal antibodies against MCF. Thus it seems likely that it will become possible to use in vitro produced and purified MCF preparations to protect an organism from infections and probably also from some tumors. Preliminary data along this line have already shown that mice can be protected against salmonella typhimurium infections by pretreatment with MCF (Lang and Lohmann-Matthes, in preparation).

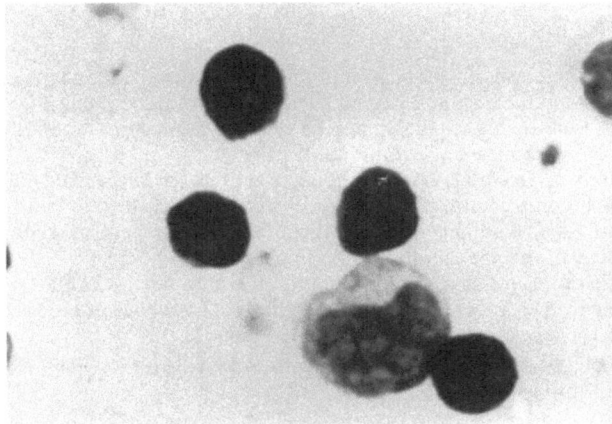

Fig. 5. Human peripheral blood cells were centrifuged through Ficoll Hypaque and then passed over several consecutive adherence columns in order to completely · remove mature macrophages. Shown are small lymphocytes and one cell which resembles the macrophage precursor cell in the mouse

The third cytotoxic activity of cells of the macrophage lineage has been a matter of continuous debate, since the opinions on the nature of the so-called natural killer cell were and still are controverse. In the meantime there is now agreement on the question of the morphology of the effector cell. It is a cell of 10–14 µ diameter with a usually bean-shaped nucleus, and it is nonadherent and nonphagocytic. It has recently been characterized by three antimacrophage-monoclonal antibodies [1, 2, 32] to be of macrophage origin. Other groups present evidence that this cell belongs to the T cell lineage because they show a rather low amount of Thy 1.2 [10] and other T cell markers like Quat 4 and Quat 5 on these cells. However, for Thy 1.2 it is now clear that this marker, which is shared with epithelial and brain cells, is also present on macrophages (unpublished observation). Many of the other T cell markers have not yet been carefully checked through all maturation stages of the macrophage lineage. Our data with a rat anti-mouse macrophage monoclonal antibody (M 102) seem to indicate that the NK cells represent a subpopulation of macrophages, since monoclonal antibody M 102 is strictly macrophage specific but does not inhibit other macrophage functions like antigen-presentation or lymphokine-induced cytotoxicity [32].

Thus in conclusion cells of the macrophage lineage have three cytotoxic effector mechanisms and can act at different maturation stages. Cells of the monocyte-macrophage lineage are released as an immature precursor cell (promonocyte) from the bone marrow to the periphery of an organism, where they fulfill various functions. From these data it appears likely that the macrophage system plays a critical role in the body's cytotoxic defense against infectious and tumor diseases.

7. Summary

Cells of the macrophage lineage possess three different mechanisms allowing them to act as cytotoxic effector cells: lymphokine-activated macrophages kill extracellularly proliferating target cells, preferentially tumor targets, and intracellularly parasitic microorganism like Salmonella, Toxoplasma, Leishmania, and others. Young macrophages and macrophage precursor cells kill antibody coated nucleated target cells. Finally nonadherent, nonphagocytic macrophage precursor cells, released from the bone marrow to sites of inflammation and to the peripheral blood show strong natural killer activity. For the various macrophage functions different subpopulations seem to be responsible as can be seen from data with antimacrophage monoclonal antibodies recognizing and eliminating functionally active subpopulations. The cells of the macrophage system, with their three cytotoxic effector mechanisms which cooperate and enhance each other, are very likely to play an important role in the cytotoxic defense system of the body.

References

1. Ault KA, Springer TA (1981) Cross-reaction of a rat-anti-mouse phagocyte specific monoclonal antibody (Anti MAC-1) with human monocytes and natural killer cells. J Immunol 126:359
2. Breard J, Reinharz EL, Kung PC, Goldstein G, Schlossmann FJ (1980) A monoclonal antibody reaction with human peripheral blood monocytes. J Immunol 124:1943
3. Churchill WH, Piessens WF, Subis CA, David J (1975) Macrophages activated as suspension cultures with lymphocyte mediators become cytotoxic for tumor cells. J Immunol 115:781
4. Domzig W, Lohmann-Matthes M-L (1979) Antibody-dependent cellular cytotoxicity against tumor cells. II. The granulocyte identified as effector cell. Eur J Immunol 9:267
5. Evans R, Alexander P (1972) Role of macrophage in tumor immunity. Immunology 23:615
6. Evans R, Grant CK, Cox H, Steele K, Alexander P (1972) Thymus-derived lymphocytes produce an immunologically specific macrophage arming factor. J Exp Med 136:1318
7. Fidler IJ (1975) Activation in vitro of mouse macrophages by syngeneic, allogeneic or xenogeneic lymphocyte supernatants. J Natl Cancer Inst 55:1159
8. Fidler IJ (1980) Science 208:1469

9. Geczy CL, Friedrich W, de Weck A (1975) Production and in vivo effect of antibodies against guinea pig lymphokines. Cell Immunol 19:65
10. Herberman RB, Nunn MF, Holden HT (1978) Low density of Thy 1 antigen on mouse effector cells mediating natural cytotoxicity. J Immunol 121:304
11. Hibbs JB, Lambert CH, Renington JS (1972) Proc Soc Exp Biol Med 139:1049
12. Hibbs JB, Tainton RR, Chapman HA, Weinberg JB (1977) Macrophage tumor killing. Influence of local environment. Science 197:279
13. Holtermann CA, Klein E, Casler GP (1973) Selective cytotoxicity of peritoneal lymphocytes for moplantic cells. Cell Immunol 9:339
14. Jones TC, Len L, Hirsch JG (1975) Assessment in vitro of immunity against toxoplasma gondii. J Exp Med 141:466
15. Keller R (1976) Susceptibility of normal and transformed cell lines to cytotoxic and cytocidal effects exerted by macrophages. J Natl Cancer Inst 56:369
16. Kniep EM, Kickhöfen B, Fischer H (1980) Macrophage cytotoxicity factor. Purification and separation from MIF. In: de Weck A, Kristensen F, Landy M (eds) Biochemical characterization of lymphokines. Academic Press, New York, p 149
17. Kniep EM, Lohmann-Matthes M-L, Domzig W, Kickhöfen B (1981) Chemical characterization of the macrophage cytotoxicity factor (MCF). J Immunol 127:417
18. Lang H, Lohmann-Matthes M-L (1980) Cooperation effects between antibody dependent and lymphokine-induced macrophage mediated cytotoxicity. Immunobiology 157:109
19. Lang H, Lohmann-Matthes M-L (to be published) In vivo activation of macrophages to cytotoxicity by application of the lymphokine MCF (MAF)
20. Lang H, Lohmann-Matthes M-L (to be published) Specific macrophage mediated cytotoxicity revisited. Eur J Immunol
21. Lohmann-Matthes M-L, Domzig W (1979) Antibody dependent cellular cytotoxicity against tumor targets. I. Cultivated bone-marrow cells kill tumor cells. Eur J Immunol 9:261
22. Lohmann-Matthes M-L, Zähringer M (1981) Zerstörung von Tumorzellen durch Makrophagen-Vorstufen bei Maus und Mensch. Blut 42:283
23. Lohmann-Matthes M-L, Schipper H, Fischer H (1972) Macrophage mediated cytotoxicity against allogeneic target cells in vitro. Eur J Immunol 2:45
24. Lohmann-Matthes M-L, Ziegler FG, Fischer H (1973) Macrophage cytotoxicity factor. A product of in vitro sensitized thymus-dependent cells. Eur J Immunol 3:56
25. Lohmann-Matthes M-L, Kolb B, Meerpohl HG (1978) Susceptibility of malignant and normal target cells to the cytotoxic action of bone-marrow macrophages activated in vitro with MCF. Cell Immunol 41:231
26. Lohmann-Matthes M-L, Domzig W, Roder J (1979) Promonocytes have the functional characteristics of natural killer cells. J Immunol 123:1883
27. Meltzer MS, Wahl LM, Leonard EJ, Nacy LA (1980) Macrophage activation by lymphokines. In: de Weck A, Kristensen F, Landy M (eds) Biochemical characterization of lymphokines. Academic Press, New York, p 161
28. Ojo E, Wigzell H (1978) Corynebacterium parvum induced peritoneal exsudate cells with cytolytic activity against tumor cells are nonphagocytic cells with characteristics of natural killer cells. Scand J Immunol 215
29. Piessens W, Churchill H, David J (1975) Macrophage activated in vitro with lymphocyte mediators kill neoplastic but not normal cells. J Immunol 114:293
30. Ruco LP, Meltzer MS (1978) Macrophage activation for tumor cytotoxicity: Tumoricidal activity by macrophages from C_3H/HEJ mice requires at least two activation stimuli. Cell Immunol 41:35
31. Ruco LP, Meltzer MS, Rosenstreich DL (1978) Macrophage activation for tumor cytotoxicity: control of macrophage tumoricidal capacity by the LPS gene. J Immunol 121:543
32. Sun D, Lohmann-Matthes M-L (1981) Functionally active subpopulations of mouse macrophaces recognized by monoclonal antibodies. Eur J Immunol (in press)
33. Ziegler FG, Lohmann-Matthes M-L, Fischer H (1975) Studies on the mechanism of macrophage mediated cytotoxicity. Int Arch Allergy Appl Immunol 48:182

Haematology and Blood Transfusion Vol 27
Disorders of the Monocyte Macrophage System
Edited by F. Schmalzl, D. Huhn, H.E. Schaefer
© Springer-Verlag Berlin Heidelberg New York 1981

The Role of Complement in the Function of the Monocyte – Macrophage System

H.U. Schorlemmer

1. Abstract

In parallel to the behavior of many agents that activate complement via the alternative pathway and can stimulate macrophages to secrete lysosomal enzymes we investigated the interaction of mouse peritoneal macrophages cultured in a serum-free medium with various stimuli such as zymosan, polyanions, collagen type II, and immune complexes prepared from tetanus toxoid and pooled human anti-tetanus toxoid $F(ab)_2$. All these stimuli induced the release of hydrolytic enzymes from macrophages in culture. The release was time and dose dependent and is not associated with loss of the cytoplasmic enzyme lactate dehydrogenase or any other sign of cell death. The mechanism of macrophage activation by these various agents is unknown. Macrophages have surface receptors for Fc and C3b with the capacity to bind immune complexes or C3b, respectively, and this is followed by activation of the cells. Activation via the Fc part can be excluded in these experiments. The possibility therefore arose that macrophages might be stimulated by endogenous C3 via the C3b receptor, since it is known that all the substances mentioned above can activate C3. To confirm this hypothesis we tried to inhibit this reaction by using an anti-C3-Fab preparation. There was hardly any detectably enzyme release after adding the anti-C3-Fab (dose dependent) together with the various stimuli to the macrophages. An unrelated Fab preparation showed no inhibitory effect. Furthermore, incubation of macrophages and the stimuli together with β1H and C3bINA abolished the effect to activate the macrophages. The observations now presented focus attention to the possibility that endogenous C3 could play a role in the stimulation of mouse peritoneal macrophages by various activators of the alternative pathway.

2. Introduction

It has long been established that the complement protein sequence is a major mediator of the inflammatory response. These proteins which exist in an inactive precursor state can be activated specifically by antibodies on recognition of antigen. Several activated complement components have the potential to mediate inflammatory tissue injury either directly or indirectly. Since the complement fragments such as C3a and C5a can induce histamine release from basophils and mast cells and can cause increased vascular permeability and vasodilation, it is tempting to suggest that direct activation of the complement system may be of pathogenetic significance in mediating the inflammatory response. In addition macrophages are thought to play an important role in the production or release, or both, of several types of mediators of the inflammatory response. It would be of great clinical value if the mechanisms underlying inflammatory reactions were understood and individual stages of the process could be controlled pharmacologically. Because the reactions in vivo are highly complex, we have attempted to isolate components of the reactions for in vitro analysis. We will summarize here some of the basic principles and experimental approaches that we have used to determine the responses of cells of the mononuclear phagocyte lineage and the effect on the complement system in chronic inflammatory responses. We have chosen to study this cell type and the humoral defense system since they both participate to one extent or another in all inflammatory responses regardless of their etiology.

3. Material and Methods

3.1 Experimental Animals

Male mice of the outbred strain NMRI/Bom were purchased from G.L. Bomholtgaard, Ry, Denmark.

3.2 Tissue Culture Materials

Tissue culture-grade Petri dishes were obtained from Nunc GmbH, Wiesbaden, Germany; TC 199 and fetal bovine serum were from Flow Laboratories GmbH, Bonn, Germany.

3.3 Biochemical Reagents

Penicillin and streptomycin were from Flow Laboratories; Pyruvate and NADH were from Boehringer GmbH, Mannheim; heparin, preservative free, was from Nordmark Werke GmbH, Hamburg, Germany.

3.4 Macrophage Collection and Culture

Mouse macrophages were obtained by peritoneal lavage of NMRI mice with 5 ml of TC 199 containing penicillin and streptomycin (100 units/ml) and heparin (10 I.U./ml). Four-ml aliquots of the peritoneal exudate cell suspension containing 0.5 to 1.0×10^6 cells/ml were distributed into 35-mm Petri dishes and incubated in a humidified atmosphere of 5% carbon dioxide and air at 37 °C for 1 to 2 h to allow attachment of adherent cells. Nonadherent cells were removed by four washes with PBS. After the washing, the cells were cultured in TC 199 without serum. Cultures prepared in this way give a sheet of well-spread cells within 24 h.

At the end of each incubation period the medium was removed and the adherent cells were released by adding saline containing 0.1% (w/v) Triton X-100 and scraping with sterile silicone rubber bungs. The activities of various enzymes were assayed in both the media- and cell-containing fractions.

3.5 Enzyme Assays

All assays were conducted under conditions giving linear release of the product in relation to the amount of sample used and the time of incubation. Lactate dehydrogenase was assayed by determining the rate of oxidation of reduced nicotinamide adenine dinucleotide at 340 nm using 2.5-mM pyruvate in 0.05-M phosphate buffer, ph 7.5 [3]; β-glucuronidase was assayed by the method of Talalay et al. [25]. β-galactosidase was assayed by the method of Conchie et al. [6] using p-nitrophenyl-β-D-galactopyranoside as substrate. N-acetyl-β-D-glucosaminidase was assayed by the method of Woolen et al. [29] using p-nitro-phenyl-2-acetamido-2-β-D-glucopyranoside as substrate dissolved in 0.1-M acetate buffer, pH 4.5.

3.6 Method for Cytotoxicity Assay

As the release of lactate dehydrogenase (LDH) into the culture supernate is an indication for cell death [27], the ^{51}Cr release assay was replaced by the determination of LDH. The assay for the cytoplasmic LDH was conducted as described recently [22]. In positive experiments, up to 90% of the cellular LDH content appeared in the supernate within 2 h of incubation which is why this time interval was chosen as test period. Control supernatants showed very low spontaneous background release of LDH even after 6 h.

3.7 Incubation of C3 Activation and Hemolytic Assay for C3 Activity

Standard incubation mixtures for C3 activation and the assay for hemolytic C3 measurement were performed as detailed in [5]. Equal volumes of either pooled normal or C4-deficient guinea pig serum were incubated with appropriate dilutions of the test materials for 30 min at 37 °C and cooled to 4 °C in an ice bath; the concentrations referred to in the text represent final concentrations in this standard incubation mixture (µg/ml). Controls were run with PBS instead of test material. From the standard incubation mixture, samples were withdrawn and subdiluted to a final serum concentration of 1 : 100, 1 : 500, and 1 : 1000. The amount of residual hemolytically active C3 [site forming units (SFU)] was determined and calculated as previously described [5]. The values obtained were expressed as percentage of the hemolytically active C3 in the controls. The C3 content of normal and C4-deficient guinea pig serum in the controls was in the range of $1-2 \times 10^{11}$ SFU/ml.

4. The Reactivity of the Complement System

The term "complement" is applied to a system of factors occurring in normal serum or tissue fluid that are activated characteristically by antigen–antibody interaction and subsequently mediate a number of biologically significant consequences. Complement was originally recognized as a heat labile co-factor in immune hemolytic and bacteriocidal reactions and the complement sequence was originally defined and its components isolated through study of lytic reactions. It was shown that activation of complement by antigen–antibody complexes resulted in sequential activation of 11 well-characterized plasma proteins in the form of nine numbered components, C1 to C9. Antigen–antibody complexes are the usual mechanism of activation of C1, the initial classical pathway protein which is a macromolecule composed of C1q, C1r, and C1s. The natural substrates of C1 esterase are two other classical pathway components, C4 and C2. Activation of these proteins generates a C3 splitting enzyme, the classical pathway C3 convertase. In addition to the classical pathway, the complement system can be activated by the alternative pathway. This reaction sequence has been delineated and as well as certain newly recognized factors, the components involved include those of the earlier described properdin system, namely, factor B, factor D, and properdin. Briefly, alternative pathway activation occurs once C3b, a split product of C3 has been generated. It combines with factor B, which has been activated to Bb by factor D. The resulting complex C3bBb has enzymatic activity to cleave additional C3 molecules and is known as the alternative pathway C3 convertase. Properdin, for which the alternative pathway was originally named, stabilizes the C3bBb complex. A feature of C3 activation, whether by the classical pathway or the alternative pathway, is the recruitment of a positive feedback system (the C3b feedback loop) in which the primary breakdown product of C3, C3b, generates more C3 convertase activity. The extent to which the feedback is activated depends upon the balance between the generation of C3b and its interaction by the regulatory proteins C3bINA and β1H.

C3 is the most abundant complement component and that which carries the most important biological activities. The recent findings of C3b-dependent generation of a chemotactic lymphokine from B lymphocytes [13] and secretion of lysosomal enzymes from macrophages [14] and the demonstration of a C3 requirement for in vivo "activation" of mouse peritoneal macrophages [4] indicate that products of the complement reaction can modulate inflammatory cell functions.

The cleavage of C3 to C3b is thus the principal event in complement activation. It can be brought about by a variety of enzymes including trypsin, plasmin, thrombin, collagenase, and elastase. The enzymes split C3 to C3b and a small fragment C3a. C3a itself has important biological activities as an anaphylatoxin reacting with receptors particularly on mast cells, basophils, and platelets, causing exocytosis of their pharmacological mediators. The nascent C3b has a briefly activated binding site which allows it to bind firmly to a wide variety of surface structures. C3 even entirely on its own could thus serve as a mediator of nonspecific immunity in a simple system where phagocytic cells are present and where C3 activation can be produced by proteolytic enzymes derived either from the invading organisms or from the phagocytic cells themselves. Many of the complement components and factors of the alternative pathway are also synthesized by the normal macrophages, and since C3b has the property of combining with factor B in the presence of Mg^{++} ions to give a powerful C3 splitting enzyme (C3bBb), this reaction also provides a positive feedback amplification loop for enhancing phagocytosis, exocytosis, and cell activation.

It seems plausible to assume that some such positive feedback amplification of inflammatory reactions was the original function of the complement system. It is however equally possible that the original functions of the C3b, factor B loop were concerned with cell activation and that the two components for this purpose were produced by the cell for its own use rather than for export to the extracellular. Thus it has been demonstrated by us that macrophage activation involves the cleavage of endogenously formed C3 by endogenous proteases and the resulting C3b reacting with the cell's own C3 receptors. This reaction is amplified by factor B which is also synthesized by macrophages. It is therefore quite possible that the complement system in the plasma may have developed from an earlier existence as an endogenous mechanism for cell regulation.

5. Origin of Mononuclear Phagocytes and Recruitment into Lesions

Macrophages are distributed widely throughout the body and assume a variety of morphological guises depending or their stage of maturation and on tissue localization. These various forms were collectively designated the mononuclear phagocyte system. Mononuclear phagocytes are derived from bone marrow, where they differentiate from a primitive precursor stem cell. The most immature recognizable form is the monoblast, which matures into the promonocyte and finally into the monocyte. Monocytes then form a nondividing pool of cells that are recruited into the circulation and finally into the various tissues, where they differentiate further into macrophages. Macrophages are thought to play an important role in both the afferent and efferent limbs of immune responses. These responses include the production or release, or both, of several types of mediators of the inflammatory response. Under normal conditions macrophages mediate essential functions of host defense at their various sites. When an inflammatory stimulus localizes to any particular tissue or organ and provokes a chronic inflammatory response, these cells accumulate at these sites and persist until the lesion is either resolved and repaired or, alternatively, until local cell proliferation, chronic tissue damage, and destruction occurs. The majority of mononuclear phagocytes accumulating in inflammatory sites originate from the circulating blood monocytes. The accumulation of monocytes in inflammatory sites may depend on the local generation of monocyte chemotactic factors. A number of products generated at inflammatory sites attract monocytes. These include products of complement activation. Phagocytes participate in inflammation and immunity by their activity of mediator secretion: the regulatory function of these cells could, in some phases, be even more determinant than their phagocytic action. Macrophages secrete the following substances: enzymes (acid hydrolases, neutral proteases, lysozyme), products or factors with biological activity on other cells in culture (lymphocyte stimulatory factors, colony formation supporting factors, DNA synthesis inhibitors, cytotoxins), and such classical mediators of inflammation as complement components and prostaglandins as well as interferons and pyrogens.

6. Effects of Complement Components on Macrophages

Many agents eliciting inflammatory responses in vivo are activators of the alternative pathway of the complement system. We have found that mouse peritoneal macrophages in culture can be activated to secrete lysosomal enzymes [17, 18] and to become cytolytic for various target cell types by many different materials [19, 20]. Looking for a common factor among the variety of agents that induce inflammation in vivo and macrophage stimulation in vitro our attention has been drawn to the fact that they all activate the complement system [16]. An early consequence of the activation is cleavage of the complement component C3 into the small fragment C3a and the large fragment C3b. Highly purified guinea pig C3b was incubated with mouse peritoneal macrophages in a serum-free medium and found to elicit a dose-dependent release of several hydrolases but not of lactate dehydrogenase. These macrophages stimulated by C3b also acquire the capacity to lyse added tumor cells (Fig. 1). This activation of macrophages was inhibited in the presence of antibodies which react with C3b; an unrelated antiserum had no inhibitory effect.

Supernatants of activated but not of unactivated macrophages were able to cleave C3, presumably because of the secretion of a neutral proteinase with appropriate specificity [14]. Macrophages secrete complement components, including C3, and stimulated macrophages can cleave C3 into C3a and C3b which are released into the extracellular medium. The C3b becomes attached to specific receptors on the plasma membrane of the macrophages and the C3b so formed can activate other macrophages, so that serial activation can result.

It has been shown that C3a of guinea pig origin is highly lytic for tumor cells or transformed lymphocytes. Untransformed normal cells are relatively resistant to the lytic effect of C3a [7]. C3a also lyses mycoplasmas and L-forms of bacteria but has no effect on intact

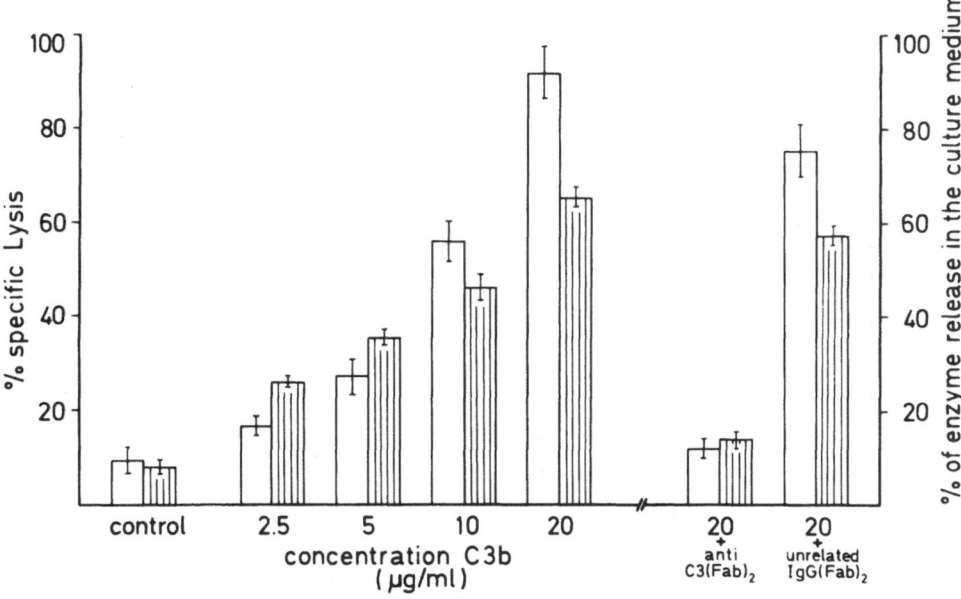

Fig. 1. Dose-dependence of C3b stimulation of macrophage cytotoxicity (□) and lysosomal enzyme secretion (▥)

bacteria [26]; erythrocytes also are highly resistant. As already mentioned, macrophages activated by C3b release into the culture medium a material like C3a with anaphylatoxic and cytolytic activity. When added to a guinea pig ileum preparation, it produced contraction, tachyphylaxis, and specific desensitization towards a reference sample of C3a [8, 20]. This suggest that the active agent in the media of stimulated macrophages was C3a. Further evidence that the lytic activity is due to C3a comes from experiments using specific antiserum against C3a. The lytic activity of culture media was virtually abolished by anti-C3a but not by normal serum. In a partial purification by PEG precipitation, ion exchange, and gel chromatography of the lytic activity in supernatants derived from stimulated macrophages, it turned out to correspond to a low molecular weight peptide with physicochemical characteristics similar to purified C3a. The molecular weight, assessed by 10% SDS polyacrylamide gel electrophoresis, was 8500 daltons. This macrophage-generated cytolytic factor turned out to be a peptide with basic character and is resistant to heat (100 °C for 10 min) and acid treatment (pH 2.0). Its smooth muscle contracting activity can be destroyed by carboxypeptidase B ("C3a inactivator"). These findings indicate that this cytolytic activity is quite similar to guinea pig or human C3a [20]. The main conclusion from these experiments is that macrophages stimulated by various agents secrete a factor indistinguishable pharmacologically and immunologically from C3a.

7. Participation of Endogenous C3 in Macrophage Activation by Various Stimuli

Unstimulated peritoneal macrophages do not secrete enzymes other than lysozyme; they do synthesize native complement components, but they do not release complement cleavage products. However, when macrophages are stimulated in vivo or in vitro they can become activated to perform various biologically important functions. In parallel to the behavior of many agents that activate complement via the alternative pathway and can stimulate macrophages to secrete lysosomal enzymes we investigated the interaction of mouse peritoneal macrophages cultured in a serum-free medium with various stimuli, like zymosan, collagen type II, and immune complexes prepared of tetanus toxoid and pooled human anti-tetanus-toxoid-F(ab)$_2$. All these stimuli induced the release of lysosomal en-

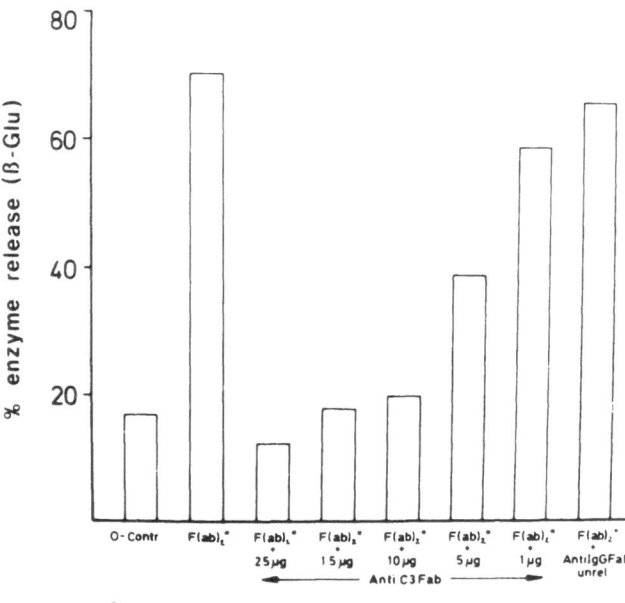

Fig. 2. Dose-dependent inhibition of enzyme secretion from macrophages induced by immune complexes with an anti C3-Fab preparation

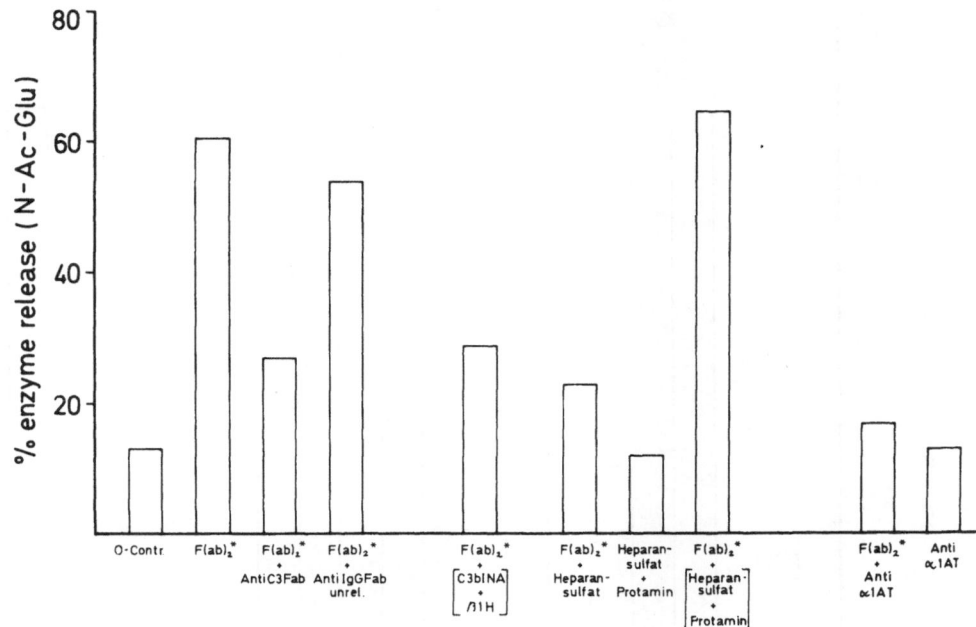

a   * Immune complexes

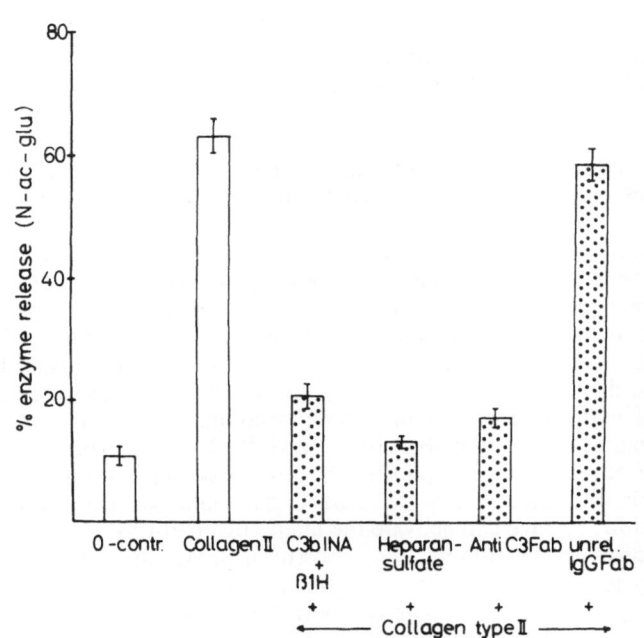

b

65

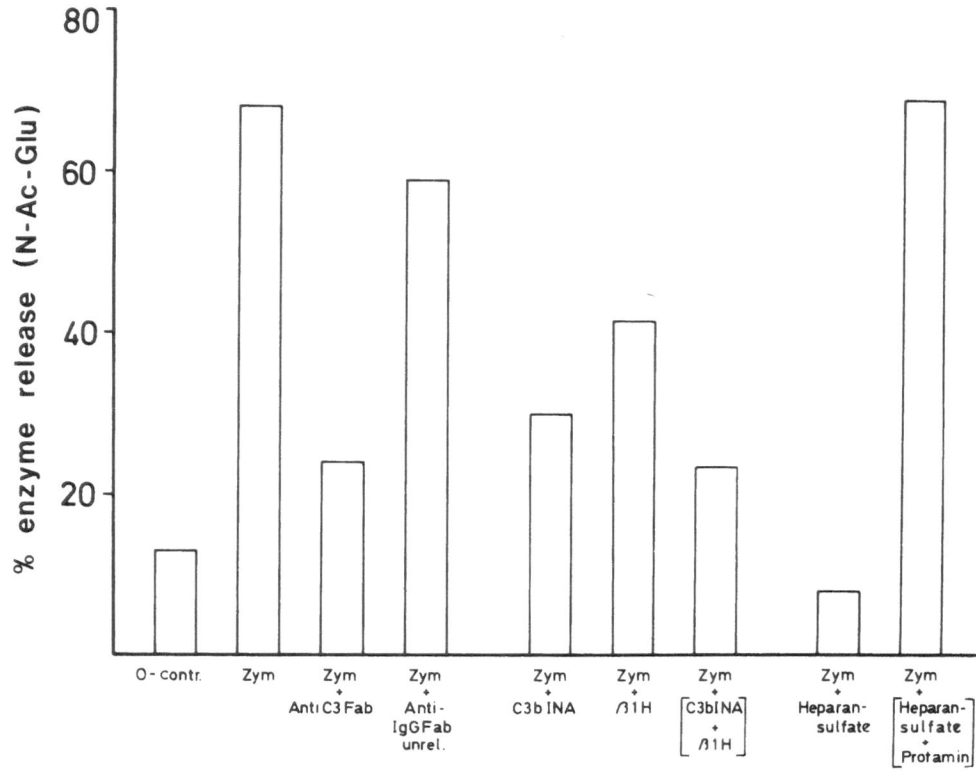

c

Fig. 3a, b, c. The effect of various inhibitors of the alternative pathway activation on lysosomal enzyme secretion from cultured macrophages induced by immune complexes, collagen type II, and zymosan

zymes from macrophages in culture. The release was time and dose dependent and was not associated with loss of the cytoplasmic enzyme lactate dehydrogenase or any other sign of cell death.

The mechanism of macrophage activation by these various agents is unknown. But we know that macrophages have considerable capacity to synthesize proteins like complement components and factors of the alternative pathway. We have shown that purified C3b is able to activate mouse peritoneal macrophages in culture to secrete hydrolytic enzymes [15]. There is also a parallelism in the capacity of various agents to stimulate macrophages to release lysosomal enzymes and to activate the alternative pathway. Macrophages have surface receptors for Fc and C3b with the capacity to bind immune complexes or C3b, respectively, and this is followed by activation of the cells. Activation via the Fc part can be excluded in our experiments with F(ab)₂-immune complexes. The possibility therefore arose that macrophages might be stimulated by endogenous C3 via the C3b receptor, since it is known that all the substances mentioned above can activate the alternative pathway.

To confirm this hypothesis we tried to inhibit this reaction by using an anti-C3-Fab preparation. There was hardly any detectable enzyme release after adding the anti-C3-Fab to the macrophage cultures (Fig. 2). An unrelated Fab preparation showed absolutely no inhibitory effect. This inhibition of macrophage activities was dose dependent and could be diluted out, so that optimal activation was observed again. Also incubation with β1H and C3bINA abolished the ability of immune complexes (F[ab]₂), collagen, or zymosan to activate macrophages (Fig. 3). These observations focus attention to the possibility that addition of stimuli to serum-free macrophage cultures generates a factor which is then able

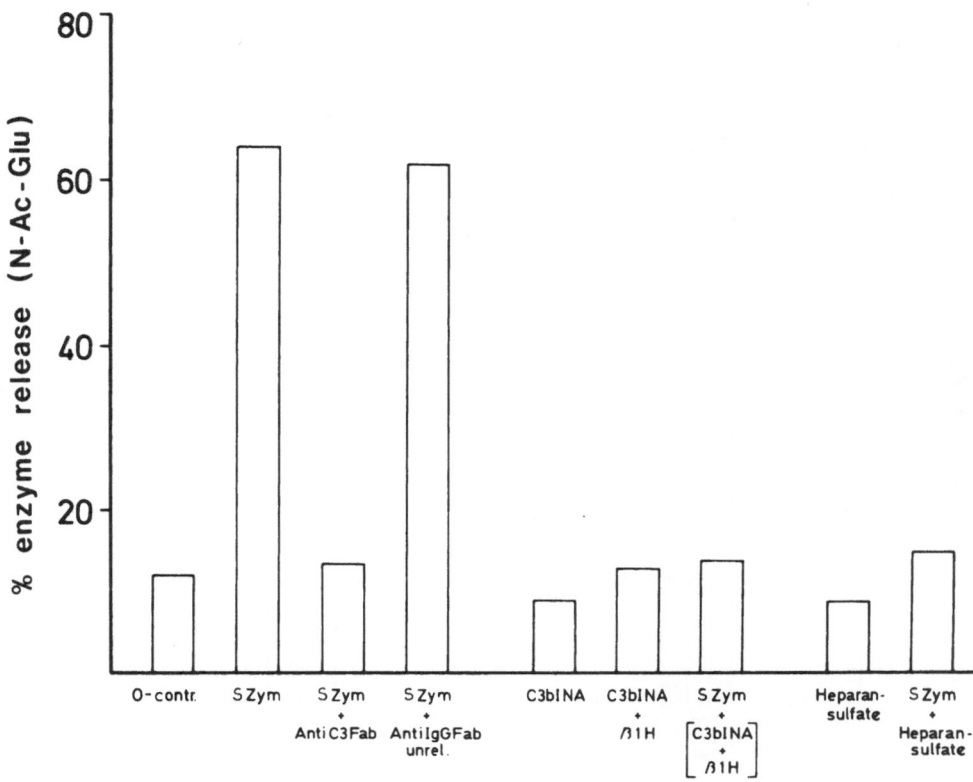

Fig. 4. There is a factor present in macrophage cultures originally stimulated by various agents [e.g., by zymosan (S Zym)] which is able to activate new macrophages

to activate macrophages. This factor seems to be very similar to C3b, indicated, as shown, by the inhibition experiments. If it is true that there is a generation of an activating factor in supernatants of stimulated macrophages we should be able to transfer this factor to new macrophages and to test its activating potency. After adding a cell and particle-free supernatant of originally activated macrophages (e.g., stimulated by zymosan) to new macrophage cultures we get an activation of these macrophages to secrete lysosomal enzymes. The activating potency of this factor is inhibited either by an anti-C3-Fab preparation, by β1H and C3bINA, or by low molecular weight polyanions, indicating again that this factor is similar to C3b (Fig. 4).

Such a factor is not present in supernatants of normal untreated macrophages. But when you add a stimulus like zymosan to a cell-free culture medium of normal macrophages, incubate the mixture for 1 h, then centrifuge and make it particle free again and add this supernatant to fresh macrophage cultures, you get a stimulation of the macrophages (Fig. 5). By using anti-C3-Fab, C3bINA and β1H, or low molecular weight polyanions you either can inhibit the induction phase, which means the generation of this factor (the white columns), or when the factor is already generated you can block the effector phase with these inhibitors, which means the activating potency of the factor to stimulate the macrophages to secrete lysosomal enzymes (these are the dark columns). The observations now presented focus attention to the possibility that endogenous C3 (synthesized by the macrophages themselves) could play a role in the stimulation of mouse peritoneal macrophages by various activators of the alternative pathway.

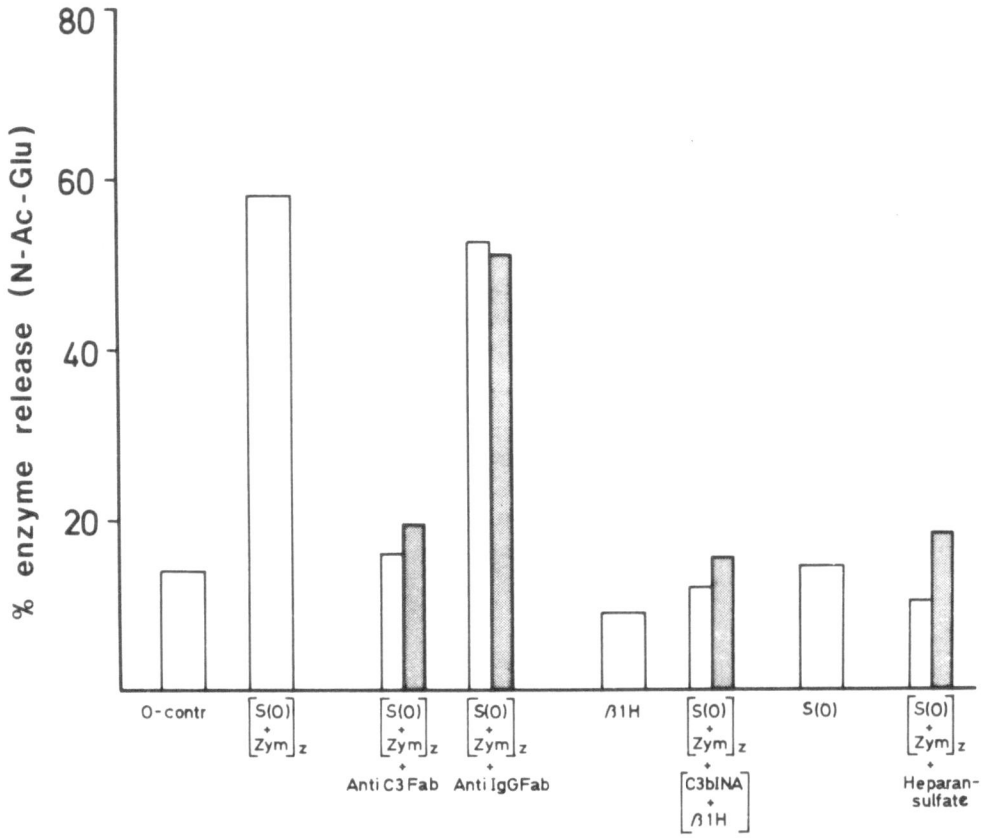

Fig. 5. Inhibition of the activating potency of a factor in macrophage supernatants by regulatory proteins of the alternative pathway. S (O), supernatant of normal macrophages; [S (O) + Zym]$_z$, supernatant of normal macrophages incubated with zymosan and then centrifugated

8. Discussion

Mackaness [11] introduced the term "macrophage activation" to describe morphological changes in mononuclear phagocytes. Subsequently the terms "activation" and "stimulation" have been used to describe the responses of macrophages to a large number of stimuli both in vivo and in vitro. As indicated elsewhere [1] the use of these terms should always be made with reference to both the stimulus being studied and the cellular response which is measured. The analysis of responses of mononuclear phagocytes to various agents in vivo is difficult since inflammatory lesions in which macrophages are present also contain other cells. Extensive use has been made of in vitro culture methods for studying biochemical responses of macrophages to an endocytic stimulus is characteristic of many other responses of macrophages to agents brought into their environment. The simplest way to characterize these responses is by direct morphological observations. But macrophage responsiveness also can be defined by a large number of functional criteria which develop or are enhanced after a given stimulation. These include increase in pinocytosis and the development of the capacity to kill viruses, infectious agents, and tumor cells. We have reviewed elsewhere various functional and biochemical responses of macrophages as well as their secretory activities [21, 23].

To illustrate how the biochemical responses of mononuclear phagocytes vary according to the state of stimulation of the cells and the nature of the stimulus, we will discuss here only the selected example of the complement components as a stimulus. As shown by us [14, 15] and others [9, 24], there is a conversion of the resting macrophage to an activated state by complement components and factors of the alternative pathway which constitutes a highly effective surveillance and potent antitumor mechanism. Once activated, macrophages can regulate cell proliferation and destroy cells with abnormal growth properties. The mechanisms for macrophage activation expressed as selective enzyme secretion and target cell lysis are still far from being understood.

There is growing evidence that the agents which induce chronic inflammation share the capacity to activate macrophages [16, 17]. Looking for a common factor among the variety of agents that induce chronic inflammation in vivo and hydrolase secretion in vitro, our attention has been drawn to the fact that they all activate complement by the alternative pathway [16]. An early consequence of the activation is cleavage of the complement component C3 into a smaller fragment C3a and a large fragment C3b. Normal macrophages are known to synthesize and secrete several intact complement components, including C3 [2, 10] and factor B of the alternative pathway [2, 28]. However, when the macrophages are activated by C3b [15] or in other ways [12, 16] they release the cleavage products C3a and C3b. The C3b so formed can activate other macrophages, so that serial activation can result. Incubation of macrophages with C3b in a serum-free medium resulted in the appearance of a substance with anaphylatoxic activity [20]. When added to a guinea pig terminal ileum preparation, it produced contraction and tachyphylaxis, and specific desensitization toward a reference sample of C3a identified the material in the supernatant as C3a. We suggest that a common factor in macrophage stimulation is activation of complement by the alternative pathway. Attachment of C3b, which is formed to the complement receptor on macrophages, is a powerful activator resulting in acquisition of the capacity to secrete lysosomal enzymes and to lyse tumor cells in culture [21].

It is known that the complement system is a group of potent biologically active mediators, synthesized in significant amounts by mononuclear phagocytes obtained from various sources. A major product of these cells incude C3b which has been shown to bind to receptors on various cell types with resultant stimulation. Thus C3b appears to be a common and important agent activating macrophages to perform various functions. It seems that three mechanisms for activating complement must be distinguished: 1. activation of the classical pathway through C1, C4, and C2 to cleave C3, 2. activation of the alternative pathway, including the C3b amplification mechanism, and 3. activating macrophages to cleave C3. Moreover, since C3b can activate macrophages to release further C3b, a cellular amplification also exists. The observations now presented focus attention to the possibility that endogenous C3 could play a role in the stimulation of macrophages by various activators of the alternative pathway; once complement activation is initiated there is the possibility of continuing activation. Under normal conditions the establishment of this interaction may be forestalled by the effective control of the alternative pathway via the regulatory proteins β1H and C3bINA. There should be a balance between serial activation and inhibition and that could be important in determining whether chronic inflammatory reactions persist or are terminated. The initiation of macrophage activation and alternative pathway activation may delineate one of the molecular feedback mechanisms that contributes to chronic inflammatory diseases.

Acknowledgments

We greatly appreciate the outstanding technical assistance of Miss A. Börner and we greatfully acknowledge the excellent secretarial help rendered by Mrs. I. Gimbel in the preparation of this manuscript.

References

1. Allison AC, Ferluga J, Prydz H, Schorlemmer HU (1978) The role of macrophage activation in chronic inflammation. Agents Actions 8:27
2. Bentley C, Fries W, Brade V (1978) Synthesis of factors D, B and P of the alternative pathway of complement activation as well as of C3 by guinea-pig peritoneal macrophages in vitro. Immunology 35:971
3. Bergmeyer HU, Bernt E (1970) Lactat-Dehydrogenase. In: Bergmeyer HU (ed) Methoden der enzymatischen Analyse. Chemie, Weinheim, p 533
4. Bianco C, Eden A, Cohn ZA (1967) The induction of macrophage spreading: role of coagulation factors and the complement system. J Exp Med 144:1531
5. Bitter-Suermann D, Hadding U, Melchers F, Wellensiek HJ (1970) Independent and consecutive activation of C5, C6 and C7 in immune hemolysis. I. Preparation of EAC1-5 with purified guinea pig C3 and C5. Immunochemistry 17:955
6. Conchie J, Findlay J, Levy GA (1959) Mamalian glycosidases. Distribution in the body. Biochem J 71:318
7. Ferluga J, Schorlemmer HU, Baptisda LC, Allison AC (1976) Cytolytic effects of the complement cleavage product C3a. Br J Cancer 34:626
8. Ferluga J, Schorlemmer HU, Baptisda LC, Allison AC (1978) Production of the complement cleavage product C3a by activated macrophages and its tumorlytic effects. Clin Exp Immunol 31:512
9. Götze O, Bianco C, Cohn ZA (1979) The induction of macrophage spreading by factor B of the porperdin system. J Exp Med 149:372
10. Lai A Fat RFM, van Furth R (1975) In vitro synthesis of some complement components (C1q, C3 and C4) by lymphoid tissues and circulating leucocytes in man. Immunology 28:359
11. Mackaness GB (1970) Cellular immunity. In: van Furth R (ed) Mononuclear phagocytes. Blackwell, Oxford, p 461
12. Pontz BF, Hanauske-Abel H, Schorlemmer HU (1979) Activation of the alternative pathway of the complement system by different collagen types. Immunobiology 156:239
13. Sandberg AL, Wahl SM, Mergenhagen SE (1975) Lymphokine production by C3b-stimulated B-cells. J Immunol 115:139
14. Schorlemmer HU, Allison AC (1976) Effects of activated complement components on enzyme secretion by macrophages. Immunology 31:781
15. Schorlemmer HU, Davies P, Allison AC (1976) Ability of activated complement components to induce lysosomal enzyme release from macrophages. Nature 261:48
16. Schorlemmer HU, Bitter-Suermann D, Allison AC (1977) Complement activation by the alternative pathway and macrophage enzyme secretion in the pathogenesis of chronic inflammation. Immunology 32:929
17. Schorlemmer HU, Davies P, Hylton W, Gugig M, Allison AC (1977) The selective release of lysosomal acid hydrolases from mouse peritoneal macrophages by stimuli of chronic inflammation. Br J Exp Pathol 58:315
18. Schorlemmer HU, Edwards JH, Davies P, Allison AC (1977) Macrophage responses to mouldy haydust, Micropolyspora faeni and zymosan, activators of complement by the alternative pathway. Clin Exp Immunol 27:198
19. Schorlemmer HU, Ferluga J, Allison AC (1977) Interactions of macrophages and complement components in the pathogenesis of chronic inflammation. In: Willoughby DA, Giroud JP, Velo GP (eds) Perspectives in inflammation. MTP, Lancaster, p 191
20. Schorlemmer HU, Hadding U, Bitter-Suermann D, Allison AC (1977) The role of complement cleavage products in killing of tumor cells by macrophages. In: James K, McBride B, Stuart A (eds) The macrophage and cancer. Econoprint, Edinburgh
21. Schorlemmer HU, Hadding U, Bitter-Suermann D (1979) Effects of complement cleavage products released from stimulated macrophages in allergic diseases. Eur J Rheum Inflammation 2:130
22. Schorlemmer HU, Opitz W, Etschenberg E, Bitter-Suermann D, Hadding U (1979) Killing of tumour cells in vitro by macrophages from mice treated with synthetic dehydrodipeptides. Cancer Res 39:1847
23. Schorlemmer HU, Hanauske-Abel H, Pontz B (1980) Influence of different collagen types on complement and macrophage activation. J Immunol 124:1539
24. Sundsmo JS, Götze O (1980) Human monocyte spreading induced by factor B of the alternative pathway of complement activation. Cell Immunol 52:1

25. Talalay P, Fishman WM, Huggins (1946) Chromogenic substances II. Phenolphthalein glucuronic acid as substrate for the assay of glucuronidase activity. J Biol Chem 166:757
26. Taylor-Robinson D, Schorlemmer HU, Furr PM, Allison AC (1978) Macrophage secretion and the complement cleavage product C3a in the pathogenesis of infections by mycoplasmas and L-forms of bacteria and in immunity to these organisms. Clin Exp Immunol 33:486
27. Weissmann G, Dukor P, Zurier RB (1971) Effect of cyclic AMP on release of lysosomal enzymes from phagocytes. Nature [New Biol]. 231:131
28. Whaley K (1980) Biosynthesis of the complement components and the regulatory proteins of the alternative complement pathway by human peripheral blood monocytes. J Exp Med 151:501
29. Woolen JW, Heyworth R, Walker PG (1961) Studies on glucosaminidase and N-acetyl-β-galactosaminidase. Biochem J 78:111

Haematology and Blood Transfusion Vol 27
Disorders of the Monocyte Macrophage System
Edited by F. Schmalzl, D. Huhn, H.E. Schaefer
© Springer-Verlag Berlin Heidelberg New York 1981

Macrophage-Dependent Production of Erythropoietin and Colony-Stimulating Factor

W. Heit, I.N. Rich and B. Kubanek

Monocytes and macrophages play an essential role in the thymus-dependent immune response by taking up an antigen and presenting it to the T-cell. They phagocyte and kill infectious agents during secondary immune response. Furthermore, macrophages secrete a wide range of biologically active molecules which influence the development of the immune-responsive lymphocytes and even play a role in the control of hemopoiesis. Proliferation and differentiation of hemopoietic precursor cells in vitro require specific growth factors such as colony-stimulating factor (CSF) for granulopoiesis and erythropoietin (Epo) for erythropoiesis. In the presence of these glycoproteins committed hemopoietic stem cells form distinct granulocytic–monocytic cell aggregates (colony-forming unit in culture, CFU-c) or erythroid colonies (colony-forming unit - erythroid, CFU-e), when incubated in semisolid culture medium. CSF is ubiquitously distributed in rodent and human tissues [3-5, 10, 14, 17, 18, 26, 28] and monocytes and macrophages are of the major sources of it [6, 12, 13, 17]. The role of macrophages in controlling erythropoiesis was until very recently only suggestive.

Erythropoietin (Epo) is produced by the kidney and regulates red cell production. Erythropoietin can also be produced extrarenally; the primary organ which has been implicated in this function has been the liver [9, 29]. The cell thought to be responsible for Epo production in the liver is the Kupffer cell or liver macrophage [20, 30]. Since macrophages are extremely versatile cells, producing and releasing many substances which affect the functioning of other cell types [8, 16], we initiated a series of studies to investigate whether macrophages from various sources could release Epo. Using silica as a specific cytotoxic agent [2, 15, 19], macrophages could be shown to release an erythropoietic stimulating factor (ESF) into the extracellular fluid. This ESF was identified to be Epo by in vivo and in vitro bioassays and, in addition, by the neutralization of the ESF activity by antierythropoietin [7, 25]. The important question as to whether Epo can be produced by the macrophage was investigated in vitro. We report a new culture technique whereby Epo can be shown to be generated for at least 4 weeks by spleen cells in suspension on gas-permeable hydrophobic Teflon surfaces. Macrophages were defined by morphological (including electronmicroscopy) and cytochemical criteria and by their ability to ingest latex particles.

Crystalline silica, although chemically inert, can be extremely toxic to biological membranes. In contrast to so-called rapid cytotoxicity, delayed cytotoxicity occurs at low silica concentrations and in the presence of serum [1]. Silica is ingested and reacts with the lysosomal membrane, causing its rupture and death to the cell [1]. The process whereby Epo is released from macrophages by silica treatment is absolutely dependent on phagocytosis of the silica particles (Rich and Kubanek, in preparation). Epo activity released from different macrophage-containing cell suspensions after 30-min exposure to silica varies in the following manner: spleen > lung > bone marrow > peritoneal macrophages > fetal liver > adult liver > kidney [23, 24]. Since the spleen released the highest amount of Epo activity, the cells from this organ were used in the following experiments.

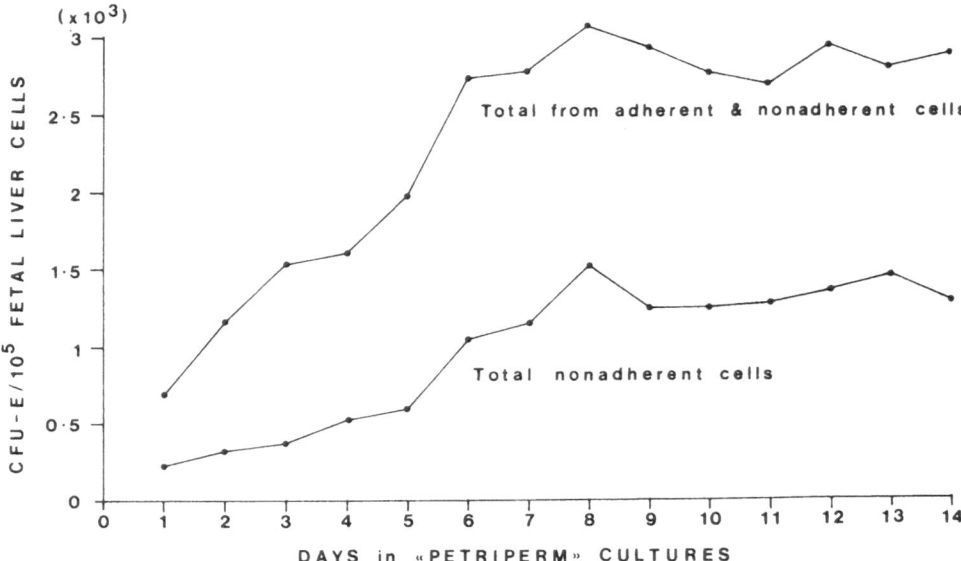

Fig. 1. Erythropoietin production from separated mouse spleen cells cultured in suspension in Petriperm dishes. Spleens from 10 CBA/Ca virgin female mice 10–12 weeks of age were removed and a single cell suspension prepared in Hank's balanced salt solution (Hank's BSS) containing 3% fetal calf serum (FCS) as described previously [21]. Prior to use, the fetal calf serum was treated with 1 mg/ml Dextran T-40 (Pharmacia Fine Chemicals) and 10 mg/ml Norit A, activated charcoal (Serva Chemicals, West Germany) for 30 min. This procedure consistently increased CFU-E growth regardless of serum batch. The single cell spleen suspension was underlayered with 2 ml FCS/10 ml suspension and left on ice for 30 min. After withdrawing the cell suspension, the latter was washed twice with Hank's BSS containing FCS, centrifuging between each washing (1300 rpm; 10 min; 4 °C). The cells were resuspended in 20 ml Hank's BSS containing FCS and the nucleated cells counted using a Coulter Counter Model ZF. The cells were then incubated at a concentration of 10×10^6 cells/ml in a total volume of 30 ml in 250 ml (75 cm^2) tissue culture flasks (Costar, Cambridge, Mass., USA) for 2 h at 37 °C and 5% CO_2. After this time, the contents of the flasks were swirled and the nonadherent cells withdrawn and transferred to 50 ml plastic tubes (Falcon Plastics, Oxnard, Cal., USA). The remaining adherent cells were removed with a rubber policeman and also transferred to plastic tubes which were centrifuged as described above. The cell pellets were resuspended in 10 ml Iscove's Modified Dulbecco's Medium (IMDM) containing 5% FCS and the nucleated cells counted again. Nonadherent and adherent cell suspensions were transferred to Petriperm dishes (Haereus, Hanau, West Germany) at a concentration of 24×10^6 cells/3 ml volume diluted in IMDM + 5% FCS.

Four groups were prepared; nonadherent and adherent cells were cultured in the presence or absence of alpha-thioglycerol at a final concentration of 1×10^{-4} M. On each of the days shown in the above diagram, one culture group was terminated. The original medium contained in the dishes was removed carefully and transferred to a plastic tube. The contents were centrifuged (see above) and the supernatants collected, stored at –20 °C, and tested together with the other samples at the end of the experiment. The cells remaining in the dishes were then treated with silica as described in the legend for Fig. 2. All supernatants were tested in the 12–13 day fetal liver erythroid colony-forming technique to detect erythrocytic colony-forming units or CFU-E as described previously [22]. In this bioassay, 1×10^5 cells/ml from 12–13 day CBA/Ca fetal livers prepared as described elsewhere [21] were stimulated with 10% v/v of the supernatant. In addition pure mouse transferrin [27] was added to cultures at an end concentration of $1.3 \times 10^{-12} M$

The detailed culture procedure is explained in the legend to Fig. 1. In essence, nonadherent and adherent cells were obtained by incubating washed CBA/Ca mouse spleen cell suspensions in tissue culture flasks. Thereafter, the separated cells were transferred to "Petriperm" dishes having a gas-permeable hydrophobic culture surface in order to prevent

74

Table 1. Activity of Epo and CSF in spleen cell suspension cultures[a]

Colonies/10^5	Days of Culture					
	1	3	7	10	14	21
CFU-e	507	647	807	767	840	747
CFU-c	44	90	127	143	206	202

[a] CFU-e originated from fetal mouse livers (13/14 days of gestation): CFU-c grew in cultures of adult mouse bone marrow cells containing IMDM supplemented with fetal calf serum (25% v/v for CFU-e, 20% v/v for CFU-c). The supernatants of mouse spleen cell suspension cultures harvested between day 1 and 21 served as source of Epo and CSF (10% v/v)

adherence of the cells to this surface. For each of the two groups containing adherent or nonadherent spleen cells, supplemented either with or without alpha-thioglycerol, the original supernatants were withdrawn at daily intervals for a period of 2 weeks, and the remaining cells treated with silica in order to determine the Epo activity released during the time of incubation as well as that present residually in the cell. The Epo activity was assayed in the in vitro bioassay using 12–13 day fetal liver cells plated in the presence of supernatant and assayed for CFU-E growth in which colony incidence above 15–20 CFU-E/10^5 cells plated can be attributed to Epo activity in the supernatant.

A biphasic production curve was obtained in all groups over the first 14 days of culture, the plateau beginning at about day 7 and continuing until day 14. In a separate experiment,

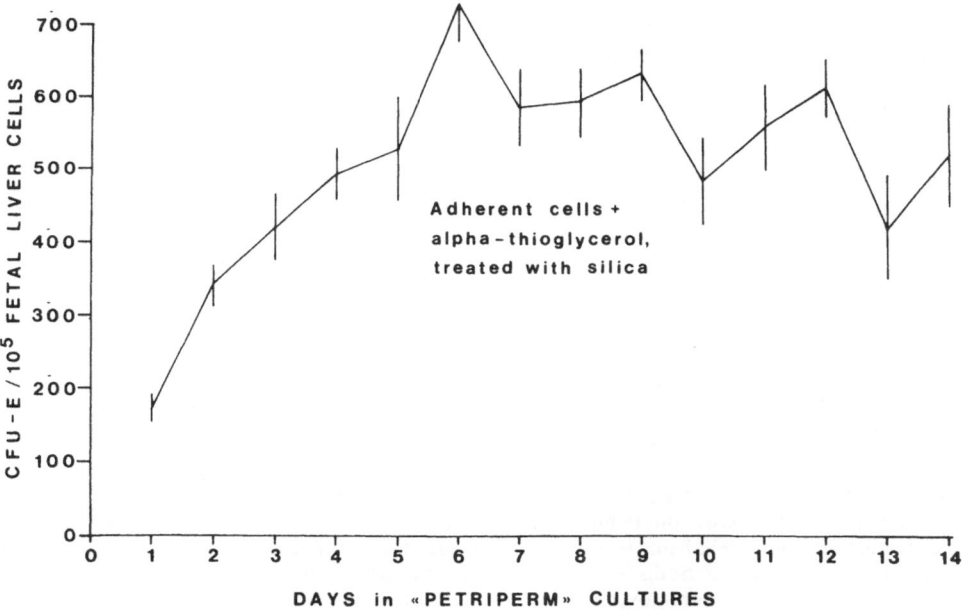

Fig. 2. Release of residual erythropoietin by treatment of Petriperm cultured spleen cells with crystalline silica. After the original supernatants had been withdrawn, the remaining cells were incubated with 10^{-4} g/ml final concentration of crystalline silica (min-U-Sil, 2–5 um; whittaker, Clarke & Daniels, New Jersey, USA) suspended in Hank's BSS + 3% FCS for 30 min at 37 °C in 5% CO_2. Thereafter, the cells and silica suspension was withdrawn, transferred to a plastic tube, and the contents centrifuged at 1300 rpm for 10 min at 4 °C. The supernatant was collected and stored at –20 °C until tested at the end of the experiment together with the original supernatants for Epo activity in the 12–13 day fetal liver CFU-E bioassay. Results given as the mean of two experiments ± S.E.M

Table 2. Erythropoietic production in macrophages generated by CFC in vitro[a]

	Conditioned medium from:					
	CFC-derived cells			CFC-derived adherent cells		
	−	Silica	+	−	Silica	+
Fetal liver CFU-E/10^5 cells	195		220	535		265

[a] Mouse bone marrow cells at a concentration of 1×10^5 cells/ml were prepared for culture in 0.8% methyl cellulose containing 20% horse serum and 10% heart conditioned medium (CSF) [30] which was inactivated by heating to 56 °C in a water bath for 30 min. After 10 days of culture, macrophage colonies predominated (155 colonies/10^5 cells). 1×10^7 cells or 5570 colonies were then resuspended in 1 ml serum-free IMDM and incubated for 30 min at 37 °C in 5% CO_2 or in 5 ml IMDM supplemented with 20% horse serum and cultured for 24 h at 37 °C in 5% CO_2 [24, 25]. Thereafter, the supernatant was decanted and the adherent cells overlayered with 5 ml of serum-free IMDM for another 30 min. In both groups, part of the cultures were exposed to 1×10^{-4} gm/ml silica. The conditioned media were then used as a source of Epo tested in the in vitro bioassay

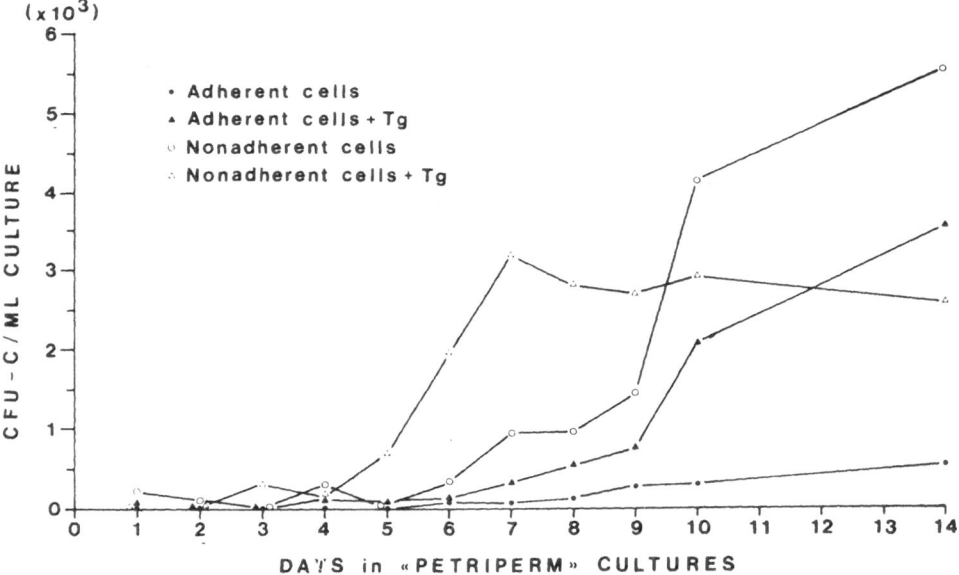

Fig. 3. Incidence of granulocyte–macrophage colony-forming units (CFC) per ml using Petriperm suspension cultures. The basic culture procedure was the same as that described for Fig. 1. However, instead of the remaining cells being treated with silica, the cells were brought into complete suspension by swirling the contents of the dishes and, if necessary, by washing the culture surface with the supernatant. The suspension was removed from the dish using a syringe with a 0.5×25 mm needle. In this way, aggregates of cells (many of which were present) could be disaggregated and made into a single cell suspension. This was controlled by microscopic examination of the cell suspension. More than 95% of the cells in the dish could be recovered. The cells were centrifuged to a pellet in plastic tubes and the supernatant withdrawn (and stored for later testing) and finally resuspended in 1 ml IMDM + 3% FCS. The number of nucleated cells was counted as described in the legend to Fig. 1 and GM-CFC cultures prepared in 0.8% methyl cellulose containing alpha thioglycerol and 10% v/v of FCS and horse serum. In addition, 10% v/v of heart conditioned medium was used as a source of CSF [5]. Cultures were counted after 7 days of incubation under the same conditions as CFU-E cultures. The number of cells plated was 0.5×10^5 cells/ml in all cases

plateau levels were observed over a period of 4 weeks. The total Epo content in the culture was calculated from the combined activity of the original supernatants and the residual activity after silica exposure of the cells (Fig. 1, upper curve). Erythropoietin production was identical in both adherent and nonadherent cell suspensions. Based on Epo sensitivity curves of the in vitro bioassay, Epo activity at plateau levels corresponded to approximately 25 mU/ml/day for adherent and nonadherent cells. The indication that besides Epo production spleen cells may generate significant amounts of CSF is given in Table 1. The activity over the 21 day period implies that both CSF and Epo are produced rather than released from an intracellular storage pool in culture. This hypothesis is substantiated by the release of Epo activity after silica exposure (Fig. 2), since the same biphasic production curve is obtained.

Since we could show that nonadherent as well as adherent spleen cells could produce and release Epo, it was of interest to see whether under the culture conditions employed macrophages were being produced, thereby indicating that the system could support proliferation and differentiation.

Experiments suggesting that this was the case demonstrated that macrophages derived from granulocyte–macrophage colony-forming cells (CFU-C) in the presence of mouse heart conditioned medium (CSF) were able to release Epo in vitro (Table 2).

An attempt was then made at correlating the kinetics of Epo production in Petriperm cultures with macrophage (M-CFC) production in these same cultures. In Fig. 3 the GM-CFC content per ml culture is shown assessed daily over a 14-day period. All colonies consisted of macrophages, the growth curve of which was comparable to the kinetics of Epo production by these cells. This indicates that Epo production may be ascribed to de novo formed macrophages.

In conclusion, macrophages possess the capacity to produce, store, and secrete Epo independently of renal production, a suggestion in keeping with earlier findings on erythropoietic differentiation in nephrectomized rats [11]. Although kidney-produced Epo is the main regulating principle of red cell production in the adult, extrarenal macrophage-produced Epo which is also an oxygen-dependent regulating mechanism [23] appears to play an important role in the anephric individual, maintaining red cell production at a lower level.

It is now possible to add Epo which is produced simultaneously with CSF to the list of biologically active molecules produced and secreted by macrophages. However, the role this cell plays in the physiological control of cellular growth and differentiation remains to be elucidated.

Acknowledgments

This work was supported by the Deutsche Forschungsgemeinschaft (SFB 112/A2). The authors would like to thank Mrs. U. Schnappauf, Mrs. I. Brackmann, and Miss N. Higgins for their excellent technical assistance. We would also like to thank Dr. Siegwald Pentz for his stimulating discussions.

References

1. Allison AC (1971) Arch Intern Med 128:131–139
2. Allison AC, Harrington SS, Birbach M (1966) J Exp Med 124:141–153
3. Bradley TR, Metcalf D (1966) Aust J Exp Biol Med Sci 44:287–299
4. Bradley TR, Stanley ER, Sumner MA (1971) Aust J Exp Biol Med Sci 49:595–603
5. Byrne P, Heit W, Kubanek B (1978) Br J Haematol 40:187–204
6. Chervenick PA, Lo Buglio AF (1972) Science 178:164–166
7. Cotes PM, Bangham DR (1961) Nature 191:1065–1067
8. Davies P, Bonney RJ (1979) J Reticuloendothel Soc 26:37–47
9. Fisher JW (1979) J Lab Clin Med 93:695–699
10. Golde DW, Finley TN, Cline MJ (1972) Lancet 2:1397–1399

11. Gordon AS, Kaplan SM (1977) In: Fisher JW (ed) Kidney hormones-erythropoietin. Academic Press, New York, pp 187–229
12. Heit W, Kern P, Kubanek B, Heimpel H (1974) Blood 44:511–515
13. Heit W, Kern P, Heimpel H, Kubanek B (1977) Scand J Haematol 18:105–112
14. Iscove NN, Senn JS, Till JE, McCulloch EA (1971) Blood 37:1–5
15. Kessel RWI, Monaco L, Maraschisio MA (1963) J Exp Pathol 44:351–366
16. Lejeune FJ, Vercommen-Grandjean A (1979) Front Biol 48:149–159
17. Moore MAS, Williams N (1972) J Cell Physiol 80:195–206
18. Paran M, Sachs L, Barak Y, Resnitzki P (1970) Proc Natl Acad Sci USA 67:1542–1549
19. Pearsall MM, Weiser R (1968) J Reticuloendothel Soc 5:107–120
20. Peschle C, Marone G, Genovese A, Rapport IA, Condorelli M (1976) Blood 47:325–337
21. Rich IN, Kubanek B (1979) J Embryol Exp Morphol 50:57–74
22. Rich IN, Kubanek B (1980) J Embryol Exp Morphol 58:143–155
23. Rich IN, Heit W, Kubanek B (1980) Blut 40:297–303
24. Rich IN, Anselstetter V, Heit W, Kubanek B (1980) In: Lucarelli G, Fliedner TM, Gale P (eds) Proceedings of the first international symposium on foetal liver transplantation. Elsevier/North Holland, Amsterdam
25. Rich IN, Anselstetter V, Heit W, Kubanek B (to be published) J Supramol Struct
26. Robinson WA, Pike BL (1970) In: Stohlmann F Jr (ed) Symposium on hemopoietic cellular differentiation. Grune & Stratton, New York, pp 249–259
27. Sawatzky G, Anselstetter V, Kubanek B (to be published) Biochim Biophys Acta
28. Sheridan W, Stanley ER (1971) J Cell Physiol 78:451–460
29. Zanjani ED, Peterson EN, Gordon AS, Wasserman LR (1974) J Lab Clin Med 83:281–287
30. Zucali JR, Stevens V, Mirand EA (1975) Blood 46:85–90

Haematology and Blood Transfusion Vol 27
Disorders of the Monocyte Macrophage System
Edited by F. Schmalzl, D. Huhn, H.E. Schaefer
© Springer-Verlag Berlin Heidelberg New York 1981

The Relation Between Monocytes and Resident (Tissue) Macrophages

J.M. de Bakker, A.W. de Wit and W.T. Daems

1. Introduction

Consensus has been reached concerning the bone marrow origin of peritoneal exudate macrophages [19] but not about the nature of precursor cells for the normal peritoneal resident macrophage [11]. From functional, biochemical, and EM cytochemical studies it is known that there are differences between macrophages obtained from unstimulated and stimulated peritoneal cavities [11]. These differences might either be due to the existence of two peritoneal macrophage populations or reflect the differentiation of monocytes into mature macrophages.

The concept of the Mononuclear Phagocyte System [20] considers all macrophages – whether from a stimulated or unstimulated peritoneal cavity – to be monocyte derived. Another view advocates a dual origin for macrophages, acknowledging that under inflammatory and experimental conditions monocytes become macrophages but considering the peritoneal resident macrophages as a self-sustaining macrophage population [11, 32, 35]. According to this latter view, resident macrophages can be assumed to be able to synthesize DNA in the steady-state peritoneal cavity.

From LM autoradiographical studies it is known that after the administration of ^{3}H-thymidine, silver grains are present over the blood monocytes and also over the macrophages in the peritoneal cavity, the liver, the lung, etc. [18]. Induction of inflammation leads to an increase in the number of labeled cells at the site of the inflammation, and these cells have been considered, on the basis of morphological, cytochemical, and/or receptor studies, to be monocytes or exudate (monocyte-derived) macrophages. Such studies contributed to the conclusion that resident macrophages might be derived from monocytes [18].

Other studies, however, shed doubt on such a relationship between monocytes and resident macrophages [11, 35]. It was found, for instance, that both types of cells could be distinguished by EM cytochemistry, resident macrophages having peroxidatic (PO) activity in the rough endoplasmic reticulum (RER) and the nuclear envelope and monocytes with PO activity in granules [15]. These findings and the results of kinetic studies [11, 35] led to the assumption that monocytes and resident macrophages are two separate types of cell with a different origin. Furthermore, resident macrophages in mitosis have been found in the unstimulated peritoneal cavity [11], which supports the assumption that, for example, peritoneal resident macrophages are capable of proliferation and form a self-sustaining population of cells [12].

To add new data to those already available, the proliferative capacity of the peritoneal macrophages in vivo was investigated by EM autoradiography of ^{3}H-thymidine-labeled cells. In addition, studies were performed with iron as a lysosomal and cytoplasmic marker to find out how long the macrophages retain the marker in the peritoneal cavity under steady-state conditions. To collect more information about the origin of the steady-state peritoneal macrophages, experiments were performed in which the peritoneal cavity was depleted of macrophages and the pattern of re-establishment of a normal population was studied. The results of these experiments will be briefly summarized here.

2. Results

Distinction between monocytes and peritoneal resident macrophages was made on the basis of a different localization of the peroxidatic activity [12, 13] (Fig. 1).

2.1 ^{3}H-Thymidine Labeling

Mice received one intravenous injection of ^{3}H-thymidine, and peritoneal and blood cells were collected at various intervals from 15 min to 8 days after the injection. EM autoradiography was used to examine the cells with respect to the presence of label over the nucleus. Labeled resident macrophages (Fig. 2) occurred from 15 min onward after ^{3}H-thymidine administration. About 4% of the peritoneal resident macrophages were labeled. Monocytes with incorporated ^{3}H-thymidine were first found in the peripheral blood 24 h after administration of the label.

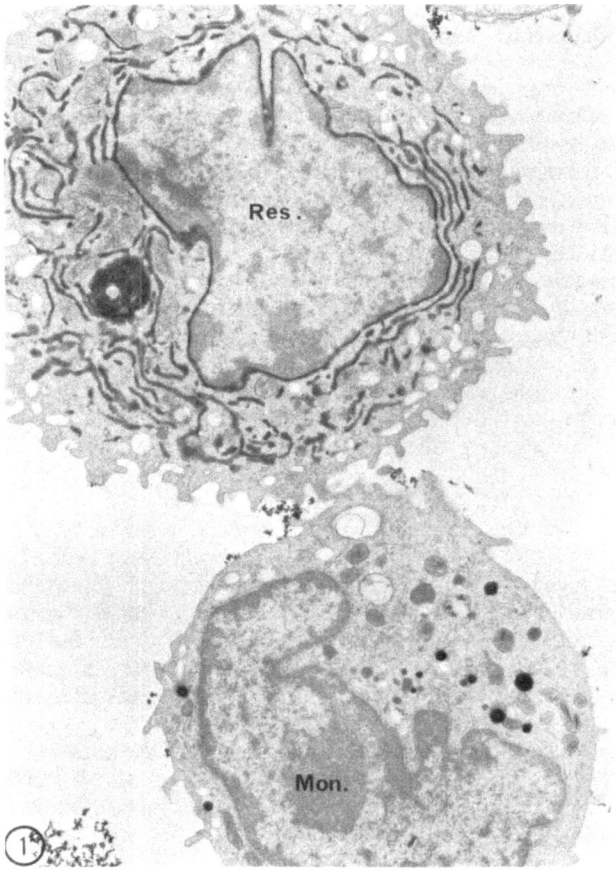

Fig. 1. Peroxidatic activity in a peritoneal resident macrophages (Res.) and a peritoneal monocyte (Mon.). In the former this activity is located in the rough endoplasmic reticulum and the nuclear envelope, in the latter in granules. × 7700

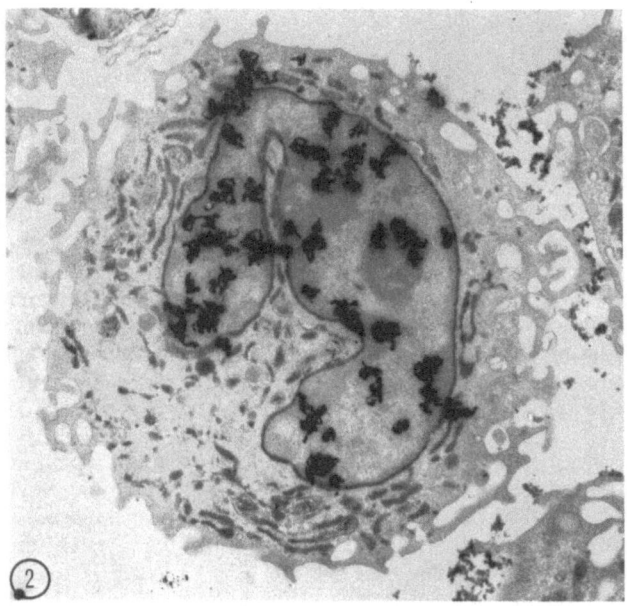

Fig. 2. EM autoradiograph of a peritoneal resident macrophage isolated from an unstimulated mouse peritoneal cavity 24 h after one intravenous injection of ^{3}H-thymidine. Silver grains, indicating the presence of ^{3}H-thymidine, lie over the nucleus. Peroxidatic activity is present in the RER and the nuclear envelope. × 7700

In addition, LM cytospin preparations of unstimulated peritoneal cells were Feulgen-stained, and the absorption of each cell was determined at 559 nm as a measure of the DNA content. This cytophotometric investigation of peritoneal cells revealed that about 5% of the steady-state peritoneal macrophages had an amount of DNA higher than 2 N.

2.2 Imferon Marking

Mice were given 2 ml of an iron-dextran solution (Imferon^R) intraperitoneally as a lysosomal and cytoplasmic marker. At various intervals from 3 min to 11 months after the injection of Imferon, the peritoneal cells were isolated. The iron-dextran was taken up by the peritoneal resident macrophages by micropinocytosis, metabolized by the cells, and converted into ferritin (Fig. 3). Imferon induced an inflammatory reaction in the peritoneal cavity. From 6 h on, blood monocytes entered the peritoneal cavity, and after that the exudate macrophages too contained Imferon. The number of peritoneal cells reached normal

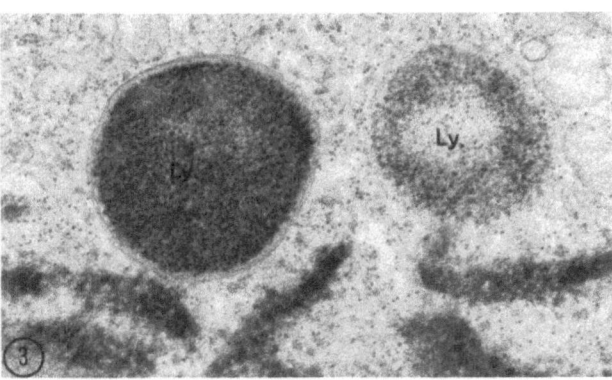

Fig. 3. A peritoneal resident macrophage 6 months after intraperitoneal administration of iron-dextran. At a high magnification ferritin particles are visible in the cytoplasm and the lysosomes (Ly). Peroxidatic activity is present in the RER. × 67 000

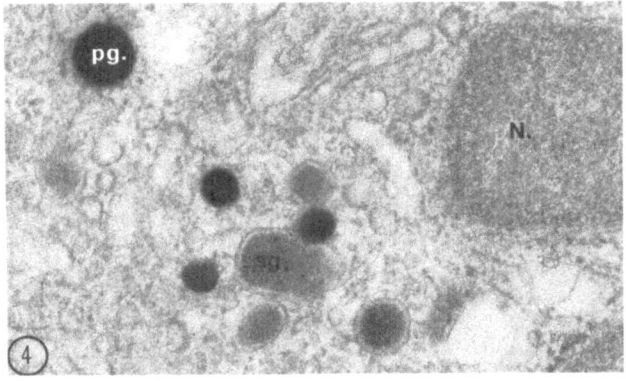

Fig. 4. A peritoneal monocyte collected 6 months after intraperitoneal administration of iron-dextran. The primary (p.g.) and secondary (s.g.) granules and cytoplasm are devoid of ferritin. Peroxidatic activity is present in part of the primary granules N., nucleus. × 67 000

values on the 5th day after intraperitoneal injection. As soon as the steady state was restored in the peritoneal cavity, the monocytes no longer showed iron in the lysosomes or cytoplasm (Fig. 4). Monocytes in the circulation did not have iron either. Following the presence of iron in the peritoneal macrophages it was found that up to 4 months after Imferon administration, 100% of the resident macrophages were marked (Fig. 5). After 11 months, resident macrophages containing ferritin in lysosomes and the cytoplasm could still be observed.

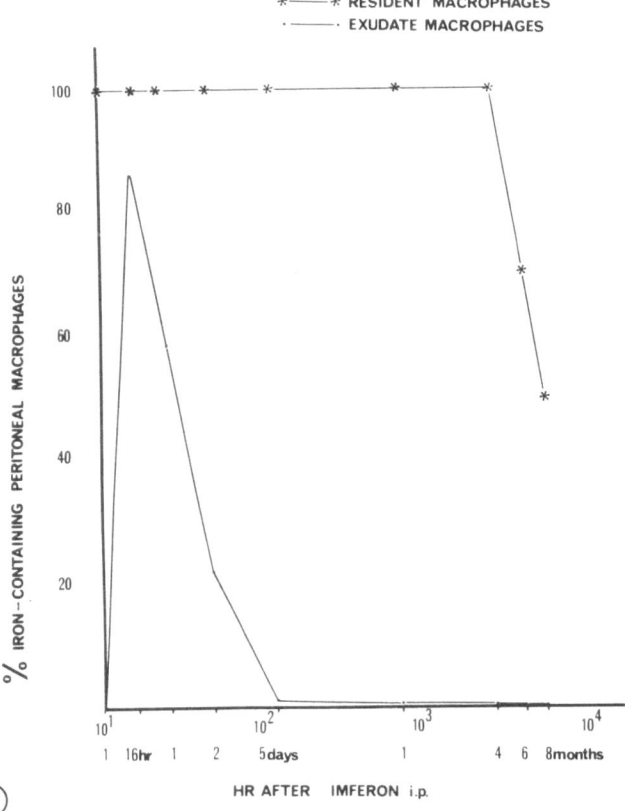

Fig. 5. At intervals between 3 min and 8 months after intraperitoneal administration of iron-dextran peritoneal macrophages were examined for the presence of iron in the cytoplasm and lysosomes

In a preliminary experiment mice were given an intraperitoneal injection of thiogly-colate 1 month after receiving Imferon intraperitoneally. One month later 82.6% of the peritoneal resident macrophages were iron positive. In the controls without thioglycolate, 100% of the resident macrophages were iron positive.

2.3 Repopulation Experiments

In an attempt to establish the origin of the peritoneal macrophage population, the peritoneal cavity of mice was depleted of cells by peritoneal washing. The peritoneal cells were removed under general anaesthesia by injecting four times 5 ml 0.9% NaCl (37 °C). After depletion, the restoration of the macrophage population was studied by isolating peritoneal cells at various times. The cells were studied by EM-PO cytochemistry and DNA-Feulgen cytophotometry. Similar experiments were performed with lethally irradiated mice.

The results showed restoration of the peritoneal resident macrophage population after depletion, together with enhanced DNA synthesis − as shown by LM cytophotometry − by the peritoneal macrophages. Whole body irradiation experiments showed that the restoration of the resident macrophage population occurred independently of the circulating pool of blood monocytes.

3. Discussion

It was found in this and earlier studies that, unlike monocytes and monocyte-derived macrophages, resident (= tissue) macrophages have peroxidatic activity in the RER and nuclear envelope [12, 13, 26, 33]. In contrast to monocytes, the lysosomes of resident macrophages are devoid of endogeneous peroxidase activity [15]. Under certain conditions, monocytes may acquire PO activity in the RER [7, 9, 16].

The view that these cells are transitional between monocyte-derived macrophages on the one hand and resident macrophages on the other is based on the assumption that the PO activity in the RER in both types of cell is identical. There is, however, no firm evidence that this is the case. Cytochemically, this enzyme activity can be distinguished from that in resident macrophages [15]. This, together with the fact that monocytes end as PO-negative cells, both in vitro and in vivo [23], whereas resident macrophages have the PO-positive RER and nuclear envelope as a permanent feature, points to the conclusion that the transient PO activity of monocytes is different from that in resident macrophages.

This view is strengthened by the observation that even in man, where PO activity in RER could not be demonstrated in normal resident macrophages, the monocytes acquire PO-positive RER when cultured in vitro. It should be emphasized, in addition, that the occurrence of the so-called exudate-resident macrophage [7] is restricted to a few cases: in vitro experiments and a limited number of situations in vivo.

Other in vivo experiments failed to reveal PO activity in the RER [25]. Even under conditions in which a large number of monocytes are present in the peritoneal cavity, e.g., after thioglycolate stimulation or after removal of the entire population of peritoneal cells [4], monocytes with both PO-positive granules and PO-positive RER are absent. Under steady-state conditions, too, cells that are, with respect to the distribution of PO activity, transitional between monocytes and resident macrophages are absent [7, 15].

It is of importance to emphasize here that the differences in PO activity between peritoneal resident macrophages and monocytes–exudate macrophages appear to be correlated with differences in cell surface characteristics (vide infra).

To obtain further information about the origin and life-span of resident macrophages, these cells were studied in two types of labeling experiments. EM autoradiography of ^{3}H-thymidine-labeled cells made it possible to establish that peritoneal resident macrophages are a proliferating population of cells [5]. The presence of labeled cells as early as 15 min after intravenous administration of ^{3}H-thymidine demonstrated that peritoneal resident macrophages are capable of synthesizing DNA in the peritoneal cavity and excludes the possibility that these cells took up the label in the bone marrow. Additional proof of DNA synthesis was obtained by cytophotometry of Feulgen-stained peritoneal macrophages under steady-state conditions. The results of our labeling experiments with ^{3}H-thymidine agree well quantitatively with those reported by other authors [2, 19] but point to the conclusion that LM studies cannot be used for the elucidation of the origins of monocytes and resident macrophages, because none of these studies permitted the conclusion that the cells labeled in the first 48 h were resident macrophages and not, as was assumed, monocytes or monocyte-derived macrophages.

Monocyte-derived macrophages and resident macrophages have many functions in common [19], but in a number of respects macrophages from unstimulated peritoneal cavities differ from those from peritoneal exudates [8, 11, 22, 24, 34–36]. This is the case, for instance, for the activity of 5'-nucleotidase and the release of plasminogen activator (Schnyder, this volume p. 27), for C3b-mediated endocytosis [24], and for the affinity for wheat germ agglutinin (WGA) [1, 21]. In view of this, particularly the macrophage population at the site of an inflammation must considered to be a heterogeneous one, with the degree of heterogeneity depending on the proportion of resident versus monocyte-derived macrophages and this in turn depending on the type and age of the inflammation.

In fact, the question should be posed whether the functional differences between the macrophage populations from stimulated and unstimulated peritoneal cavities reflect differences in the degree of activation of all macrophages or are the result of the entry into the peritoneal cavity of monocytes, giving rise to exudate macrophages with properties differing essentially from those of the pre-existing resident population. Among these properties are the binding of WGA, a nonmitogenic lectin which binds to a considerably lower degree to macrophages from an unstimulated peritoneal cavity compared with the binding to

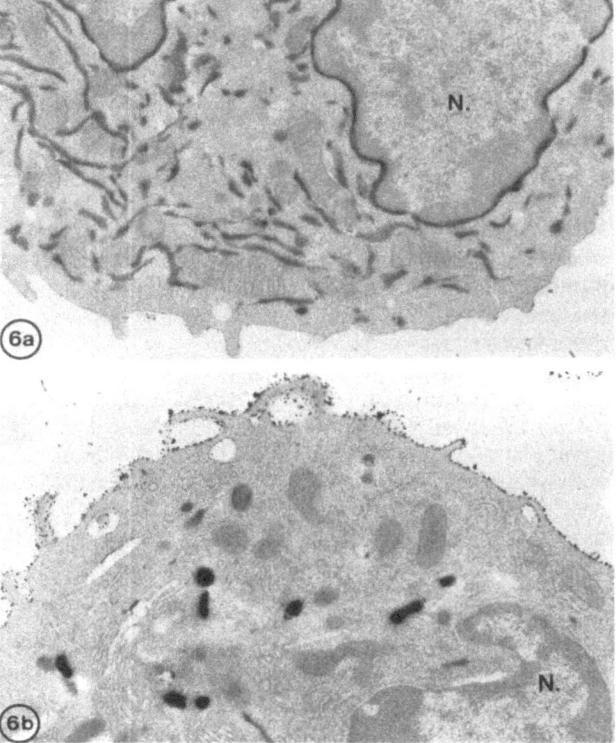

Fig. 6a, b. Electron micrographs showing a difference between a peritoneal resident macrophage and a peritoneal monocyte with respect to the affinity for wheat germ agglutinin. The binding of WGA, shown with the gold-labeling method, is negligible to the resident macrophage (**a**) and strong to the monocyte (**b**). N, nucleus. Photographs reproduced with permission of Dr. R. de Water. × 11 500

macrophages from a thioglycolate-induced peritoneal exudate [21]. In cytochemical experiments [37] it was found that in both the stimulated and unstimulated peritoneal cavity the binding of WGA, determinated with the gold-labeling method, was negligible in the case of resident macrophages (Fig. 6a) and strong to monocytes or monocyte-derived macrophages (Fig. 6b). This is in agreement with the finding that WGA binding to bone marrow macrophage was lower than that to bone marrow monocytes [1]. In addition, 5'-nucleotidase activity, known to be high in a macrophage population from an unstimulated peritoneal cavity compared with that of an exudate macrophage population, was found in EM cytochemical experiments to be present on peritoneal resident macrophages (Ginsel et al., unpublished observations).

Another important functional difference between resident macrophages and monocyte-derived macrophages is shown by studies on the formation of epithelioid and giant cells [14, 30, 31]. This work demonstrated that resident macrophages are not involved in the formation of epithelioid cells and multinucleated giant cells: both types of cell are PO negative in their mature state and on other grounds as well can be assumed not to derive from the resident macrophages. Similar findings have been reported by Papadimitriou [27, 28] and Naito et al. [25]. The observation that monocytes and not resident macrophages participate in the formation of multinucleated giant cells is further supported by the finding that in experiments in which the resident macrophages are labeled with iron dextran, the giant cells do not show the label. Under these conditions the monocytes do not have iron in either the cytoplasm or lysosomes (van der Rhee and Cambier, personal communication).

These results indicate, in our opinion, that the question as to the functional heterogeneity of macrophage populations should preferably be answered on the basis of studies done at the level of the single cell and combining various methods for characterization with the determination of functional differences.

The use of a cytoplasm–lysosome label in the form of ferritin showed that under these experimental conditions peritoneal resident macrophages form a long-lived population of cells [6]. In the Imferon studies all peritoneal resident macrophages were marked for at least 4 months. However, blood monocytes and peritoneal monocytes, or monocyte-derived macrophages, were iron negative. From these findings it is concluded that during the period of 4 months studied, the (iron-positive) peritoneal macrophages could not have derived from (iron-negative) monocytes. In addition, the experiment showed that part of the peritoneal resident macrophage population can reside in the peritoneal cavity for periods of up to 11 months. These results contradict a postulated turn-over time of 5–6 weeks (van Furth, personal communication). Recirculation of iron in the peritoneal cavity could be excluded. The induction of a peritoneal inflammatory reaction 30 days after Imferon administration led to an increased monocyte influx and subsequent differentiation into peritoneal exudate macrophages. The fact that none of these exudate macrophages contained iron shows that no pool of iron was present to give rise to newly formed iron-positive resident macrophages. It could be established with other methods as well that the transformations of monocytes into peritoneal resident macrophages is not obligatory [29].

The results of our studies lead to several conclusions. Peritoneal resident macrophages show the capacity for local proliferation. Blood monocytes form a transient (traveling) population of cells in the peritoneal cavity. Circulating blood monocytes do not transform into peritoneal resident macrophages.

Induction of a peritoneal exudate leads, among other things, to an increased influx of monocytes from the peripheral blood into the peritoneal cavity. These cells differentiate locally and become exudate macrophages, which can be distinguished from resident macrophages.

Thus, our findings point to a dual origin of peritoneal macrophages, under both steady-state and inflammatory conditions: one population of (resident) macrophages that proliferates locally and another macrophage population that originates from bone marrow derived monocytes. A similar situation was recently described for alveolar macrophages [2, 10] and for Kupffer cells [17].

We postulate the existence of a precursor cell of the peritoneal resident macrophage, most probably being located in the milky spots. This precursor cell can be assumed to give rise by mitosis to immature "proresident" macrophages which have a PO pattern similar to that of the resident cells and can be isolated from the peritoneal cavity. These proresident cells give rise by mitosis to the mature resident macrophages.

Support for the existence of a precursor cell in the omentum is found in the observation that a considerable number of macrophages are formed in the isolated omentum in vitro [3]. The existence of the precursor is also supported by the finding that in the Imferon experiments intraperitoneal stimulation leads to replacement of part of the pre-existing resident macrophage population. This replacement occurs from an iron-negative cell, indicating that the population of resident macrophages can be maintained from a precursor differing from the pre-existing resident macrophages. Our results further show that this replenishment is independent of the monocyte pool in the blood and support the hypothesis that the precursor cell of the resident macrophage is not the circulating monocyte.

Acknowledgment

The investigations were supported in part by the Foundation for Medical Research (FUNGO), which is subsidized by the Netherlands Organisation for the Advancement of Pure Research (ZWO).

References

1. Ackerman GA (1979) Distribution of wheat germ agglutinin binding sites on normal human and guinea pig bone marrow cells: An ultrastructural histochemical study. Anat Rec 195:641
2. Adamson IYR, Bowden DH (1980) Role of monocytes and interstitial cells in the generation of alveolar macrophages. II. Kinetic studies after carbon loading. Lab Invest 42:518
3. Aronson M, Shahar M (1965) Formation of histiocytes by the omentum in vitro. Exp Cell Res 38:133
4. Bakker JM de, Daems WT (to be published) The heterogeneity of mouse peritoneal macrophages. In: Proceedings of the International Workshop on the heterogeneity of mononuclear phagocytes, Vienna, Austria, July 15–19
5. Bakker JM de, Brederoo P, Onderwater JJM, Wit AW de, Daems WT (1980) DNA synthesis in mouse peritoneal macrophages: an autoradiographic and cytophotometrical study. In: Brederoo P, Priester W de (eds) Biology. Electron microscopy 1980, vol 2, The Hague p 194
6. Bakker JM de, Noordende JM van't, Daems WT (1980) The existence of a peritoneal resident macrophage population as studied by Imferon-treated mice. In: Brederoo P, Priester W de (eds) Electron microscopy, vol 2, Biology. The Hague p 192
7. Beelen RHJ, Fluitsma DM, Meer JWM van der, Hoefsmit ECM (1980) Development of exudate-resident macrophages, on the basis of the pattern of peroxidatic activity in vivo and in vitro. In: Furth R van (ed) Mononuclear phagocytes, functional aspects, vol 1. Nijhoff, The Hague Boston London, p 87
8. Bianco C, Edelson PJ (1978) Plasma membrane expressions of macrophage differentiation. In: Lorner R (ed) The molecular basis of cell-cell interaction, vol 14. Liss, New York, p 119
9. Bodel PT, Nichols BA, Bainton DF (1977) Appearance of peroxidase reactivity within the rough endoplasmic reticulum of blood monocytes after surface adherence. J Exp Med 145:264
10. Bowden DH, Adamson IYR (1980) Role of monocytes and interstitial cells in the generation of alveolar macrophages. I. Kinetic studies of normal mice. Lab Invest 42:511
11. Daems WT (1980) Peritoneal macrophages. In: Carr I, Daems WT (ed) The reticuloendothelial system, vol 1, Morphology. Plenum, New York London p 57
12. Daems WT, Brederoo P (1973) Electronmicroscopical studies on the structure, phagocytic properties, and peroxidatic activity of resident and exudate peritoneal macrophages in the guinea pig. Z Zellforsch 144:247
13. Daems WT, Koerten HK (1978) The effects of various stimuli on the cellular composition of peritoneal exudates in the mouse. Cell Tissue Res 190:47
14. Daems WT, Rhee HJ van der (1980) Peroxidase and catalase in monocytes, macrophages, epithelioid cells and giant cells of the rat. In: Furth R van (ed) Mononuclear phagocytes, functional aspects, vol 1. Nijhoff, The Hague Boston London, p 43

15. Daems WT, Roos D, Berkel ThJC van, Rhee HJ van der (1979) The subcellular distribution and biochemical properties of peroxidase in monocytes and macrophages. In: Dingle JT, Jacques PJ, Shaw IH (eds) Lysosomes in applied biology and therapeutics, vol 6. North Holland, Amsterdam New York Oxford, p 463
16. Deimann W, Fahimi HD (1977) The ontogeny of mononuclear phagocytes in fetal rat liver using endogenous peroxidase as a marker. In: Wisse E, Knook DL (eds) Kupffer cells and other sinusoidal cells. Elsevier/North Holland Biomedical Press, Amsterdam, p 487
17. Deimann W, Fahimi HD (1980) Hepatic granulomas induced by glucan. An ultrastructural and peroxidase-cytochemical study. Lab Invest 43:172
18. Furth R van (1976) Origin and kinetics of mononuclear phagocytes. Ann NY Acad Sci 278:161
19. Furth R van (1980) Cells of the mononuclear phagocyte system. Nomenclature in terms of sites and conditions. In: Furth R van (ed) Mononuclear phagocytes, functional aspects, vol 1. Nijhoff, The Hague Boston London, p 1
20. Furth R van, Cohn ZA, Hirsch JG, Humphry JH, Spector WG, Langevoort HL (1972) The mononuclear phagocyte system: A new classification of macrophages, monocytes and their precursor. Bull W H Org 46:845
21. Geoghegan WD, Ackerman GA (1977) Differential binding of wheat germ agglutinin to murine lymphocytes and macrophages (Abstr). J Reticuloendothel Soc 22:33a
22. Griffin FM, Bianco C, Silverstein SC (1975) Characterization of the macrophage: Receptor for complement and demonstration of its functional independence from the receptor for the Fc portion of immunoglobulin G. J Exp Med 141:1269
23. Meer JWM van der (1980) Characteristics of mononuclear phagocytes in culture. In: Carr I, Daems WT (eds) The reticuloendothelial system, vol 1, Morphology. Plenum, New York London p 735
24. Munthe Kaas AC, Kaplan G (1980) Endocytosis by macrophages. In: Carr I, Daems WT (eds) The reticuloendothelial system, vol 1, Morphology. Plenum, New York London p 19 Morphology (A comprehensive treatise). Plenum, New York London (The reticuloendothelial system, vol 1, p 19)
25. Naito M, Oka K, Miyazaki M, Sato T, Kojima M (1978) Ultrastructural and cytochemical changes of monocytes and macrophages in vivo and in vitro. Recent Adv RES Res 18:66
26. Ogawa T, Koerten HK, Daems WT (1978) Peroxidatic activity in monocytes and tissue macrophages of mice. Cell Tissue Res 188:361
27. Papadimitriou JM (1978) Macrophage fusion in vivo and in vitro: A review. In: Poste G, Nicolson GL (eds) Membrane fusion. Elsevier/North Holland Biomedical Press, Amsterdam New York, p 181
28. Papadimitriou JM (1979) The role of resident and exudate macrophages in multinucleated giant cell formation. J Pathol 128:93
29. Rhee HJ van der, Winter CPM de, Daems WT (1977) Fine structure and peroxidatic activity of rat blood monocytes. Cell Tissue Res 185:1
30. Rhee HJ van der, Burgh-de Winter CPM van der, Daems WT (1979) The differentiation of monocytes into macrophages, epithelioid cells, and multinucleated giant cells in subcutaneous granulomas. I. Fine structure. Cell Tissue Res 197:355
31. Rhee HJ van der, Burgh-de Winter CPM van der, Daems WT (1979) The differentiation of monocytes into macrophages, epithelioid cells, and multinucleated giant cells in subcutaneous granulomas. II. Peroxidatic activity. Cell Tissue Res 197:379
32. Shands JW, Axelrod BJ (1977) Mouse peritoneal macrophages: tritiated thymidine labeling and cell kinetics. J Reticuloendothel Soc 21:69
33. Soranzo MR, Koerten HK, Daems WT (1978) Peroxidatic activity and morphometric analysis of alveolar macrophages in the guinea pig. J Reticuloendothel Soc 23:343
24. Unkeless JC (1980) Fc receptors on mouse macrophages. In: Furth R van (ed) Mononuclear phagocytes, functional aspects, vol 1. Nijhoff, The Hague Boston London, p 29
35. Volkman A (1976) Disparity in origin of mononuclear phagocyte populations. J Reticuloendothel Soc 19:249
36. Walker WS (1976) Functional heterogeneity of macrophages. In: Nelsen DS (ed) Immunobiology of the macrophage. Academic Press, New York San Francisco London, p 91
37. Water R de, Noordende JM van 't, Ginsel LA, Daems WT (1981) Heterogeneity in wheat germ agglutinin binding by mouse peritoneal macrophages. Histochemistry (in press)

Haematology and Blood Transfusion Vol 27
Disorders of the Monocyte Macrophage System
Edited by F. Schmalzl, D. Huhn, H.E. Schaefer
© Springer-Verlag Berlin Heidelberg New York 1981

Development of Exudate-Resident Macrophages In Vivo and In Vitro

R.H.J. Beelen

1. Summary

The development of the pattern of peroxidatic activity (PA) has been studied in vivo and in vitro in macrophages and other mononuclear phagocytes. In vivo in the animal model the macrophages show four different PA patterns characterizing exudate macrophages, exudate-resident macrophages, resident macrophages and PA-negative macrophages. Cultured in vitro, rat and human blood monocytes and rat, mouse and human exudate macrophages acquire the characteristic PA pattern of resident macrophages via the transitional stage of cells with the characteristics of exudate-resident macrophages. The cytochemistry, the occurrence of this cell type in vivo and the kinetics in vitro indicate that exudate-resident macrophages represent a transitional form between the exudate and the resident macrophages. The results obtained in vivo and in vitro strongly suggest that divergent PA patterns of mononuclear phagocytes represent differences in the stages of development of these cells in the sequence monoblast – promonocyte – monocyte – exudate macrophage – exudate-resident macrophage – resident macrophage. Moreover, PA-negative cells may in fact represent, firstly, another transitional stage between exudate and resident cells and, secondly, an end-stage macrophage in vitro.

2. Introduction

For the study of the localization of PA[1] in various types of cell, including macrophages, the diaminobenzidine (DAB) technique developed by Graham and Karnovsky [17] has widely been used. On the basis of the intracellular distribution of PA two types of macrophage have been distinguished, viz. resident macrophages and exudate macrophages.

Peritoneal resident macrophages show PA exclusively in the nuclear envelope and rough endoplasmic reticulum (RER); they have been described in the rat [2, 28], the guinea pig [12, 13], the mouse [20, 25] and the rabbit [9]. Tissue macrophages, such as guinea pig alveolar macrophages [30], rat tingible body macrophages and medullary lymph node macrophages [18], Kupffer cells [35, 36] and thyroid follicular cells [32] show this same PA pattern. Macrophages with the PA pattern of resident macrophages have not been found in vivo in man [1, 3, 34]; however, they were seen in vitro after culture of blood monocytes [3, 8].

Peritoneal exudate macrophages and blood monocytes show PA only in some of the lysosomal granules. This PA pattern has been described in the rat [2, 27], the guinea pig [13] and the mouse [20, 25]. Human blood monocytes [24] and pleural exudate macrophages [3] also show this PA localization. Rabbit blood monocytes, however, have no PA in the lysosomal granules [24].

---

[1]The term PA refers to the enzymatic oxidation product of DAB with H_2O_2 which becomes electron opaque with OsO_4

According to the concept of the mononuclear phagocyte system (MPS) [16, 19], all macrophages present in serous cavities (pleural and peritoneal macrophages) and tissue macrophages (e.g. Kupffer cells) are derived from blood monocytes originating from a common precursor in the bone marrow. Although kinetic studies support the view that resident macrophages too are monocyte derived [15], doubts have remained [13, 14, 26, 29, 33], mainly because of the differences in PA pattern between blood monocytes and resident macrophages.

The PA pattern has been used to support the theory that some macrophages are derived from reticulo-endothelial precursors. This so-called RES theory of macrophage development is obviously in conflict with the previously mentioned MPS concept. However, promonocytes of the bone marrow, which have been shown to be the precursors of monocytes [15], show yet another PA pattern, viz. PA in the nuclear envelope, the Golgi system and the lysosomal granules [7]. This shows that differences in the PA pattern do not necessarily distinguish between types of cell but in this case do distinguish between different developmental stages of one cell type.

The studies in his paper refer to the development of various PA patterns in rat peritoneal macrophages during acute and chronic inflammations in vivo and after culture in vitro. Macrophages from other species, including man, were also studied both in vivo and in vitro. Finally the PA pattern of mononuclear phagocytes developing in vitro in bone marrow cultures and in vivo in milky spots were studied.

3. Material and Methods

Male Wistar rats weighing 180–200 g were used. In addition, the peritoneal cells of male guinea pigs, Swiss mice and rabbits were studied. Peritoneal exudates were induced by intraperitoneal administration of sterile newborn calf serum (NBCS), which causes an acute inflammation, or paraffin oil, which provokes a chronic inflammation. The cells from the peritoneal cavities were isolated at various intervals as described in detail elsewhere [2, 4]. Fluid from the cavities of patients suffering from chronic inflammatory diseases was collected via a pleural puncture or during a peritoneal laparoscopy [3]. Blood monocytes were obtained according to Boyum [10, 11]. The mononuclear cells were fixed immediately (in vivo) or processed further for culture (in vitro) on plastic surfaces, to which cells adhere, according to Bodel et al. [9] or on Teflon dishes according to van der Meer [21], where surface adherence does not occur. Milky spots were obtained from the omentum of mice and rats as described in detail elsewhere [5, 6].

3.1 Cytochemical Procedures

Cells were fixed for 10 min in 1.5% glutaraldehyde, washed three times in 0.1-M Na-cacodylate (pH 7.4), brought into an Eppendorf tube and reacted for PA in diaminobenzidine (DAB) as described previously [2, 3], i.e. pH 6.5; preincubation and incubation (0.01% H_2O_2) were for 1 h at 20 °C. The cells were then washed, postfixed in OsO_4, processed for electron microscopy as described previously and embedded in araldite. Since the oxidized DAB-polymer becomes electron opaque with OsO_4 [32], staining with lead citrate was omitted. Staining with uranyl acetate was omitted because it could have extracted or masked a reaction product [8].

In vitro incubated adherent cells were fixed, washed, reacted for PA and processed as just described while still in dishes until dehydration in absolute alcohol was complete [8]. The cells were then removed with propylene oxide as described previously [3, 8, 31], concentrated to small pellets and embedded in araldite. All electron microscopic data were derived from two to six experiments in each of which 100 to 400 cells were studied.

4. Results

4.1 Localization of Peroxidatic Activity in Rat Blood Monocytes and Peritoneal Macrophages In Vivo

After induction of an acute inflammatory reaction with NBCS or a chronic inflammatory reaction with paraffin oil, four populations of macrophages can be distinguished on the basis

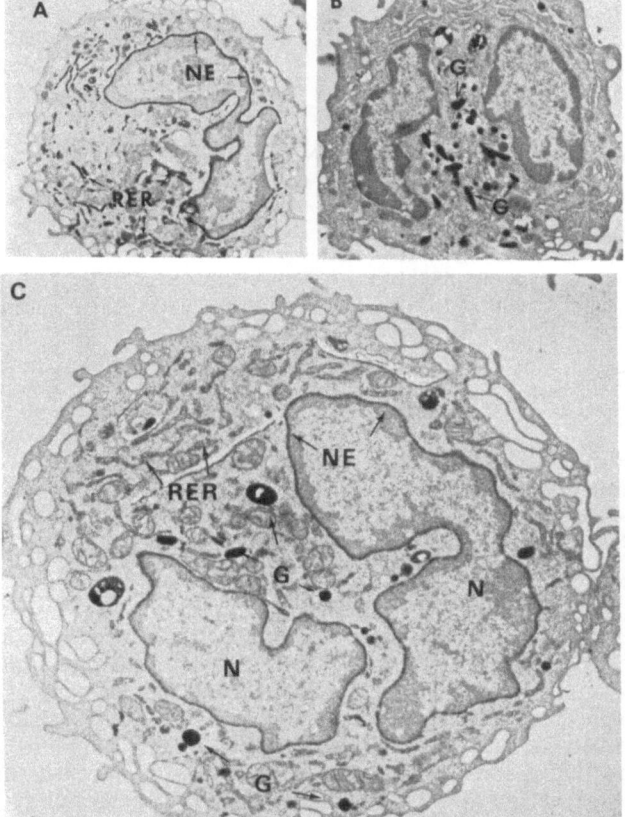

Fig. 1. Three patterns of PA representing three types of macrophage. **A** Resident macrophages with reaction product in the nuclear envelope *(NE)* and rough endoplasmic reticulum *(RER)*. **B** Exudate macrophages with reaction product in cytoplasmic granules *(G)* (lysosomal vesicles). **C** Exudate-resident macrophages with reaction product in the lysosomal vesicles as well as in the nuclear envelope and RER. *N,* nucleus. **A** × 5400; **B** × 5400; **C** × 10 000

of their different patterns of PA. Firstly some macrophages of the peritoneal exudates showed reaction product in the nuclear envelope and RER (Fig. 1A) and so show the PA pattern of normal resident macrophages. Secondly, some macrophages of the peritoneal exudates showed reaction product in a varying number of lysosomal granules (Fig. 1B) and so show the PA pattern of blood monocytes and exudate macrophages. Thirdly, some macrophages of the peritoneal exudates, however, showed reaction product both in the nuclear envelope and the RER and in a varying number of lysosomal granules (Fig. 1C). The granules of these cells showed exactly the same localization (preferentially Golgi area), dimensions (100 to 450 nm), morphological characteristics (i.e. oval or spherical and occasionally elongated) and cytochemical characteristics as those of the exudate macrophages. The reactivity of the nuclear envelope and RER in these cells behaved just like those in resident macrophages. Since these cells have the combined morphological and cytochemical features of both the exudate and the resident macrophages, we have called them exudate-resident macrophages [2]. In addition, there was a fourth population of macrophages in the peritoneal exudates; these cells showed no reaction product at all and so are called PA-negative macrophages.

4.2 Quantitative Aspects of These Four Types of Macrophage

During the normal steady state, resident macrophages usually predominate in the peritoneal cavity (> 90%). After induction of an acute inflammation the total number of resident

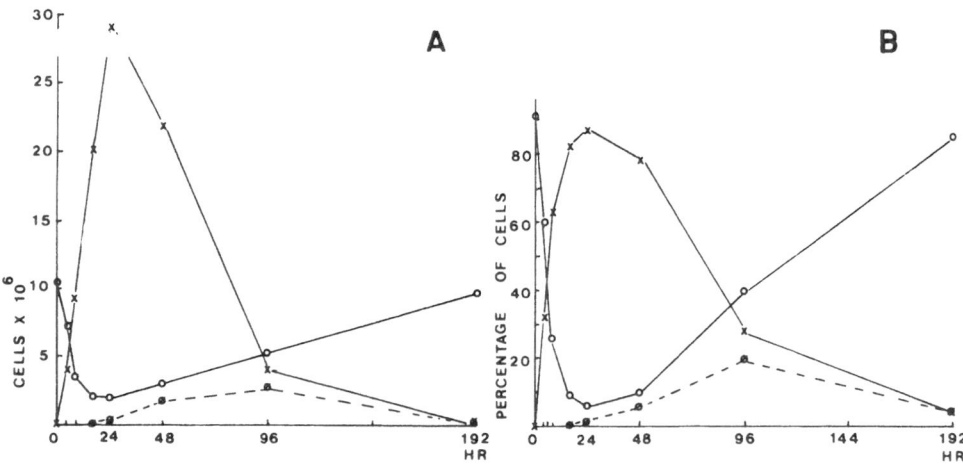

Fig. 2. Numerical course of the exudate macrophages (— x —), resident macrophages (— o —) and exudate-resident macrophages (---- ∅ ----) in the inflammatory peritoneal exudates 0 to 8 days after an injection of NBCS. **A** shows the total number of each cell type, and **B**, the percentages

macrophages decreased to 2×10^6 cells at 24 h and then increased until the normal level was regained 8 days after induction. The number of exudate macrophages rose sharply, with a peak at 24 h and then decreased, returning to the normal level after 8 days. The exudate-resident macrophages, which first appeared 24 h after induction of the inflammation, reached a maximum of about 2×10^6 cells at 48–96 h and then decreased (Fig. 2A). The kinetics of these three types of macrophages during acute inflammation (Fig. 2B) clearly show that the maxima of exudate-resident cells coincide with the sharp decrease in the number of exudate macrophages and the rapid increase in the number of resident macrophages, which indicates that exudate-resident macrophages are indeed a transitional form between exudate and resident macrophages. PA-negative macrophages were present only in low and fairly constant numbers during an acute inflammation.

After the induction of a chronic inflammation, the number of exudate macrophages and PA-negative macrophages rose sharply to about 12×10^6 cells at 2 days and remained at that level up to 8 days. The number of resident macrophages, which decreased after induction of the inflammation, remained at a constant level of about 2×10^6 cells. Exudate-resident macrophages, which first appeared after induction of the inflammation, also remained at a fairly constant level of 2×10^6 cells for at least 8 days [4]. The kinetics of these four types of macrophages during chronic inflammation suggest a constant influx of monocyte-derived exudate macrophages due to the chronic inflammatory conditions. These exudate macrophages may subsequently lose the PA in their lysosomal granules and become PA negative.

4.3 Localization of Peroxidatic Activity in Rat Blood Monocytes and Peritoneal Macrophages In Vitro

Cultured blood monocytes and cultured peritoneal macrophages from a peritoneal exudate, elicited 24 h previously with NBCS, develop in vitro the four PA patterns of exudate macrophages, exudate-resident macrophages, resident macrophages and PA-negative macrophages (Fig. 3).

The relative numbers of these mononuclear phagocytes and the four types of localization of PA at different time points are given in Table 1. The relative number of exudate ma-

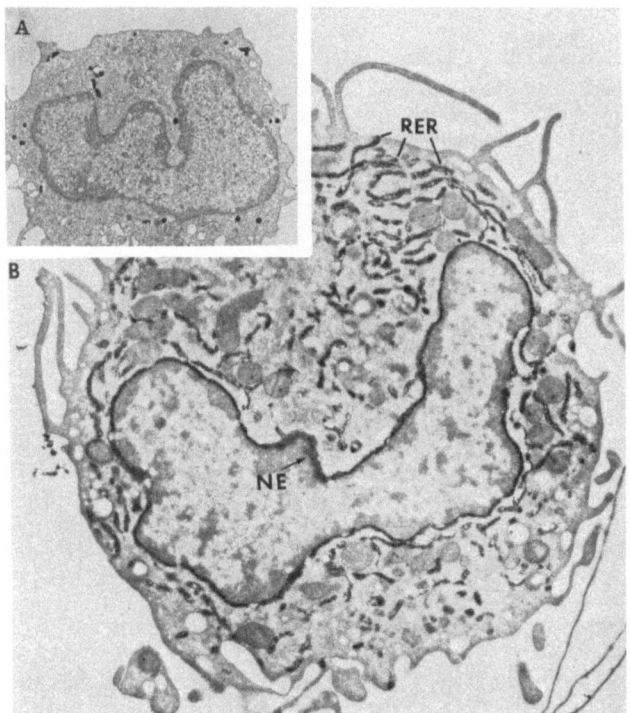

Fig. 3 A, B. Two patterns of PA observed in vitro after culturing rat blood monocytes and rat NBCS induced rat PM. **A** Exudate macrophage at the start of culture (0 days). This cell shows PA only in the lysosomes *(G)*. **B** Blood monocyte after 8 days of culture. This cell has the cytochemical characteristics of a resident macrophage, i.e. it shows PA in the nuclear envelope (NE) and rough endoplasmic reticulum (RER). **A** × 4700; **B** × 9000

crophages declines during incubation in vitro to less than 5% at 8 days. Exudate-resident macrophages first appear in high numbers after incubation in vitro at day 2. The relative number of resident macrophages increased in vitro throughout culture and this cell becomes the predominant type after 8 days. The results indicate that just as in vivo, exudate macrophages in vitro also acquire the characteristic PA pattern of resident macrophages via the intermediate stage of exudate-resident macrophages. Finally, the relative number of PA-negative macrophages also increases at the end of the culture period. The same was found, however, for the control culture of resident macrophages obtained from an animal in the normal

Table 1. The four different types of localization of PA in cultured rat blood monocytes and rat peritoneal macrophages

Origin	Days in culture	Localization of peroxidatic activity in			
		Granules (%)	Granules + RER (%)	RER (%)	None (%)
Blood monocytes	0	90			10
	2–4	30	30	20	20
	8	< 5	< 5	60	30
Peritoneal macrophages in	0	90			5	5
acute inflammation[a]	2–4	30	30	20	20
	8	< 5	< 5	60	30
Peritoneal macrophages in the	0			95	5
normal steady state	2–4			> 90	
	8			> 60	

[a] 24 h after NBCS, cultured on plastic and on Teflon

Table 2. The various types of macrophage occurring in the peritoneal cavity in the normal steady state and during inflammation in different species

Origin	Exudate macrophages	Exudate-resident macrophage	Resident macrophage	PA-negative macrophages
	%	%	%	%
Rat				
Normal steady state			95	5
Acute inflammation[a]	90		5	5
Chronic inflammation[b]	45	5	45	5
Mouse				
Normal steady state[b]			98	2
Acute inflammation[b]	65		5	30
Man				
Chronic inflammation[c]	50			50

[a] 2–8 days after paraffin oil
[b] 24 h after NBCS
[c] Pleural effusions

steady state (Table 1), and this finding is in agreement with the findings of others in vitro [1, 8, 9, 22].

In the rabbit the appearance of PA in the RER after surface adherence in vitro has also been shown but was temporary in nature [9]. Table 1 shows our results obtained in the cultures on plastic and refers to the adherent cells. Moreover, macrophages cultured on a Teflon surface, which prevents real surface adherence from the start [21], also develop these same PA patterns (Table 1).

Table 3. The four different types of localization of PA in cultured mouse peritoneal macrophages, cultured human blood monocytes and pleural macrophages

Origin	Days of culture	Localization of PA in			
		Granules (%)	Granules + RER (%)	RER (%)	None (%)
Mouse					
Peritoneal macrophages,	0			98	2
normal steady state	2			>98	
	4			>98	
Peritoneal macrophages,	0	65		5	30
acute inflammation[a]	2		20	75	5
	4		5	90	5
Man					
Blood monocytes	0	+	−	−	+
	1–2	+	+	+	+
	6			+	(>90)
Pleural macrophages	0	+	−	−	+
	1	+	+	+	+
	2–6			+	(>90)

[a] 24 h after NBCS

4.4 Localization of Peroxidatic Activity in Macrophages of Different Species In Vivo

In peritoneal macrophages deriving from normal unstimulated rats, mice, guinea pigs and rabbits we found predominantly (> 90%) resident macrophages, which agrees with the findings of others [9, 14, 20, 30]. After the induction of an acute or chronic inflammation in the mouse and guinea pig, approximately the same results were obtained as those described for the rat, viz. the development of exudate, exudate-resident, resident and PA-negative macrophages, although we found more PA-negative cells in the mouse and the guinea pig. In the rabbit after acute and chronic inflammations we found predominantly PA-negative cells, since rabbit blood monocytes have no PA in the lysosomal granules [23]. In human pleural effusions under chronic inflammatory conditions we found exclusively exudate and PA-negative macrophages [3]. Some of the results obtained in the different species are summarized in Table 2. These data suggest that especially during chronic inflammation (high turn-over) the loss of PA in the lysosomal granules may be accelerated, possibly due to the high phagocytic activity, and the appearance of the PA in the RER may be inhibited, possibly due to the presence of some inflammatory factors in vivo; as a result, monocyte-derived exudate macrophages might transform into PA-negative macrophages directly.

Cultures of rat and mouse peritoneal macrophages gave fairly similar results (Table 3). The appearance of cells with PA localization in the RER (as in resident macrophages and exudate-resident macrophages), however, occurs earlier in the mouse. On the other hand, the percentage of PA-negative cells at the start is about 30% and falls to about 5% after 2–4 days in culture. Since the cell loss during the first 4 days of culture was less than 10%, this means that at least some of the PA-negative macrophages may acquire the cytochemical characteristics of resident macrophages in vitro.

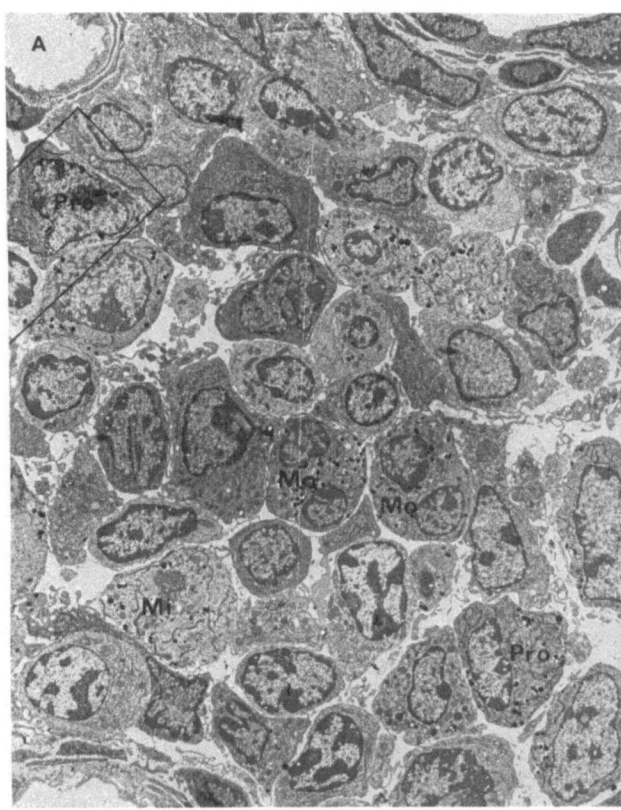

Fig. 4A–C. Part of milky spot on mouse omentum, obtained after induction of acute inflammation with NBCS, showing promonocytes *(Pro)* and surrounded by monocytes *(Mo)*. The promonocyte shows reaction product in the nuclear envelope *(NE)*, rough endoplasmic reticulum *(RER)*, Golgi system *(GS)* and a number of lysosomal granules *(G)*. A × 1700; B × 7200; C × 14400

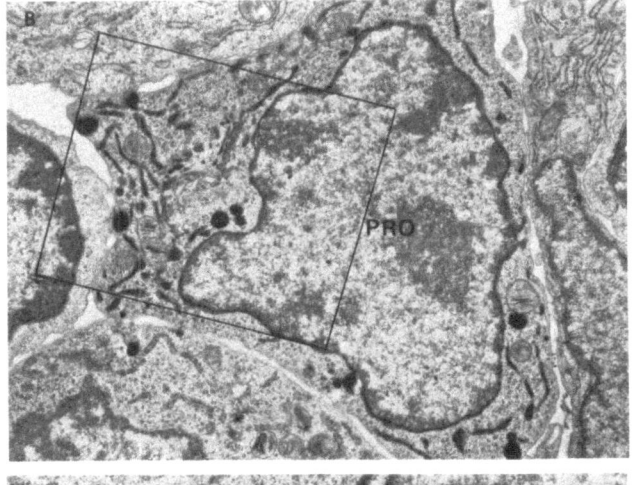

Fig. 4B

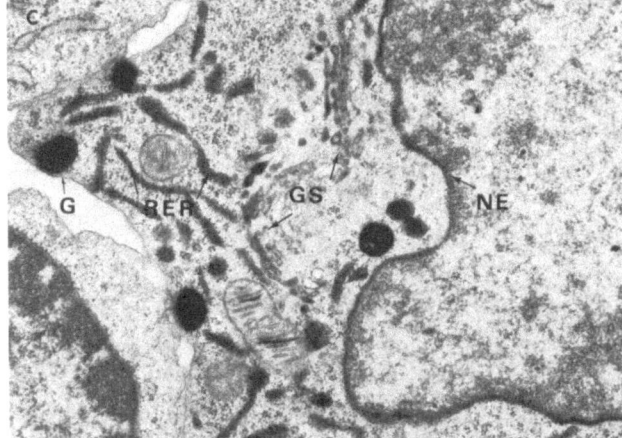

Fig. 4C

Cultured human blood monocytes and pleural macrophages also develop the four patterns of PA localization described in the rat. As shown in Table 3, all four patterns of PA localization are present simultaneously at 6–24 h. However, in these human cells the PA has disappeared after 2–6 days of culture, which is in agreement with the findings of Bodel et al. [8]. Most strikingly, human resident macrophages did develop in vitro, whereas in vivo they were not found at all.

4.5 Development of Macrophages in Bone Marrow Cultures

In long-term cultures of mouse bone marrow, mononuclear phagocytes proliferate, and the various stages of development of these cells were recognized [22]. In the first few days of culture the majority of the cells were monoblasts and promonocytes. Later in the culture period macrophages with all four PA patterns were found: exudate macrophages, exudate-resident macrophages, resident macrophages and PA-negative macrophages.

4.6 Development of Mononuclear Phagocytes in Milky Spots

After induction of an acute inflammation with NBCS exudate, exudate-resident and resident macrophages were found not only in the peritoneal exudates as described but also pre-

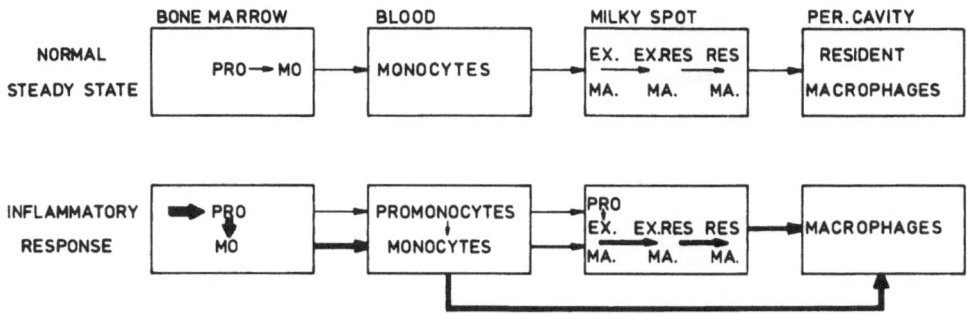

Fig. 5. Schematical representation of the mononuclear phagocytes in four compartments under steady state conditions and during acute inflammation. During the inflammatory responde there is an increase in the formation of monocytes in the bone marrow and an accelerated influx of these cells into the milky spots and the circulation, which may contribute to the population of macrophages into the peritoneal cavity. Moreover, during inflammation even promonocytes may move to the milky spots and serve locally as a pool of monocytes and macrophages. *pro*, promonocyte; *mo*, monocyte; *ex. ma*, exudate macrophage; *ex. res. ma*, exudate-resident macrophage; *res. ma*, resident macrophage

ceedingly in the milky spots in the omentum. Moreover, after induction of an acute inflammation in a state of cell-mediated immunity, promonocytes (Fig. 4A) were found perivascularly in milky spots. Some of these cells were in mitosis and were surrounded by exudate macrophages (Fig. 4A). These promonocytes (Fig. 4B) show PA in the nuclear envelope, RER, the Golgi system and the lysosomal granules (Fig. 4C). The localization and the kinetics of the three types of macrophage in the milky spots thus indicate that at least part of the free peritoneal macrophages are derived from the milky spots both during normal steady state and during inflammation (Fig. 5).

5. Discussion

In our studies in which the pattern of PA was used as criterion to distinguish various types of mononuclear phagocytes in the rat we found not only the well-known types of PA patterns [14] but also the intermediate stage of exudate-resident macrophages present both during acute inflammation [2] and chronic inflammation in vivo and in cultures of blood monocytes and exudate macrophages in vitro. Moreover, we also found this type of cell in vivo in the mouse and guinea pig as well as in vitro in human blood and cell cultures [3], mouse macrophage cultures [4] and mouse bone marrow cultures [22].

The in vivo results confirm the existence of exudate-resident macrophages as a commonly occurring cell in inflammatory processes. Both the PA pattern and the kinetics of the appearance and the disappearance of these cells in vivo and in vitro indicate that it is a transitional form, intermediate between exudate and resident macrophages. Bodel et al. [8, 9] did not find such a transitional cell in the rabbit, but since rabbit mononuclear phagocytes never have PA in the lysosomes [23], only PA-negative macrophages and resident macrophages could be expected on the basis of the PA in this species.

Our studies show that in macrophages surface adherence may not be essential for the appearance of PA in the RER in vitro as postulated by Bodel et al. [8, 9]. Our results seem to suggest that this transformation is the in vitro equivalent of an in vivo developmental process but does not explain the comparatively rapid and somewhat transitory appearance of the PA in the RER in vitro in rat, mouse, rabbit [9] and human macrophages.

A remarkably high number of PA-negative macrophages were found during chronic inflammation in man and also in the experimental animal model. The occurrence of this PA-negative cell in vivo strongly suggests a derivation directly from exudate macrophages

Table 4. Localization of PA in mononuclear phagocytes

	Nuclear envelope and RER	Golgi system	Lysosomal granules
Monoblast	+	?[a]	+
Promonocyte	+	+	+
Monocyte	−	−	+
Exudate macrophage	−	−	+
Exudate-resident macrophage	+	−	+
Resident macrophage	+	−	−
PA-negative macrophage	−	−	−

[a] In the monoblast the Golgi system is hardly developed, which makes it impossible to determine the PA of this organelle

and blood monocytes. Since the kinetics of this PA-negative cell in vitro show that at least a number of such cells transform into resident macrophages, PA-negative cells may also constitute a population of macrophages transitional between exudate and resident macrophages. Moreover, the transitory nature of PA acquired in vitro may ultimately result in an increasing number of PA-negative macrophages, so that PA-negative cells may include a population of end-stage macrophages in vitro.

The different PA patterns seen in mononuclear phagocytes in vivo and in vitro and described here are summarized in Table 4 as representing different stages of development of

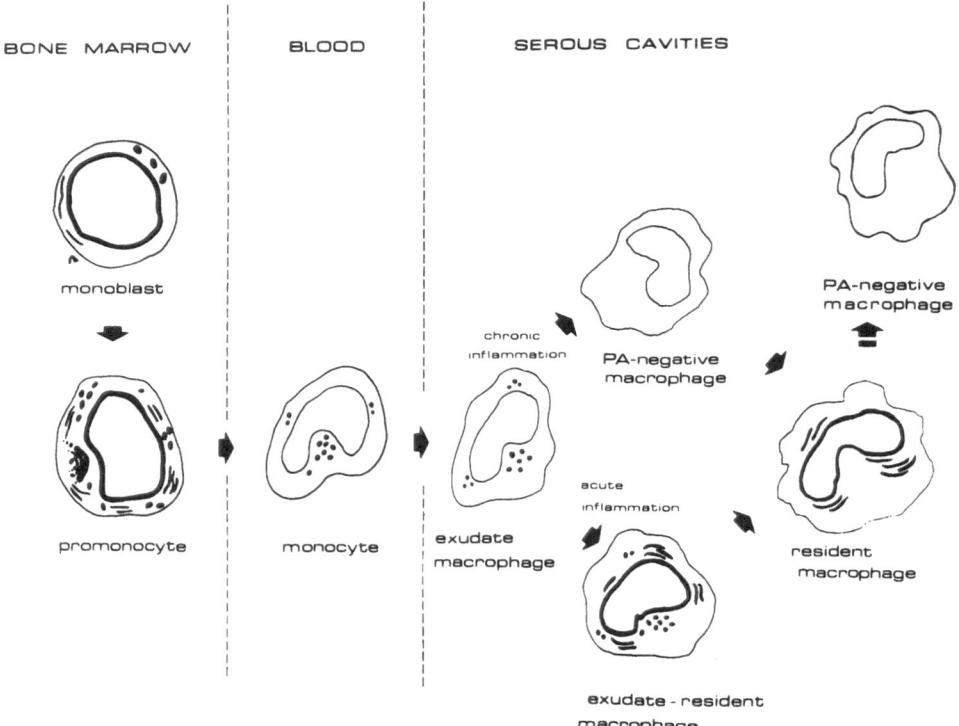

Fig. 6. Schematical representation of the different localizations and stages of development of mononuclear phagocytes on the basis of the PA pattern

these cells. The results of our in vivo and in vitro experiments lead to the hypothesis illustrated in Fig. 5. In the normal steady state, only resident macrophages are present in peripheral regions. In an acute inflammation, monocyte-derived exudate macrophages can become transformed into exudate-resident and resident macrophages in the serous cavities. In a chronic inflammation, monocyte-derived exudate macrophages may loose their PA-containing lysosomes and become PA-negative cells. These cells may subsequently acquire the characteristics of resident macrophages in the absence of inflammatory factors, as has been shown in vitro. Bone marrow cultures show monoblasts and promonocytes as well as the PA pattern of exudate, exudate-resident and resident macrophages [22]. In this process PA-negative cells may again be an end-stage cell in vitro. Moreover, during an acute inflammation in mice we found promonocytes perivascularly in milky spots. This finding shows that promonocytes may even move out of the bone marrow to a perivascular region and give rise locally to a pool of monocytes and macrophages.

In conclusion it may be said that different PA patterns (Table 4, Fig. 6) in macrophages and other mononuclear phagocytes of different species, both in vivo and in vitro, represent different stages of development in the sequence monoblast – promonocyte – monocyte – exudate macrophage – exudate-resident macrophage (PA-negative macrophage) – resident macrophage (PA-negative macrophage). The differences in PA of mononuclear phagocytes thus offer a useful tool for the study of the development of these cells under different circumstances. The exact enzymatic nature of the PA has still to be identified precisely; however, our studies have clearly shown that differences in the PA patterns of macrophages cannot be used to challenge the concept of the RES versus the concept of the MPS, as has been suggested in the literature [13, 14, 35, 36].

Acknowledgments

The authors wish to thank Mrs. Donna Fluitsma for technical assistance, Mr. Gerben Welling for preparing the photographs and Mrs. Anjo Steenvoorden-Bosma for typing the manuscript.

References

1. Bainton DF, Golde DW (1978) Differentiation of macrophages from normal human bone marrow in liquid culture – electron microscopy and cytochemistry. J Clin Invest 61:1555
2. Beelen RHJ, Broekhuis-Fluitsma DM, Korn C, Hoefsmit ECM (1978a) Identification of exudate-resident macrophages on the basis of peroxidatic activity. J Reticuloendothel Soc 23:103
3. Beelen RHJ, van't Veer M, Fluitsma DM, Hoefsmit ECM (1978b) Identification of different peroxidatic activity patterns in human macrophages in vivo and in vitro. J Reticuloendothel Soc 24:355
4. Beelen RHI, Fluitsma DM, van der Meer JWM, Hoefsmit ECM (1979) Development of the different peroxidatic activity patterns in peritoneal macrophages in vivo and in vitro. J Reticuloendothel Soc 25:315
5. Beelen RHJ, Fluitsma DM, Hoefsmit ECM (1980a) The cellular composition of omentum milky spots and the ultrastructure of milky spot macrophages and reticulum cells. J Reticuloendothel Soc 28:585
6. Beelen RHJ, Fluitsma DM, Hoefsmit ECM (1980b) Peroxidatic activity of mononuclear phagocytes developing in omentum milky spots. J Reticuloendothel Soc 28:601
7. Bentfield ME, Nichols BA, Bainton DF (1977) Ultrastructural localization of peroxidase in leukocytes of rat bone marrow and blood. Anat Rec 187:219
8. Bodel PT, Nichols BA, Bainton DF (1977) Appearance of peroxidatic reactivity within the rough endoplasmic reticulum of blood monocytes after surface adherence. J Exp Med 145:264
9. Bodel PT, Nichols BA, Bainton DF (1978) Differences in peroxidase localization of rabbit peritoneal macrophages after surface adherence. Am J Pathol 91:107
10. Boyum A (1964) Separation of white blood cells. Nature 204:793
11. Boyum A (1968) Separation of leukocytes from blood and bone marrow. Scand J Clin Lab Invest 97:21

12. Cotran RS, Litt M (1970) Ultrastructural localization of horseradish peroxidase activity and endogenous peroxidase activity in guinea pig peritoneal macrophages. J Immunol 105:136
13. Daems WT, Brederoo P (1971) The fine structure and peroxidatic activity of resident and exudate peritoneal macrophages in the guinea pig. In: DiLuzio NR, Flemming K (eds) The reticuloendothelial system and immune phenomena. Plenum Press, New York, p 19
14. Daems WT, Brederoo P (1973) Electron microscopical studies on the structure, phagocytic properties and peroxidatic activity of resident and exudate macrophages in the guinea pig. Z Zellforsch mikrosk Anat 144:247
15. Furth R van, Cohn ZA (1968) The origin and kinetics of mononuclear cells. J Exp Med 128:415
16. Furth R van, Cohn ZA, Hirsch JG, Humphrey JH, Spector WG, Langevoort HL (1972) The mononuclear phagocyte system: a new classification of macrophages, monocytes and their precursor cells. Bull WHO 46:845
17. Graham RC, Karnovsky MJ (1966) The early stages of absorption of injected horseradish peroxidase in the proximal tubules of mouse kidney: ultrastructural cytochemistry by a new technique. J Histochem Cytochem 14:291
18. Hoefsmit ECM (1975) Mononuclear phagocytes, reticulum cells and dendritic cells in lymphoid tissues. In: Furth R van (ed) Mononuclear phagocytes in immunity, infection and pathology. Blackwell Scientific Publications, Oxford Edinburgh, p 17
19. Langevoort HL, Cohn ZA, Hirsch JG, Humphrey JA, Spector WG, Furth R van (1970) The nomenclature of mononuclear phagocytes: a proposal for a new classification. In: Furth R van (ed) Mononuclear phagocytes. Blackwell Scientific Publications, Oxford, p 1
20. Lepper AWD, D'Arcy Hart P (1976) Peroxidase staining in elicited and non-elicited mononuclear cells from BCG-sensitized and non-sensitized mice. Infect Immun 14:522
21. Meer JWM van der, Bulterman D, Zwet TL van, Elzenga-Claasen I, Furth R van (1978) Culture of mononuclear phagocytes on a teflon surface to prevent adherence. J Exp Med 147:271
22. Meer JWM van der, Beelen RHJ, Fluitsma DM, Furth R van (1979) Ultrastructure of mononuclear phagocytes developing in liquid bone marrow cultures – a study on peroxidatic activity. J Exp Med 149:17
23. Nichols BA, Bainton DF, Farguhar MG (1971) Differentiation of monocytes. Origin, nature and fate of their azurophil granules. J Cell Biol 50:498
24. Nichols BA, Bainton DF (1975) Ultrastructure and cytochemistry of mononuclear phagocytes. In: Furth R van (ed) Mononuclear Phagocytes in Immunity, Infection and Pathology. Blackwell Scientific Publications, Edinburgh Melbourne, p 17
25. Ogawa T, Koerten HK, Daems WT (1978) Peroxidatic activity in monocytes and tissue macrophages of mice. Cell Tissue Res 188:361
26. Ohta H, Kamiya O, Nugata H (1973) Blood monocytes and macrophages. Kinetic study with radioisotopes and cytochemical methods. J Histochem Cytochem 17:675
27. Rhee HJ van der, Winter CPM de, Daems WT (1977) Fine structure and peroxidatic activity of rat blood monocytes. Cell Tissue Res 185:1
28. Robbins D, Fahimi HD, Cotran RS (1971) Fine structural cytochemical localization of peroxidase activity in rat peritoneal cells: mononuclear cells, eosinophils and mast cells. J Histochem Cytochem 19:571
29. Shands JW, Axelrod BJ (1977) Mouse peritoneal macrophages: traited thymidine and cell kinetics. J Reticuloendothel Soc 21:64
30. Soranzo MR, Koerten HK, Daems WT (1978) Peroxidatic activity and morphometric analysis of alveolar macrophages in the guinea pig. J Reticuloendothel Soc 23:343
31. Steinman RM, Cohn ZA (1972) The interaction of soluble horseradish peroxidase with mouse peritoneal macrophages in vitro. J Cell Biol 55:186
32. Tice LW (1974) Effect of hypophysectomy and TSH replacement on the ultrastructural localization of thyroidperoxidase. Endocrinology 95:421
33. Volkman A (1976) Disparity of the origin of mononuclear phagocyte populations. J Reticuloendothel Soc 19:249
34. Watanabe N, Masubuchi S, Kageyama K (1973) Ultrastructural localization of peroxidase in mononuclear phagocytes from the peritoneal cavity, blood, bone marrow and omentum, with a special reference to the origin of peritoneal macrophages. Rec Adv RES Res 11:120
35. Wisse E (1974a) Observations on the fine structure and peroxidase cytochemistry of normal rat liver Kupffer cells. J Ultrastruct Res 46:343
36. Wisse E (1974b) Kupffer cell reactions in rat liver under various conditions is observed in the electron microscope. J Ultrastruct Res 46:499

Haematology and Blood Transfusion Vol 27
Disorders of the Monocyte Macrophage System
Edited by F. Schmalzl, D. Huhn, H.E. Schaefer
© Springer-Verlag Berlin Heidelberg New York 1981

The Monocyte–Macrophage System In Granulomatous Inflammation

J.L. Turk and R.B. Narayanan

1. Introduction

A granuloma may be defined as a collection of cells of the monocyte–macrophage system with or without the additional presence of other inflammatory cells [14]. Granulomas may be classified according to whether they are of immunological or non-immunological origin.
1. Immunological e.g. tuberculosis, tuberculoid leprosy, syphilis, schistosomiasis, sarcoidosis, berillium, zirconium.
2. Non-immunological, a) non-toxic, e.g. plastic beads, carbon particles, non-toxic metals (Fe); b) toxic, e.g. silica, talc, asbestos; c) activation of C3, e.g. carageenan, kaolin, aluminium hydroxide, lepromatous leprosy.

Toxic granulomas are mainly caused by substances that non-specifically cause increased permeability of lysosomal membranes. Substances that activate C3 through the alternative pathway or through plasminogen-plasmin could also indirectly result in a similar process on cell membranes.

In this communication we shall illustrate the various mechanisms of granuloma formation by our recent studies on granulomas formed in the skin and lymph nodes in response to metal compounds containing aluminium and zirconium and in mycobacterial infections. In immunological granulomas cells of the monocyte–macrophage series characteristically take on the features described since the time of Metchnikoff [5] as "epithelioid cells". These granulomas are often therefore referred to as "epithelioid cell granulomas". In most non-immunological granulomas cells of the monocyte–macrophage series take on a less well differentiated appearance and often contain large amounts of ingested foreign material. Immunological granulomas are also classically surrounded by other inflammatory cells-lymphocytes and granulocytes brought in as a result of the immunological reaction–and also may be surrounded by fibroblasts actively secreting collagen. Fibrosis is a frequent end result of the evolution of granulomas of immunological origin. This is a point of intense clinical importance.

2. What Is an Epithelioid Cell?

The term epithelioid cell is derived from observations using the light microscope. Classically these are large polygonal cells with pale oval nuclei and abundant eosinophilic cytoplasm whose borders blend imperceptibly with those of their neighbours. The origin of these cells from other cells of the monocyte–macrophage series was first recognised by Metchnikoff [5]. We have studied the ultrastructural appearance of these cells in both experimental and clinical material, comparing the appearance in immunological granulomas induced by sodium zirconium lactate with those induced by mycobacteria in man and in experimental animals.

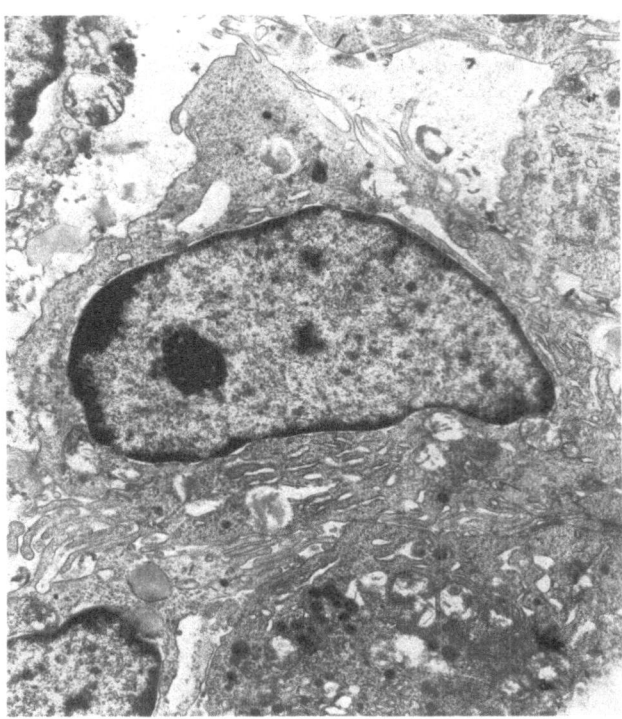

Fig. 1. Epithelioid cell from zirconium granuloma in the guinea pig with fenestrated cell periphery, fimbriated and interdigitating cell membrane and relatively abundant endoplasmic reticulum and polyribosomes. × 6300 [17]

Guinea pigs immunised with sodium zirconium lactate (Na Zr L) in Freund's complete adjuvant can be induced to develop allergic type nodular granulomas when skin tested with Na Zr L during the period of immunisation. These granulomas occur in two thirds of the animals that show delayed hypersensitivity and reach peak intensity 7 days after skin test [16, 17]. Ultrastructurally (Fig. 1) most of the cells of the monocyte–macrophage system have a distinctive appearance. The nuclei are large, spherical and reticulated. The cell periphery has a fenestrated appearance with a fimbriated cell membrane interdigitating with other cells, and there is a relatively abundant rough endoplasmic reticulum. In addition, the cytoplasm contains a number of vesicles and vacuoles. The nucleus is usually oval with finely marginated nuclear chromatin, and nucleoli are large, spherical and reticulated. Epithelioid cells of this type have been described in sarcoidosis in man [2]. Additional features of these granulomas are the multinucleate giant cells, which also contain a large number of vesicles and vacuoles, and fibroblasts surrounded by newly produced collagen.

If one studies the ultrastructural appearance of cells of the monocyte–macrophage system in these granulomas one can observe a sequence from cells resembling circulating monocytes through typical activated macrophages to the secretory epithelioid cell with its large amount of rough endoplasmic reticulum. However, many epithelioid cell granulomas such as those in tuberculoid leprosy [12] contain mainly typical activated macrophages without much rough endoplasmic reticulum (Fig. 2). Some of these go onto a highly vesicular degenerate appearance (Fig. 3) producing another cell type that may also be seen in the tuberculoid-type granulomas of sarcoidosis.

The relation between the development of epithelioid cells and delayed hypersensitivity is still controversial. It has been suggested that macrophages incubated in vitro with lymphokine take on a light microscopic appearance resembling that of epithelioid cells [6]. These changes are associated with an activation of Krebs cycle enzymes and enzymes of the hexose monophosphate shunt [8]. However, these activated macrophages do not develop an endoplasmic reticulum and thus do not develop the typical ultrastructural appearance of the secretory epithelioid cell [15].

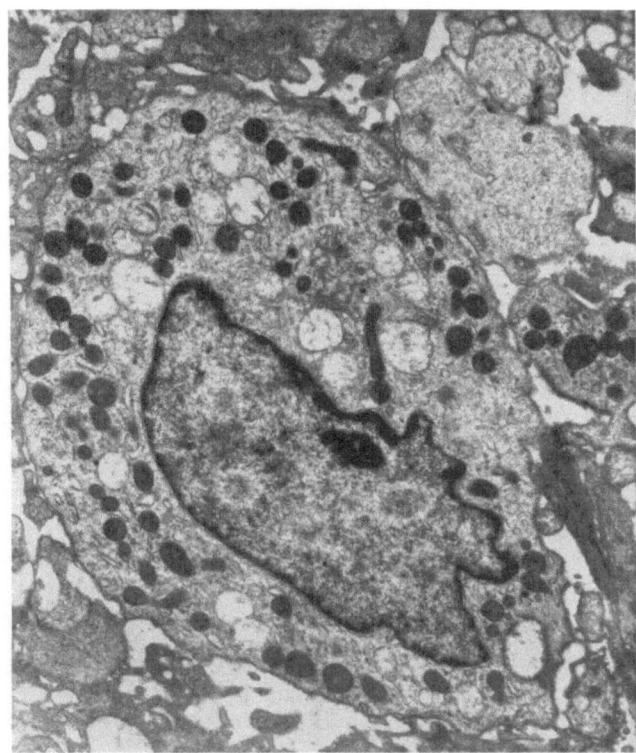

Fig. 2. Epithelioid cell from borderline tuberculoid leprosy containing large numbers of intracellular vacuoles. × 7800 [13]

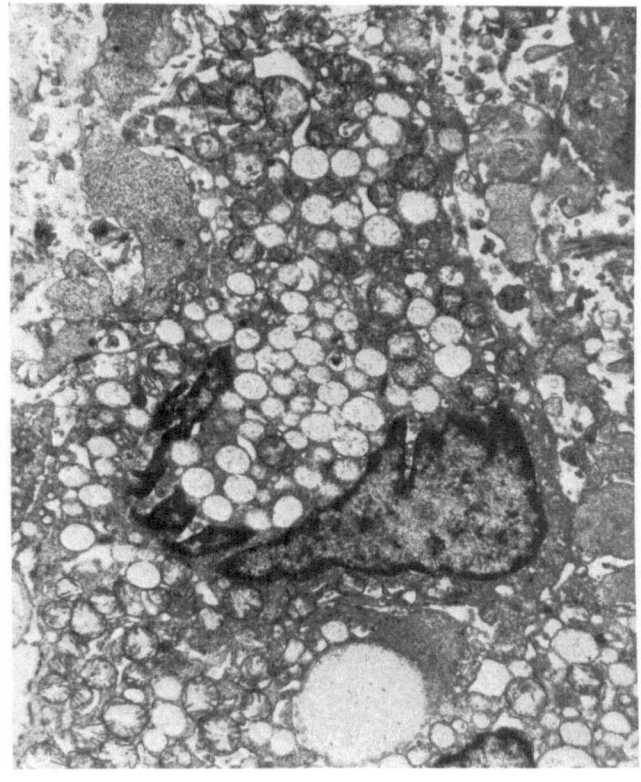

Fig. 3. Highly vacuolar epithelioid cell from borderline tuberculoid leprosy. × 7750 [13]

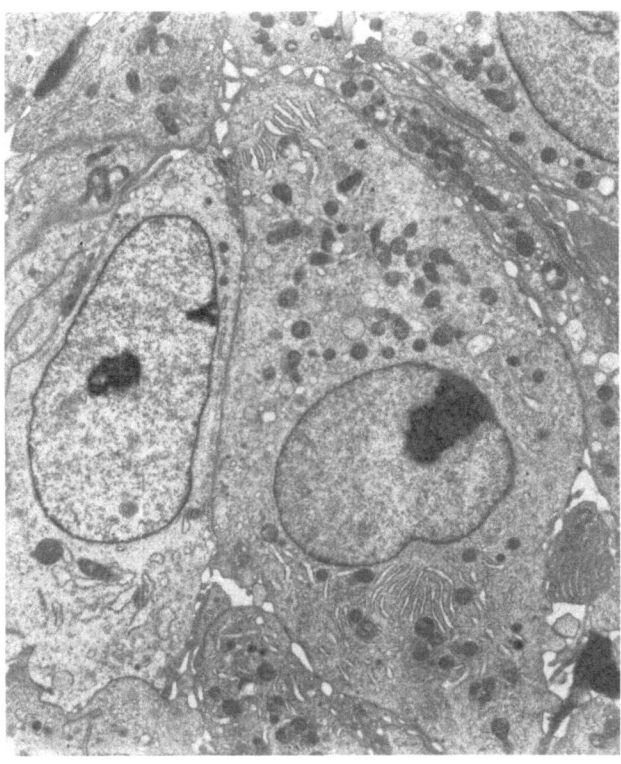

Fig. 4. Epithelioid cell containing arrays of rough endoplasmic reticulum from borderline tuberculoid leprosy *in reaction.* × 4300 [13]

The question therefore arises as to what causes circulating monocytes to develop into "secretory" epithelioid cells in one type of granuloma and what causes them to take on the highly vesicular form in another type of granuloma. The role of lymphokine in the evolution of these cells and their relation to delayed hypersensitivity needs further elucidation. In our recent study on the ultrastructure of cells of the monocyte–macrophage system in leprosy [13] we found that epithelioid cells in chronic tuberculoid lesions were mainly of the "degenerate"-looking vesicular type. However, where the lesions were the site of an acute reaction, cells were found with a more developed endoplasmic reticulum (Fig. 4), indicating a definite secretory function. In these reaction states it has been shown that increase in the lymphocyte transformation test as an indicator of allergic reactivity to specific *M.leprae* antigen was proportional to the intensity of erythema and oedema in the lesions [1]. Thus the presence of epithelioid cells with a strong endoplasmic reticulum could be taken as developing in the presence of an active allergic reaction. Similar epithelioid cells with large amounts of endoplasmic reticulum are found at the site of the nodular granulomas following the intradermal injection of lepromin (heat killed *M.leprae*) in highly sensitive individuals.

One of the most important cell types in these granulomas is the fibroblast. Fibroblasts may be found mainly in those granulomas containing secretory type epithelioid cells. It is important to determine more closely the relation between the two different types of epithelioid cells and the fibroblasts and particularly to delineate their role in the process of fibrogenesis. The end result of granuloma formation in man is frequently severe scarring, whether it is in the skin or in the lung. Thus knowledge of the inter-relationship between cells of the monocyte–macrophage series and fibroblasts is particularly vital if severe tissue damage is to be avoided.

3. Experimental Mycobacterial Granulomas in Guinea Pig Lymph Nodes: Presence of Epithelioid Cells and Their Relationship to Fibrosis

A model of granuloma formation has been used in the lymph node draining the site of intra-dermal injection of particulate antigens, e.g., bacille Calmette-Guérin (BCG) vaccine [3]. The draining lymph node has an advantage over other sites in that it can be weighed and the progress of granuloma development quantitated. In addition, it allows easy access to the infiltrating cell population for isolation and in vitro study. Guinea pigs were injected in each ear with 1×10^7 viable BCG (Pasteur), 1×10^9 cobalt irradiated BCG or 1×10^9 cobalt-irradiated *M.leprae*. The BCG vaccine, whether live or irradiated, gave strong sensitisation to 25 µg tuberculin PPD, which developed maximally at 3 weeks. In *M.leprae* injected animals sensitivity to PPD was maximal between 5–7 weeks. The increase in lymph node weight was parallelled by the area of infiltration. BCG produced its maximum granu-lomatous effect at one to two weeks. At this time the light microscopic appearance was of cells with an epithelioid appearance. Fibrosis, necrosis and tubercle formation was apparent from 7 weeks onwards and extensive by 10 weeks in animals infected with live BCG. With the irradiated BCG the granuloma had resolved by 5 weeks. Following the injection of irradiated *M.leprae*, peak granuloma formation and lymph node weight was at 5 weeks. Most of the cells in the infiltrate had the appearance of undifferentiated macrophages. Fibrosis was not a major event.

Electron micrographs of 2-week-old BCG granulomas showed a high percentage of cells characterised by (1) large nucleus with prominent nucleolus, (2) paucity of cytoplasmic organelles (vesicles, lysosomes, phagocytic vacuoles and mitochondria) and (3) character-istically swollen rough endoplasmic reticulum. These cells were similar to those seen in ex-perimental zirconium granulomas in the skin and in borderline tuberculoid leprosy in reac-tion. By contrast, most of the cells infiltrating the *M.leprae* granuloma at its peak were ma-crophages, many of which contained phagocytosed organisms. At the later stage (1 week) the characteristic cells in the live BCG granulomas were typical fibroblasts. Thus in BCG lymph nodes there was an early infiltration with secretory epithelioid cells with subsequent necrosis and tubercle formation developing between 7 and 10 weeks. In *M.leprae* granu-lomas the appearance was of macrophage infiltration only.

Further studies were made to determine the relative degrees of collagen formation in the lymph nodes. Biochemical experiments were carried out to determine ^{14}C-proline incor-poration into collagen in the lymph nodes after 24 h culture in vitro. Proline incorporation into TCA-precipitated protein was measured with and without digestion by specific collage-nase [7], the difference being due to specific incorporation into collagen. There was very lit-tle basal collagen synthesis detected in lymph nodes from *M.leprae* injected animals. In ani-mals injected with irradiated BCG increasing collagen synthesis could be detected up to 5 weeks after injection. Following live BCG this was delayed to 10–12 weeks after injection.

In these studies there seems to be a definite association between the presence of se-cretory epithelioid cells in the granuloma and fibroblast infiltration associated with biochemical evidence of collagen synthesis. Further studies are in progress to determine whether a specific factor secreted by these epithelioid cells exists which can be detected with the fibroblast as its target. To date the only positive results have been obtained with media taken from a 24-h culture of 2 week BCG lymph node cells. After dialysis against fresh medium and fivefold concentration, these stimulated proline incorporation into total protein but not proline incorporation into collagen. The nature of this stimulating factor is being investigated. Despite reports that lymphokine preparations will stimulate collagen synthesis and thymidine incorporation in fibroblast cultures [4, 18], we have been unable to get this effect using lymphokine prepared by culturing with PPD lymph node cells from guinea pigs sensitised with *M.tuberculosis*. Thus, the stimulation of total protein synthesis by granuloma-conditioned media is unlikely to be due to a lymphokine.

4. Membrane Receptors on Cells of the Mononuclear Phagocyte Series Across the Spectrum in Leprosy

Granulomas in the skin of patients across the clinical spectrum of leprosy were examined for Fc and C3 receptors [11]. Cryostat sections were layered with sheep erythrocytes coated with antibody (EA) for Fc receptors and sheep erythrocytes with antibody and complement (EAC) for C3 receptors. EAC but not EA adherence was seen at the BT end of the spectrum and was particularly dense around the epithelioid cells. EA adherence occurred over the active macrophages of BL or LL leprosy. In LL leprosy EAC adherence was poor or absent. The phenomenon seemed to be related to bacterial density. Thus, two patterns of membrane receptors can be discerned. In epithelioid cell granulomas, including sarcoidosis, there is a loss of Fc receptors, whereas in non-immunological granulomas of lepromatous leprosy there is a loss of C3 receptors.

In an attempt to elucidate the mechanism of changes in membrane receptors, peritoneal macrophages from guinea pigs were cultured for 7 days in the presence of BCG vaccine as a source of mycobacteria. An 80% decrease in the number of EAC receptors was seen at 24 h. 100% receptor activity returned after 7 days of culture [12]. Neither zymosan nor latex caused a change in EA or EAC receptors over a 7 day period. Thus, the effect was not due to phagocytosis alone. No loss of EA receptors was found after treatment with lymphokine, so this change in membrane receptors associated with epithelioid cell transformation could not be directly related to delayed hypersensitivity. The loss of C3 receptors appeared, however, to be related directly to the ingestion of mycobacteria.

Recently we have been studying the activation of C3 by mycobacteria through the alternative pathway [10]. *M.leprae, M.lepraemurium* and mycobacterial cord factor (trehalose dimycolate) but not muramyl dipeptide (MDP) were found to activate the alternative pathway of complement in normal and C4-deficient guinea pig serum. It was felt that complement activation could be one of the mechanisms by which mycobacteria induce chronic granulomatous inflammation in the absence of cell-mediated immunity as in lepromatous leprosy. The relation of C3 activation to the loss of C3 receptors is difficult to understand as ingestion of zymosan which is a potent activator of the alternative complement pathway does not have this effect. It could be, however, that mycobacteria have the added action of being able to inhibit the regeneration of the C3 receptor once it has been lost.

The relation between activation of C3 and the development of non-immunological granulomas is interesting, since in previous studies we have found that a number of metal compounds such as aluminium hydroxide, aluminium chlorhydrate, zirconium aluminium glycinate and kaolin, all of which are potent producers of "toxic" granulomas, are non-immunological activators of C3, not through the alternative pathway but through the conversion of plasminogen to plasmin which is potent in splitting C3 into its biologically active components [9]. This could be the result of activation of the Hageman factor.

5. Summary

Granulomas may be immunologically induced or non-immunologically induced. In immunologically induced granulomas cells of the monocyte–macrophage series take on the appearance of epithelioid cells. Ultrastucturally epithelioid cells may have a secretory appearance with much rough endoplasmic reticulum or take on a highly degenerate vesicular appearance. Other epithelioid cells look like activated macrophages. Secretory epithelioid cells may be found associated with acute local inflammation as in borderline tuberculoid leprosy in reaction, the lepromin reaction, following injection of BCG vaccine and in experimental zirconium granulomas. In these situations there may also be strong histological and biochemical evidence of increased fibroblast activity and collagen synthesis. It is suggested that these cells are actively secreting a fibroblast-activating factor.

Epithelioid cells may lose their Fc receptors, undifferentiated macrophages in lepromatous leprosy can lose their C3 receptors. It is suggested that in a number of situations granuloma formation may be associated with complement activation through the alternative pathway as in the case of mycobacterial granulomas. Toxic granulomas produced by metals may be caused by C3 being split by plasmin after conversion from plasminogen by activation of the Hageman factor.

References

1. Bjune G, Barnetson RSC, Ridley DS, Kronvall G (1976) Lymphocyte transformation test in leprosy: correlation of the response with inflammation. Clin Exp Immunol 25:85
2. Elias PM, Epstein WL (1968) Ultrastructural observations on experimentally induced foreign body and organised epithelioid-cell granulomas in man. Am J Pathol 52:1207
3. Gaafar SM, Turk JL (1970) Granuloma formation in lymph nodes. J Pathol 100:9
4. Johnson RL, Ziff M (1976) Lymphokine stimulation of collagen accumulation. J Clin Invest 58:240
5. Metchnikoff E (1893) Lectures on comparative pathology of inflammation. Kegan, Paul, Tench, Trubner, London, pp 159–166
6. Nath I, Poulter LW, Turk JL (1973) Effect of lymphocyte mediators on macrophages in vitro. A correlation of morphological and cytochemical changes. Clin Exp Immunol 13:455
7. Petrofsky B, Diegelman (1971) Use of a mixture of proteinase free collagenases for the specific assay of radioactive collagen in the presence of other proteins. Biochemistry 10:988
8. Poulter LW, Turk JL (1976) Studies on the effect of soluble lymphocyte products (lymphokines) on macrophage physiology. II. Cytochemical changes associated activation. Cell Immunol 20: 25
9. Ramanathan VD, Badenoch-Jones P, Turk JL (1979) Complement activation by aluminium and zirconium compounds. Immunology 37:881
10. Ramanathan VD, Curtis J, Turk JL (1980) Activation of the alternative pathway of complement by mycobacteria and cord factor. Infect Immun 29:30
11. Ridley MJ, Ridley DS, Turk JL (1978) Surface markers on lymphocytes and cells of the mononuclear phagocyte series in skin sections in leprosy. J Pathol 125:91
12. Ridley MJ, Turk JL, Badenoch-Jones P (1978) In vitro modification of membrane receptors on cells of the mononuclear phagocyte system. J Pathol 127:173
13. Ridley MJ, Badenoch-Jones P, Turk JL (1980) Ultrastructure of cells of the mononuclear phagocyte system across the spectrum of leprosy. J Pathol 30:223
14. Turk JL (1971) Granuloma formation in lymph nodes. Proc R Soc Med 64:942
15. Turk JL (1980) Immunologic and non-immunologic activation of macrophages. J Invest Dermatol 74:301
16. Turk JL, Parker D (1977) Sensitization with Cr, Ni and Zr salts and allergic type granuloma formation in the guinea pig. J Invest Dermatol 68:341
17. Turk JL, Badenoch-Jones P, Parker D (1978) Ultrastructural observations on epithelioid cell granulomas induced by zirconium in the guinea pig. J Pathol 124:45
18. Wahl SM, Wahl LM, McCarthy JB (1978) Lymphocyte–mediated activation of fibroblast proliferation and collagen production. J Immunol 121:942

Haematology and Blood Transfusion Vol 27
Disorders of the Monocyte Macrophage System
Edited by F. Schmalzl, D. Huhn, H.E. Schaefer
© Springer-Verlag Berlin Heidelberg New York 1981

Macrophages in Granulomas: Histochemical Evidence Suggesting Local Control of Heterogeneous Functions[1]

A. M. Dannenberg, Jr., M. Suga and J. E. Garcia-Gonzales

1. Introduction

Cells of the monocyte–macrophage system are produced in the bone marrow [120, 121] and enter the blood stream (as monocytes) in a rather immature, undifferentiated state. From the blood stream they enter various tissues, e.g., the liver, where they may become Kupffer cells; the lung, where they may become alveolar macrophages; and sites of inflammation where, as exudate macrophages, they may differentiate in a variety of ways [1, 27]. The various functions that macrophages perform have been identified mainly in cell culture and not in the cells' natural microenvironment, which can be seen in tissue sections. Our laboratory has studied macrophage enzyme levels in such sections by means of double-staining histochemical techniques. These techniques enable us to identify functional macrophage heterogeneity in inflammatory lesions. This report describes these studies and speculates on the control and meaning of such heterogeneity. It also outlines in vivo–in vitro experiments which should provide additional histochemical–functional correlations that further elucidate the role of macrophages in the inflammatory process.

2. Evidence for Macrophage Heterogeneity in the Literature

Table 1 lists a variety of functions or properties in which macrophage heterogeneity has been observed, mainly in vivo. Macrophage populations are heterogeneous in phagocytic abilities, in microbicidal activities and cytotoxicity for tumor cells, in many hydrolytic activities, in various regulatory functions for the immune and inflammatory responses, and in other properties. Macrophages may become specialized for one or several of these functions, but we doubt whether a given macrophage can carry out more than a few unrelated functions at one time. They seem to change as their local environment changes [22, 109, 114].

3. Macrophage Functional Heterogeneity In Vivo (Demonstrated Histochemically by Double-staining BCG Granulomas for Two Different Enzymes) [114]

Tuberculous granulomas were produced in the skin of rabbits by the intradermal injection of viable bacille Calmette-Guérin (BCG) or in their lungs by the intravenous injection of

[1] Supported by Grants ES-01879 and ES-00454 from the National Institute of Environmental Health Sciences, US Public Health Service; Grant HL-14153 from the National Heart, Lung, and Blood Institute for the Johns Hopkins Specialized Center on Lung, US Public Health Service; and Contract DAMD-17-80-C-0102 with the US Army Medical Research and Development Command

Table 1. Functional heterogeneity of macrophages[a]

Phagocytic abilities
    Phagocytosis [24, 56, 68, 88, 95, 101, 108]
    IgG or IgM (Fc) receptors [23, 55, 65, 82, 94, 101, 108, 126 (see 35, 125)]
    Complement receptors [46]
    Fibrinogen-fibrin binding [23]
Cidal capacities
    Microbicidal ability [68, 71, 132]
    Peroxidatic activity [7, 59, 69, 102 (see 38)]
    Cytostasis or cytolysis [19, 33, 60, 64, 70, 71, 102, 108, 129, 132 (see 51)]
Digestive and other hydrolytic abilities
    Activities of various hydrolases [5, 34, 40, 50, 66, 72, 89, 122, 123]
Regulatory functions in the immune response [128]
    Antigen binding or antigen presentation [42, 61, 65, 126]
    Immunostimulation and immunosuppression [59, 62]
    Ia antigen expression [14, 25, 60]
    Immunogenic RNA production [95]
    Response to lymphokines [reviewed in 110]
Regulatory functions in the inflammatory response
    Protease inhibitors [57]
    SRS-A formation [85]
    Prostaglandin production [50, 89]
    Production of the C5 component of complement [84]
Miscellaneous properties
    Susceptibility to cell death (following phagocytosis of BCG or zymosan) [67]
    Ability to cluster into foci (of 10 to 30 macrophages: "focus forming") [101]
    Expression of macrophage differentiation (or maturation) antigen (M1/70) [111]
    Spontaneous rosette formation (with sheep and chicken erythrocytes) [116, see 101]

[a] Other references are quoted in the papers listed and in the reviews [48, 127]. Differences in mouse peritoneal and pulmonary alveolar macrophages are reviewed in 46

heat-killed virulent *Mycobacterium tuberculosis*. Paraffin-embedded sections, prepared from the lesions at various times, were stained histochemically for acid phosphatase [28, 47, 114], β-galactosidase [114, 133], esterase [28, 47, 114], and cathepsin D [99, 114].

The level of each macrophage enzyme varied with the age of the BCG lesion and the site within the lesion where the macrophages were located. When the lesions were at their peak size at 21 days, they contained the highest levels of all four enzymes, i. e., more enzymatically active macrophages were present. Within the BCG lesions, macrophages containing β-galactosidase and esterase were more numerous at the edge of the caseous necrotic center and in the viable tissue nearby. (In these areas, both the bacilli (and their products) and the tissue debris are plentiful.) In contrast, macrophages that contained acid phosphatase and cathepsin D were more numerous in the peripheral parts of the lesion (where little or no necrosis is present and bacilli are rare or absent). Thus, overall macrophage (enzymatic) function is determined by the age of the BCG granuloma and the location within it. We call this *macrolocal* control.

Macrophage enzymatic function, i. e., the content of enzymes in macrophages, is also determined by *microlocal* conditions. Two macrophages, existing side by side, may stain for different enzymes. We demonstrated this phenomenon by double-staining techniques. The first enzyme produced a red-colored histochemical product; the second enzyme produced a blue product. Cathepsin D and β-galactosidase (and acid phosphatase and β-galactosidase) were usually found in different macrophages. Esterase and β-galactosidase were almost always found in the same macrophage, as were red-esterase and blue-esterase (i.e., esterase hydrolyzing naphthol-ASD acetate and 5-bromo-4-chloroindoxyl acetate, respectively). In this case, a purple colored histochemical product resulted.

4. Discussion of the Concept of Macrolocal and Microlocal Control of Macrophage Function

The following concepts about the control of macrophage function are presented as a stimulus for thought and, perhaps, future research. The data, presented in reference [114] and in this report, support these concepts but cannot conclusively prove them.

Overall controls must exist. We have demonstrated that the levels of acid phosphatase, cathepsin D, esterase, and β-galactosidase reached a peak when the BCG lesions reached their greatest size (at about 3 weeks).

Macrolocal controls must exist. The number of macrophages containing high levels of a given enzyme was apparently dependent on the location of these cells within the BCG lesion. Esterase and β-galactosidase containing macrophages were found to be more numerous in viable tissue near the caseous necrosis, whereas acid phosphatase and cathepsin D containing macrophages were found to be more numerous in the peripheral parts of the lesion.

Microlocal controls may exist. The macrophages in the different areas of the BCG lesion could also respond as independent units. Some macrophages became rich in one enzyme, some became rich in another, and some became rich in both (or probably multiple enzymes, but double-staining only detected two at a time).

Of the three concepts, *microlocal* control is the least certain, because the observed microlocal macrophage functional heterogeneity could be produced in a variety of ways. Perhaps each of them plays a role.

1. A macrophage may become rich in a given enzyme when the lesion is young and remain rich in this enzyme when the lesion is old, even though the prevailing cells had meanwhile become rich in another enzyme. In other words, two adjacent macrophages, each staining for a different enzyme, may have been activated by different stimuli (occurring simultaneously or at different times).

2. The production of macrophage enzymes (in response to a single stimulus) may occur at different times and at different rates. For example, MN in the peripheral parts of the BCG lesion are frequently recent arrivals and often are acid phosphatase positive; whereas MN near the center of the lesion are frequently older and often β-galactosidase positive. Perhaps acid phosphatase is produced sooner and more rapidly than β-galactosidase.

3. A macrophage may be activated in one part of the lesion and migrate to another part. In other words, an acid phosphatase positive macrophage may respond to a particular local chemotactic agent whereas a β-galactosidase positive macrophage may respond to a different chemotactic agent. (No data are available to support or refute this possibility.)

4. Local substrates, e.g., carbohydrates, proteins, lipids, and nucleic acids, may not be evenly distributed in the lesion but exist in microlocal foci. In this case, the enzyme content of macrophages would reflect the microlocal environment, if adaptive enzyme production took place.

5. Macrophages may differ in what they ingest because they have different surface receptors (see Table 1). In this case, their enzyme content would reflect the type of receptors they possess, if (as discussed previously) adaptive enzyme production took place.

6. Finally, a macrophage may respond to other cells in its microenvironment. Adjacent or nearby lymphocytes (and their lymphokines), plasma cells, granulocytes, capillary endothelial cells, fibroblasts, and even other macrophages may control the enzyme levels and function of a given macrophage at various times. This last possibility is supported by the microlocal macrophage heterogeneity demonstrated by double-staining enzyme histochemistry, but the identification and regulation of the many factors responsible for such heterogeneity remain to be investigated.

5. Extracellular Functions of Macrophages

We have just described some of the heterogeneous *intracellular* functions of macrophages, namely, a few of their hydrolytic enzymes which (except for the esterase) are most active at acidic pH. Numerous *extracellular* functions of macrophages have also been described (see Table 2). These functions are usually performed at neutral pH by products secreted (and not stored) by macrophages. Most of these products have, of necessity, been identified in vitro, i.e., in cell culture. In vivo they are usually inactivated or neutralized (soon after their release) by inhibitors and/or enzymes from the serum or nearby cells. Table 1 lists heterogeneity of the macrophages with respect to some of these secretory products.

In our initial experiments on macrophage secretion in vivo, we placed plastic chambers over rabbit dermal BCG granulomas with the epidermis removed [115]. Analysis of the fluids collected in the chambers showed that collagenase was present when these granulomas were at peak size. At other times the collagenase activity was apparently neutralized by the antiproteases in the serum that also exuded into the chamber.

Table 2. Secretion (or release) of enzymes and other products from macrophages[a]

Enzymes hydrolyzing tissue components
   Acid-acting lysosomal enzymes: proteinases, nucleases, lipases, acid phosphatase, and various glycosidases [11]
   Neutral-acting enzymes: collagenase, elastase, plasminogen activator, proteinase [see also 97]
Products involved in host defense
   Lysozyme [20]
   Complement components [74, 100; see 4, 106]
   Interferons [80, 81; see 4]
   Microbicidins: cationic proteins, superoxide [53, 76, 77, 86], H_2O_2 [78, 79; see 76, 77], arginase [4, 26, 58, 83, 103]
Products modulating cells
   Colony stimulating factors (CSF) that increase the production of phagocytes by the bone marrow (reviewed in 15) (A lactoferrin-like glycoprotein inhibitor from PMN reduces CSF production by monocytes [see 15, 16])
   An erythropoietic factor [43, see 96]
   Substances that stimulate the division and differentiation of T and B lymphocytes and also substances that suppress lymphocyte function [4, 30, 31, 118, 119][b]
   Factors that stimulate the growth of new capillaries
   Substances that stimulate fibroblasts
   Osteoclast activation factor [see 32, 49, 134, 135]
   Prostaglandins and related substances [11–13, 17, 18, 41, 73, 107, see 4, 30, reviewed in 113] and cAMP [see 4]
   Chemotactic factor inactivators [52]
   Insulin-like activity [39]
   Polyamine oxidase [see 4]
   Thymidine [112, see 4]
   Fibronectin [2]
Miscellaneous products
   Endogenous pyrogen, causing fever [10, 21, 45] and decreased plasma, iron, and zinc levels [54]. It seems to be identical with lymphocyte activating factor [75]
   Thromboplastin [90, 91, 98, see 36, 37, 63]
   Platelet activating factor(s) [8, 9]
   α_2-macroglobulin (the protease inhibitor)
   α_1-proteinase inhibitor (α_1-antitrypsin) [131]
   Transferrin
   Transcobalamin II (the vitamin B_{12} transport protein) [92, 93]
   Ss protein, an antigeneic euglobulin in serum that is a product of the histocompatible gene complex [104; see 100]
   Slow reacting substance of anaphylaxis (SRS-A) [85]

[a] Organized after Unanue et al. [119]. Recent references *only* are listed. Other references are available in reviews 3, 4, 6, 29, 44, 87, 105, 115, 117 and 130
[b] The recent literature on immune regulation was too extensive to include here

These results suggested the following in vivo–in vitro system, where the effect of serum inhibitors is reduced or eliminated. Specifically, we are performing biopsies on lung and skin granulomas produced by BCG in rabbits and mice. (In time, lesions produced by silica and by heat-killed pneumococci, streptococci, and staphylococci will be studied for comparative purposes.)

The biopsies are minced and cultured for 1 to 2 days in (serum-free) Dulbecco's modified Eagle medium (with 0.2% lactalbumin hydrolysate), and the supernatant fluids are collected and concentrated by ultrafiltration.

These culture fluids are being assayed *in vitro* for proteases (and their inhibitors) with ^{14}C-casein and ^{3}H-elastin and for chemotactic factors (and their inhibitors) in the Boyden chamber.

In addition, they are being assayed *in vivo* for their ability to produce or inhibit phlogistic and chemotactic responses. The culture fluids are concentrated and injected into normal skin and into cotton-induced granulomas. Because the mediators are secreted and collected over 24 h and then concentrated, they would be only *partly* inactivated, after injection, by local serum and tissue inhibitors. However, if the initial lesion was not cultured but merely homogenized, such mediators would be in low concentration and probably inactivated by inhibitors. Our results to date have detected phlogistic and chemotactic factors in culture fluids from lungs containing BCG granulomas, but we have not yet quantitated these factors or studied healing granulomas for inhibitors.

We also plan to test these culture fluids for the presence of mitogenic factors by injecting these fluids into granulomas produced by cotton balls inserted subcutaneously, injecting tritiated thymidine intravenously, and then evaluating by autoradiography the proliferation of (i.e., ^{3}H-TdR incorporation by) macrophages, lymphocytes, fibroblasts, and capillary endothelial cells in thin plastic-embedded (JB-4) tissue sections made from the granulomas (47). In this way, the targets for the mitogenic factors will be identified.

Our in vivo–in vitro system should detect the production of proteases, chemotactic factors, and mitogenic factors (and their inhibitors) by various types of granulomas as they develop and heal. Admittedly, in this system we can not tell which cells produce the inflammatory mediators and inhibitors, but histological and histochemical correlations will be made. We hope that such studies will recognize changes in the macrophage population responsible for the development and healing of chronic inflammatory lesions, but our work, unfortunately, has not progressed enough at this time to provide you with definitive results.

6. Summary

Monocytes seem to enter sies of inflammation in a rather immature, undifferentiated state. In such sites they are called macrophages and they differentiate for a variety of functions, e.g., (a) host defense: phagocytosis, microbicidal activities, and digestion; (b) regulation of the immune response by antigen presentation and interaction with lymphocytes; (c) secretion of a variety of substances which modulate other cells, such as capillaries and fibroblasts; (d) release of enzymes which hydrolyze tissue components at neutral (and acid) pH; (e) sequestration of indigestible foreign material; and (f) combinations of these and other functions (see Tables 1 and 2). Perhaps some macrophages are programed for certain functions in the bone marrow, but many seem to differentiate in certain directions because of local stimuli in the inflammatory site.

Double-staining histochemical techniques for a variety of enzymes suggest that both macrolocal and microlocal control of macrophage differentiation exists. In the peripheral parts of 21-day tuberculous lesions produced by BCG, macrophages rich in the proteinase cathepsin D were more numerous than macrophages rich in the sugar-splitting enzyme β-galactosidase. In the viable tissue near the caseous center, the reverse situation was present. We feel that such local differentiation was under regional or *macrolocal* control.

Within each part of the BCG lesion, there were always some cells stained for the non-prevalent enzyme. We feel that such local differentiation was under the control of individual cells nearby. In other words, there was *microlocal* control.

Because inflammatory mediators, such as neutral proteases, chemotactic factors and mitogenic factors, seem to be secreted by macrophages and then rapidly neutralized or inactivated by serum factors (and others) as soon as they escape the local site, we have begun in vivo-in vitro studies on inflammatory lesions. Specifically, we are culturing biopsies of BCG lesions in serum-free medium, concentrating the culture supernates, and assaying them for the three groups of inflammatory mediators just listed and for inhibitors of these mediators. We hope to be able to correlate various histochemically demonstrable marker enzymes with some of the secretory functions of macrophages. If successful, we should gain still further insight into the macrolocal and microlocal control of the inflammatory process.

References

1. Adams DO (1976) The granulomatous inflammatory response. A review. Am J Pathol 84:164–191
2. Alitalo K, Hovi T, Vaheri A (1980) Fibronectin is produced by human macrophages. J Exp Med 151:602–613
3. Allison AC (1978) Macrophage activation and non-specific immunity. Int Rev Exp Pathol 18: 303–343
4. Allison AC (1978) Mechanisms by which activated macrophages inhibit lymphocyte responses. Immunol Rev 40:3–27
5. Ando M, Suga M, Shima K, Sugimoto M, Higuchi S, Tsuda T, Tokuomi H (1976) Different effects of phytohemagglutinin-activated lymphocytes and their culture supernatants on macrophage function. Infect Immun 13:1442–1448
6. Baggiolini M, Schnyder J, Bretz U (1979) Lysosomal enzymes and neutral proteinases as mediators of inflammation. In: Weissmann G, Samuelsson B, Paoletti R (eds) Advances in inflammation research, vol 1. Raven, New York, pp 263–272
7. Beelen RHJ, Broekhuis-Fluitsma DM, Korn C, Hoefsmit ECM (1978) Identification of exudate-resident macrophages on the basis of peroxidatic activity. J Reticuloendothel Soc 23:103–110
8. Benveniste J, Duval D, Arnoud B, Chretien J (1979) Liberation du facteur activant les plaquettes (P.A.F.) par les macrophages alveolaires. Nouv Presse Med 7:2071–2072
9. Blumenthal K, Rourke FJ, Wilder MS (1980) Platelet activation by cultured mouse peritoneal macrophages. J Reticuloendothel Soc 27:247–257
10. Bodel P, Miller H (1977) Differences in pyrogen production by mononuclear phagocytes and by fibroblasts or HeLa cells. J Exp Med 145:607–617
11. Bonney RJ, Wightman PD, Davies P, Sadowski SJ, Kuehl FA Jr, Humes JL (1978) Regulation of prostaglandin synthesis and of the selective release of lysomal hydrolases by mouse peritoneal macrophages. Biochem J 176:433–442
12. Bonney RJ, Burger S, Davies P, Kuehl FA Jr, Humes JL (1980) Prostaglandin E_2 and prostacyclin elevate cyclic AMP levels in elicited populations of mouse peritoneal macrophages. Adv Prostaglandin Thromboxane Res 8:1691–1693
13. Bray MA, Gordon D (1978) Prostaglandin production by macrophages and the effect of anti-inflammatory drugs. Br J Pharmacol 63:635–642
14. Breard J, Reinherz EL, Kung PC, Goldstein G, Schlossman SF (1980) A monoclonal antibody reactive with human peripheral blood monocytes. J Immunol 124:1943–1948
15. Brennan JK, Lichtman MA, DiPersio JF, Abboud CN (1980) Chemical mediators of granulopoiesis: A review. Exp Hematol 8:441–464
16. Broxmeyer HE, Ralph P, Bognacki J, Kincade PW, Desousa M (1980) A subpopulation of human polymorphonuclear neutrophils contains an active form of lactoferrin capable of binding to human monocytes and inhibiting production of granulocyte-macrophage colony stimulatory activities. J Immunol 125:903–909
17. Brune K, Glatt M, Kälin H, Peskar BA (1978) Pharmacological control of prostaglandin and thromboxane release from macrophages. Nature 274:261–263
18. Brune K, Kälin H, Schmidt R, Hecker E (1979) Regulation of prostaglandin release from macrophages. Adv Inflam Res 1:467–475

19. Campbell MW, Sholley MM, Miller GA (1980) Macrophage heterogeneity in tumor resistance: cytostatic and cytotoxic activity of Corynebacterium parvum-activated and proteose peptone-elicited rat macrophages against Moloney sarcoma tumor cells. Cell Immunol 50:153–168

20. Carr I, Carr J, Lobo A, Malcolm D (1978) The secretion of lysozyme in vivo by macrophages into lymph and blood in a rat granuloma. J Reticuloendothel Soc 24:41–48

21. Cebula TA, Hanson DF, Moore DM, Murphy PA (1979) Synthesis of four endogenous pyrogens by rabbit macrophages. J Lab Clin Med 94:95–105

22. Cohn ZA, Benson B (1965) The in vitro differentiation of mononuclear phagocytes. III. The reversibility of granule and hydrolytic enzyme formation and the turnover of granule constituents. J Exp Med 122:455–466

23. Colvin RB, Dvorak HF (1975) Fibrinogen/fibrin on the surface of macrophages: Detection, distribution, binding requirements, and possible role in macrophage adherence phenomena. J Exp Med 142:1377–1379

24. Cooper GN, Houston B (1964) Effects of simple lipids on the phagocytic properties of peritoneal macrophages. II. Studies on the phagocytic potential of cell populations. Aust J Exp Biol Med Sci 42:429–442

25. Cowing C, Schwartz BD, Dickler HB (1978) Macrophage Ia antigens. I. Macrophage populations differ in their expression of Ia antigens. J Immunol 120:378–384

26. Currie GA (1978) Activated macrophages kill tumour cells by releasing arginase. Nature 273:758–759

27. Dannenberg AM Jr (1980) Macrophages and monocytes. In: Spivak JL (ed) Fundamentals of clinical hematology. Harper & Row, Hagerstown, pp 137–153

28. Dannenberg AM Jr, Burstone MS, Walter PC, Kinsley JW (1963) A histochemical study of phagocytic and enzymatic functions of rabbit mononuclear and polymorphonuclear exudate cells and alveolar macrophages. I. Survey and quantitation of enzymes, and states of cellular activation. J Cell Biol 17:465–486

29. Davies P, Allison AC (1976) Secretion of macrophage enzymes in relation to the pathogenesis of chronic inflammation. In: Nelson DS (ed) Immunobiology of the macrophage. Academic Press, New York, pp 427–461

30. Davies P, Bonney RJ, Humes JL, Kuehl FA Jr (1980) Secretion of arachidonic acid oxygenation products by mononuclear phagocytes: Their possible significance as modulators of lymphocyte function. In: Unanue ER, Rosenthal AS (eds) Macrophage regulation of immunity. Academic Press, New York, pp 347–397

31. Diamantstein T, Handschumacher RE, Oppenheim JJ, Rosenstreich DL, Unanue ER, Waksman BH, Wood DD (1979) Nonspecific "lymphocyte activating" factors produced by macrophages. J Immunol 122:2633–2635

32. Dominguez JH, Mundy GR (1980) Monocytes mediate osteoclastic bone resorption by prostaglandin production. Calcif Tissue Int 31:29–34

33. Domzig W, Lohmann-Matthes ML (1979) Antibody dependent cellular cytotoxicity against tumor cells. II. The promonocyte identified as effector cell. Eur J Immunol 9:267–272

34. Edelson PJ, Erbs C (1978) Plasma membrane localization and metabolism of alkaline phosphodiesterase I in mouse peritoneal macrophages. J Exp Med 147:77–86

35. Edelson RL, Smith RW, Frank MM, Green I (1973) Identification of subpopulations of mononuclear cells in cutaneous infiltrates. I. Differentiation between B cells, T cells, and histiocytes. J Invest Dermatol 61:82–89

36. Edwards RL, Rickles FR (1980) The role of human T cells (and T cell products) for monocyte tissue factor generation. J Immunol 125:606–609

37. Edwards RL, Rickles FR, Bobrove AM (1979) Mononuclear cell tissue factor: Cell of origin and requirements for activation. Blood 54:359–370

38. Emeis JJ, Planque B (1976) Heterogeneity of cells isolated from rat liver by pronase digestion: Ultrastructure, cytochemistry and cell culture. J Reticuloendothel Soc 20:11–29

39. Filkins JP (1980) Endotoxin-enhanced secretion of macrophage insulin-like activity. J Reticuloendothel Soc 27:507–511

40. Fishman M, Weinberg DS (1979) Functional heterogeneity among peritoneal macrophages. II. Enzyme content of macrophage subpopulations. Cell Immunol 45:437–445

41. Goldyne ME, Stobo JD (1979) Synthesis of prostaglandins by subpopulations of human peripheral blood monocytes. Prostaglandins 18:687–694

42. Gorczynski RM, MacRae S, Jennings JJ (1979) A novel role for macrophage: Antigen discrimination of distinct carbohydrate bonds. Cell Immunol 45:276–294

43. Gordon LI, Miller WJ, Branda RF, Zanjani ED, Jacob HS (1980) Regulation of erythroid colony formation by bone marrow macrophages. Blood 55:1047–1050

44. Gordon S (1976) Macrophage neutral proteinases and chronic inflammation Ann NY Acad Sci 278:176–189

45. Hanson DF, Murphy PA, Windle BE (1980) Failure of rabbit neutrophils to secrete endogenous pyrogen when stimulated with staphylococci. J Exp Med 151:1360–1371

46. Hearst JH, Warr GA, Jakab GJ (1980) Characterization of murine lung and peritoneal macrophages. J Reticuloendothel Soc 27:443–454

47. Higuchi S, Suga M, Dannenberg AM Jr (1979) Histochemical demonstration of enzyme activities in plastic- and paraffin-embedded tissue sections. Stain Technol 54:5–12

48. Hopper KE, Wood PR, Nelson DS (1979) Macrophage heterogeneity. Vox Sang 36:257–274

49. Horton JE, Koopman WJ, Farrar JJ, Fuller-Bonar J, Mergenhagen SE (1979) Partial purification of a bone-resorbing factor elaborated from human allogeneic cultures. Cell Immunol 43:1–10

50. Humes JL, Burger S, Galavage M, Kuehl FA Jr, Wightman PD, Dahlgren ME, Davies P, Bonney RJ (1980) The diminished production of arachidonic acid oxygenation products by elicited mouse peritoneal macrophages: possible mechanisms. J Immunol 124:2110–2116

51. Hunninghake GW, Fauci AS (1976) Immunological reactivity of the lung. II. Cytotoxic effector function of pulmonary mononuclear cell subpopulations. Cell Immunol 26:98–104

52. Hunt JD, Ward PA (1979) Chemotactic factor inactivator release from rat leukocytes. Inflammations 3:203–214

53. Johnston RB, Godzik CA, Cohn ZA (1978) Increased superoxide anion production by immunologically activated and chemically elicited macrophages. J Exp Med 148:115–127

54. Kampschmidt RF, Pulliam LA (1978) Effect of human monocyte pyrogen on plasma iron, plasma zinc, and blood neutrophils in rabbits and rats. Proc Soc Exp Biol Med 158:32–35

55. Kávai M, Laczkó J, Csaba B (1979) Functional heterogeneity of macrophages. Immunology 36: 729–732

56. Kavet RI, Brain JD (1977) Phagocytosis: Quantification of rates and intercellular heterogeneity. J Appl Physiol 42:432–437

57. Klimetzek V, Sorg C (1979) The production of fibrinolysis inhibitors as a parameter of the activation state in murine macrophages. Eur J Immunol 9:613–619

58. Kung JT, Brooks SB, Jakway JP, Leonard LL, Talmage DW (1977) Suppression of in vitro cytotoxic response by macrophages due to induced arginase. J Exp Med 146:665–672

59. Lee KC (1980) Macrophage heterogeneity in the stimulation of T cell proliferation. In: Unanue ER, Rosenthal AS (eds) Macrophage regulation of immunity. Academic Press, New York, pp 319–332

60. Lee KC (1980) On the origin and mode of action of functionally distinct macrophage subpopulations. Mol Cell Biochem 30:39–55

61. Lee KC, Wong M (1980) Functional heterogeneity of culture-grown bone marrow-derived macrophages. I. Antigen presenting function. J Immunol 125:86–95

62. Lee KC, Kay J, Wong M (1979) Separation of functionally distinct subpopulations of Corynebacterium parvum-activated macrophages with predominantly stimulatory or suppressive effect on the cell-mediated cytotoxic T cell response. Cell Immunol 42:28–41

63. Levy GA, Edgington TS (1980) Lymphocyte cooperation is required for amplification of macrophage procoagulant activity. J Exp Med 151:1232–1244

64. Lohmann-Matthes ML, Domzig W, Roder J (1979) Promonocytes have the functional characteristics of natural killer cells. J Immunol 123:1883–1886

65. Loor F, Roelants GE (1974) The dynamic state of the macrophage plasma membrane. Attachment and fate of immunoglobulin, antigen and lectins. Eur J Immunol 4:649–660

66. Lowrie DB, Andrew PW, Peters TJ (1979) Analytical subcellular fractionation of alveolar macrophages from normal and BCG-vaccinated rabbits with particular reference to heterogeneity of hydrolase containing granules. J Biochem 178:761–767

67. McGee MP, Myrvik QN (1979) Phagocytosis-induced injury of normal and activated alveolar macrophages. Infect Immun 26:910–915

68. McIntyre J, Rowley D, Jenkin CR (1967) The functional heterogeneity of macrophages at the single cell level. Aust J Exp Biol Med Sci 45:675–680

69. McKeever PE, Walsh GP, Storrs EE, Balentine JD (1978) Electron microscopy of peroxidase and acid phosphatase in leprous and uninfected armadillo macrophages: A macrophage subpopulation contains peroxisomes and lacks bacilli. Am J Trop Med Hyg 27:1019–1029

70. Miller GA, Campbell MW, Hudson JL (1980) Separation of rat peritoneal macrophages into functionally distinct subclasses by centrifugal elutriation. J Reticuloendothel Soc 27:167–174

71. Morahan PS, Kaplan AM (1978) Antiviral and antitumor functions of activated macrophages. In: Chirigos MA (ed) Immune modulation and control of neoplasia by adjuvant therapy. Raven, New York, pp 447–457

72. Mørland B, Mørland J (1978) Selective induction of lysosomal enzyme activities in mouse peritoneal macrophages. J Reticuloendothel Soc 23:469–477

73. Morley J, Bray MA, Jones RW, Nugteren DH, van Dorp CA (1979) Prostaglandin and thromboxane production by human and guineapig macrophages and leukocytes. Prostaglandins 17:729–736

74. Muller W, Hanauske-Abel H, Loos M (1978) Biosynthesis of the first component of complement by human and guinea pig peritoneal macrophages: Evidence for an independent production of the Cl subunits. J Immunol 121:1578–1584

75. Murphy PA, Simon PL, Willoughby WF (1980) Endogenous pyrogens made by rabbit peritoneal exudate cells are identical with lymphocyte-activating factors made by rabbit alveolar macrophages. J Immunol 124:2498–2501

76. Murray HW, Cohn ZA (1979) Macrophage oxygen-dependent antimicrobial activity. I. Susceptibility of Toxoplasma gondii to oxygen intermediates. J Exp Med 150:938–949

77. Murray HW, Juangbhanich CW, Nathan CF, Cohn ZA (1979) Macrophage oxygen-dependent antimicrobial activity. II. The role of oxygen intermediates. J Exp Med 150:950–964

78. Nathan CF, Brukner LH, Silverstein SC, Cohn ZA (1979) Extracellular cytolysis by activated macrophages and granulocytes. I. Pharmacologic triggering of effector cells and the release of hydrogen peroxide. J Exp Med 149:84–99

79. Nathan CF, Silverstein SC, Brukner LH, Cohn ZA (1979) Extracellular cytolysis by activated macrophages and granulocytes. II. Hydrogen peroxide as a mediator of cytotoxicity. J Exp Med 149:100–113

80. Neumann C, Sorg C (1977) Immune interferon. I. Production by lymphokine-activated murine macrophages. Eur J Immunol 7:719–725

81. Neumann C, Sorg C (1978) Immune interferon. II. Different cellular site for the production of murine macrophage migration inhibitory factor and interferon. Eur J Immunol 8:582–589

82. Norris DA, Morris RM, Sanderson RJ, Kohler PF (1979) Isolation of functional subsets of human peripheral blood monocytes. J Immunol 123:166–172

83. Olds GR, Ellner JJ, Kearse LA Jr, Kazura JW, Mahmoud AAF (1980) Role of arginase in killing of schistosomula of Schistosoma mansoni. J Exp Med 151:1557–1562

84. Ooi YM, Harris DE, Edelson PJ, Colten HR (1980) Post-translational control of complement (C 5) production by resident and stimulated mouse macrophages. J Immunol 124:2077–2081

85. Orange RP, Moore EG, Gelfand EW (1980) The formation and release of slow reacting substance of anaphylaxis (SRS-A) by rat and mouse peritoneal mononuclear cells induced by ionophore A23187. J Immunol 124:2264–2267

86. Pabst MJ, Johnston RB Jr (1980) Increased production of superoxide anion by macrophages exposed in vitro to muramyl dipeptide or lipopolysaccharide. J Exp Med 151:101–114

87. Page RC, Davies P, Allison AC (1978) The macrophage as a secretory cell. Int Rev Cytol 52:119–157

88. Perkins EH, Leonard MR (1963) Specificity of phagocytosis as it may relate to antibody formation. J Immunol 90:228–237

89. Picker LJ, Raff HV, Goldyne ME, Stobo JD (1980) Metabolic heterogeneity among human monocytes and its modulation by PGE$_2$. J Immunol 124:2557–2562

90. Prydz H, Allison AC (1978) Tissue thromboplastin activity of isolated human monocytes. Thromb Haemost 39:582–591

91. Prydz H, Allison AC, Schorlemmer HU (1977) Further link between complement activation and blood coagulation. Nature 270:173–174

92. Rachmilewitz B, Rachmilewitz M, Chaouat M, Schlesinger M (1978) Production of TCII (Vitamin B$_{12}$ Transport Protein) by mouse mononuclear phagocytes. Blood 52:1089–1098

93. Rachmilewitz D, Ligumsky M, Rachmilewitz B, Rachmilewitz M, Tarcic N, Schlesinger M (1980) Transcobalamin II level in peripheral blood monocytes–a biochemical marker in inflammatory diseases of the bowel. Gastroenterology 78:43–46

117

94. Rhodes J (1975) Macrophage heterogeneity in receptor activity: The activation of macrophage Fc receptor function in vivo and in vitro. J Immunol 114:976-981
95. Rice SG, Fishman M (1974) Functional and morphological heterogeneity among rabbit peritoneal macrophages. Cell Immunol 11:130-145
96. Rich IN, Heit W, Kubanek B (1980) An erythropoietic stimulating factor similar to erythropoietin released by macrophages after treatment with silica. Blut 40:297-303
97. Ridge SC, Oronsky AL, Kerwar SS (1980) Induction of the synthesis of latent collagenase and latent neutral protease in chondrocytes by a factor synthesized by activated macrophages. Arthritis Rheum 23:448-454
98. Rivers RPA, Hathaway WE, Weston WL (1975) The endotoxin-induced coagulant activity of human monocytes. Br J Haematol 30:311-316
99. Rojas-Espinosa O, Dannenberg AM Jr, Sternberger LA, Tsuda T (1974) Role of cathepsin D in the pathogenesis of tuberculosis. A histochemical study employing unlabeled antibodies and the peroxidase-antiperoxidase complex. Am J Pathol 74:1-17
100. Roos MH, Kornfeld S, Shreffler DC (1980) Characterization of the oligosaccharide units of the fourth component of complement (Ss protein) synthesized by murine macrophages. J Immunol 124:2860-2863
101. Roubin R, Zolla-Pazner S (1979) Markers of macrophage heterogeneity. I. Studies of macrophages from various organs of normal mice. Eur J Immunol 9:972-978
102. Ruco LP, Meltzer MS (1978) Macrophage activation for tumor cytotoxicity: Increased lymphokine responsiveness of peritoneal macrophages during acute inflammation. J Immunol 120:1054-1062
103. Russell AS, Ruegg UT (1980) Arginase production by peritoneal macrophages: A new assay. J Immunol Methods 32:375-382
104. Saunders D, Edidin M (1974) Sites of localization and synthesis of Ss protein in mice. J Immunol 112:2210-2218
105. Schnyder J, Baggiolini M (1978) Secretion of lysosomal hydrolases by stimulated and nonstimulated macrophages. J Exp Med 148:435-450
106. Schorlemmer HU, Bitter-Suermann D, Allison AC (1977) Complement activation by the alternative pathway and macrophage enzyme secretion in the pathogenesis of chronic inflammation. Immunology 32:929-940
107. Scott WA, Zrike JM, Hamill AL, Kempe J, Cohn ZA (1980) Regulation of arachidonic acid metabolites in macrophages. J Exp Med 152:324-335
108. Serio C, Gandour DM, Walker WS (1979) Macrophage functional heterogeneity: Evidence for different antibody-dependent effector cell activities and expression of Fc-receptors among macrophage subpopulations. J Reticuloendothel Soc 25:197-206
109. Simon LM, Robin ED, Phillips JR, Acevedo J, Axline SG, Theodore J (1977) Enzymatic basis for bioenergetic differences of alveolar versus peritoneal macrophages and enzyme regulation by molecular O_2. J Clin Invest 59:443-448
110. Sorg C (1979) The biochemistry and in vitro activity of soluble factors of activated lymphocytes. Mol Cell Biochem 28:149-167
111. Springer T, Galfre G, Secher DS, Milstein C (1979) Mac-1: a macrophage differentiation antigen identified by monoclonal antibody. Eur J Immunol 9:301-306
112. Stadecker MJ, Calderon J, Karnovsky ML, Unanue ER (1977) Synthesis and release of thymidine by macrophages. J Immunol 119:1738-1743
113. Stenson WF, Parker CW (1980) Opinion: Prostaglandins, macrophages, and immunity. J Immunol 125:1-5
114. Suga M, Dannenberg AM Jr, Higuchi S (1980) Macrophage functional heterogeneity in vivo. Macrolocal and microlocal macrophage activation, identified by double-staining tissue sections of BCG granulomas for pairs of enzymes. Am J Pathol 99:305-324
115. Sugimoto M, Dannenberg AM Jr, Wahl LM, Ettinger WH Jr, Hastie AT, Daniels DC, Thomas CR, Demoulin-Brahy L (1978) Extracellular hydrolyticenzymes of rabbit dermal tuberculous lesions and tuberculin reactions collected in skin chambers. Am J Pathol 90:583-608
116. Thomas MA, Galbraith I, MacSween RNM (1978) Heterogeneity of rat peritoneal and alveolar macrophage populations: Spontaneous rosette formation using sheep and chicken red blood cells. J Reticuloendothel Soc 23:43-52
117. Unanue ER (1976) Secretory function of mononuclear phagocytes. Am J Pathol 83:396-417
118. Unanue ER (1978) The regulation of lymphocyte functions by the macrophage. Immunol Rev 40:227-255

119. Unanue ER, Beller DI, Calderon J, Kiely JM, Stadecker MJ (1976) Regulation of immunity and inflammation by mediators from macrophages. Am J Pathol 85:465–478
120. Van Furth R (ed) (1975) Mononuclear phagocytes in immunity, infection and pathology. Blackwell, Oxford
121. Volkman A (1970) The origin and fate of the monocyte. Ser Haematol 3:62
122. Wachsmuth ED, Stoye JP (1977) Aminopeptidase on the surface of differentiating macrophages: Induction and characterization of the enzyme. J Reticuloendothel Soc 22:469–483
123. Wachsmuth ED, Stoye JP (1977) Aminopeptidase on the surface of differentiating macrophages: Concentration changes on individual cells in culture. J Reticuloendothel Soc 22:485–497
124. Waksman BH, Namba Y (1976) On soluble mediators of immunologic regulation. Cell Immunol 21:161–176
125. Walker DM (1976) Identification of subpopulations of lymphocytes and macrophages in the infiltrate of lichen planus lesions of skin and oral mucosa. Br J Dermatol 94:529–534
126. Walker WS (1974) Functional heterogeneity of macrophages: Subclasses of peritoneal macrophages with different antigen-binding activities and immune complex receptors. Immunology 26:1025–1037
127. Walker WS (1976) Functional heterogeneity of macrophages. In: Nelson DS (ed) Immunobiology of the macrophage. Academic Press, New York, pp 91–110
128. Walker WS (1976) Functional heterogeneity of macrophages in the induction and expression of acquired immunity. J Reticuloendothel Soc 20:57–65
129. Weinberg DS, Fishman M, Veit BC (1978) Functional heterogeneity among peritoneal macrophages. 1. Effector cell activity of macrophages against syngeneic and xenogeneic tumor cells. Cell Immunol 38:94–104
130. Werb Z, Dingle JT (1976) Lysosomes as modulators of cellular functions: Influence on the synthesis and secretion of nonlysosomal materials. In: Dingle JT, Dean RT (eds) Lysosomes in biology and pathology, vol 5. Elsevier/North-Holland, New York, pp 127–156
131. Wilson GB, Walker JH Jr, Watkins JH Jr, Wolgroch D (1980) Determination of subpopulations of leukocytes involved in the synthesis of α_1-antitrypsin in vitro. Proc Soc Exp Biol Med 164:105–114
132. Wing EJ, Gardner ID, Ryning FW, Remington JS (1977) Dissociation of effector functions in populations of activated macrophages. Nature 268:642–644
133. Yarborough DJ, Meyer OT, Dannenberg AM Jr, Pearson B (1967) Histochemistry of macrophage hydrolases. III. Studies on β-galactosidase, β-glucuronidase and aminopeptidase with indolyl and naphthyl substrates. J Reticuloendothel Soc 4:390–408
134. Yoneda T, Mundy GR (1979) Monocytes regulate osteoclast-activating factor production by releasing prostaglandins. J Exp Med 150:338–350
135. Yoneda T, Mundy GR (1979) Prostaglandins are necessary for osteoclast-activating factor production by activated peripheral blood leukocytes. J Exp Med 149:279–283

Haematology and Blood Transfusion Vol 27
Disorders of the Monocyte Macrophage System
Edited by F. Schmalzl, D. Huhn, H.E. Schaefer
© Springer-Verlag Berlin Heidelberg New York 1981

The Role of Macrophages in Storage Diseases[1]

H.-E. Schaefer

When dealing with macrophages in storage diseases three points of interest arise:

1. Since macrophages are involved in the first line in most types of storage diseases the observation of characteristic morphological features provides an important diagnostic tool for pathologists.

2. When macrophages become overloaded with stored material the mononuclear system has to proliferate in order to furnish additional storage capacity. Under those conditions, the recruitment of tissue macrophages from monocytes or from other precursor cells can be monitored just as in an *experimentum naturae*.

3. The process of pathological storage may reveal functional differences of macrophages or of uncertain macrophage candidates.

1. Diagnostic Aspects of Macrophages in Storage Diseases

Despite the fact that most types of inherited catabolic enzyme deficiencies afflict various cell types, the most obvious storage phenomena are to be seen in macrophages, since tissue macrophages serve as scavengers for any material that cannot be catabolized by other cells. Ultimately the indigestible substances are piled up within macrophage lysosomes, thus forming distinct structures depending on the type of the accumulated metabolic substrate.

In *Niemann-Pick's disease* sphingomyelin deposits exhibit concentrically laminated structures with a periodicity of about 50 Å [23]. To some extent the appearance of these storing lysosomes is modified by the scheme of fixation applied (Fig. 1). In most types (A, C and D) of Niemann-Pick's disease inclusions occur not only in tissue macrophages but in blood monocytes and in lymphocytes as well [20, 42, 43]. It seems likely that in part storing lysosomes become exocytosed from lymphocytes [32]. This material, together with metabolic debris derived from the physiologic phagocytotic activities, ultimately accumulates in macrophages of the spleen, bone marrow, liver, lungs, and various other organs. Within this final station, the primary storage material undergoes secondary chemical modifications resulting in lipofuscin or ceroid-like deposits due to slow peroxidative processes. These secondary inclusions exhibit the tinctorial qualities of "sea-blue histiocytes" [4, 11], become autofluorescent, and acquire tartrate-resistance of acid phosphatase [34]. This type of modification is not at all specific for Niemann-Pick's disease; it may develop in various types of inherited lipoidosis and in reactive forms of aging lipid storing cells as well.

A further type of residual bodies with a concentrical lamellar ultrastructure results from deposition of globotriaosylceramide and galabiosylceramide in *Fabry's disease* (deficiency of lysosomal α-galactosidase A). Despite some resemblance with the sphingomyelin inclu-

[1] The investigations have been supported by the Minister für Wissenschaft und Forschung des Landes Nordrhein-Westfalen

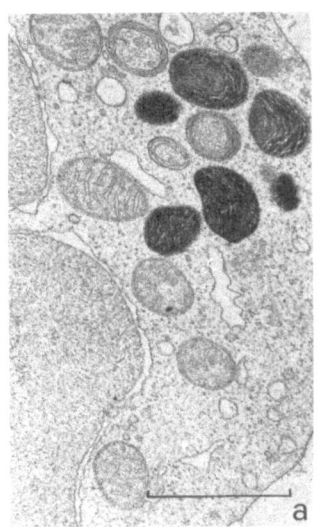
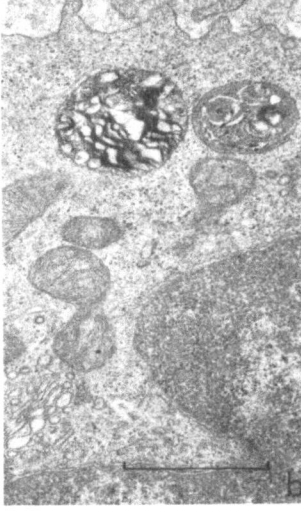

Fig. 1a, b. Blood lymphozytes in *Niemann-Pick's disease* (type C). **a** On fixation with OsO$_4$ only, the enlarged sphingomyelin-storing lysosomes display a more concentrically laminate substructure. **b** Combined fixation with OsO$_4$ and glutaraldehyde brings about lysosomes containing more irregularly arranged foliaceous deposits. *Scale bar, 1 μm*

sions in Niemann-Pick's disease, the cellular distribution of storage phenomena is quite different in Fabry's disease, e.g., with respect to the peculiar involvement of endothelial cells and of renal epithelial cells [38, 40].

The Fabry-type inclusions tend to aggregate and to fuse to birefringent globular masses measuring up to 10 μm in diameter [40]. In vital state the largest inclusions display a snail shell aspect that can be observed with ease in foam cells of an urinary sediment (Fig. 2).

From the comparison of the engorged macrophages in *Gaucher's disease* (accumulation of glucocerebroside) and in *Krabbe's disease* (accumulation of galactocerebroside) it can be concluded that accumulation of chemically related substances brings about similarities in the morphological characteristics [6, 12, 37]. In human cases of Krabbe's disease [45] as well as in dogs [19] and in mice [29] deficient for β-galactosidase, the perivascular so-called globoid cells which gather in the central nervous system and the macrophages occurring in degenerated peripheral nerves among more irregular inclusions contain typically twisted tubules. These tubules are quite similar to the tubular substructure [21] of the storing lysosomes to be seen in macrophages of the spleen or the bone marrow or in Kupffer cells in Gaucher's disease (Fig. 3a, 4).

Lymph node [25] or bone marrow macrophages engorged with rectangular and hexagonal crystalline profiles provide diagnostic evidence for *cystinosis* with the same degree of security as does a chemical tissue analysis for cystine storage (Fig. 5). *Tangier disease* gives an additional example for the notion that lysosomal storage phenomena are not at all confined to inherited deficiences of catabolic or lysosomal enzymes. The absence of high density lipoprotein (HDL) [8, 14] and the abnormal metabolism of apolipoprotein A-I [13] brings

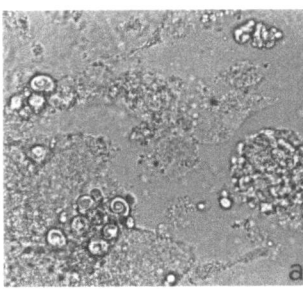
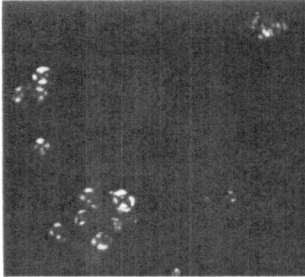

Fig. 2. **a** Urinary sediment obtained from a patient suffering from *Fabry's disease*. Unstained foam cells contain large snail shell shaped inclusions. **b** Viewed in plane-polarized light the inclusions exhibit maltese cross birefringence

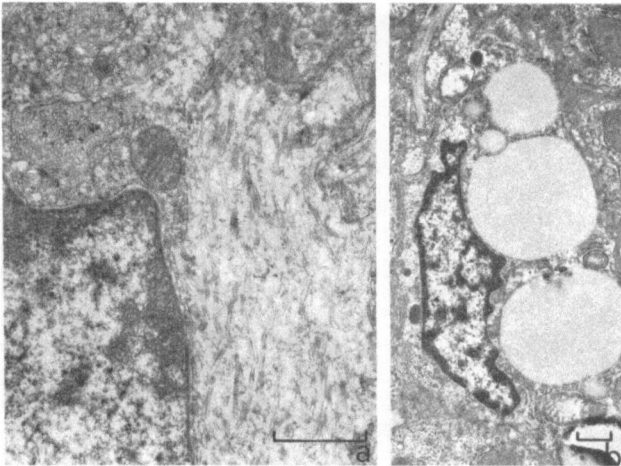

Fig. 3. a Von Kupffer cell in *Gaucher's disease* containing giant lysosomes filled with tubular glucocerebrosides deposits. **b** An Ito cell from the same liver equipped with its typical lipid droplets does not participate in the storing process. *Scale bar, 1 µm*

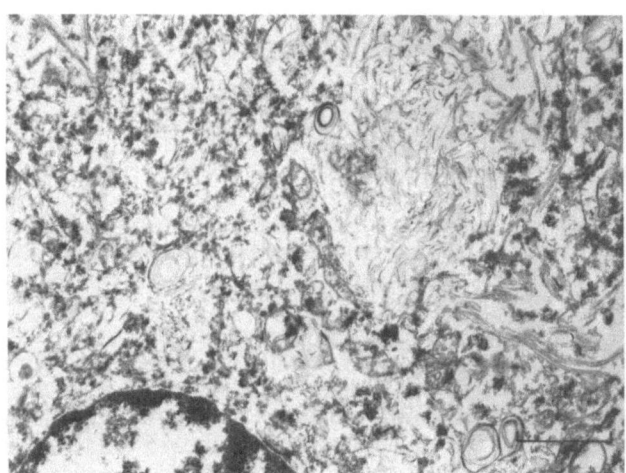

Fig. 4. *Krabbe's disease.* The cytoplasma of globoid cells contains deposits of galactocerebroside arranging themselves in tubular structures similar to the glucocerebroside deposits to be seen in Gaucher cells (compare with Fig. 3a). Specimen obtained from autopsy, embedded in paraffin prior to conventional processing for electron microscopy. *Scale bar, 1 µm*

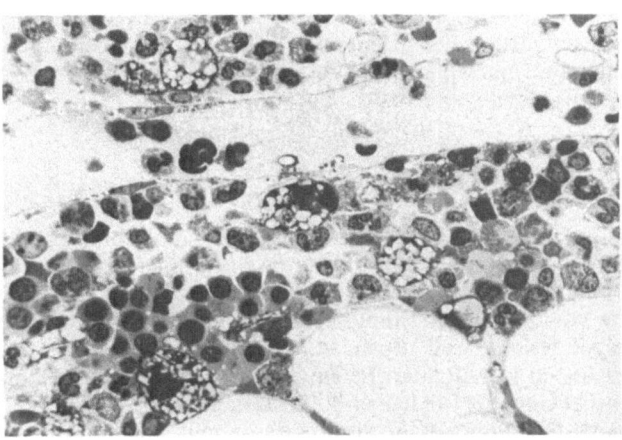

Fig. 5. Bone marrow biopsy from a patient suffering from cystinosis (adult type). Phagocytic reticulum cells are filled with polygonal (hexagonal and rectangular) inclusions of transparent cystine crystal. Semithin section, Epon-embedded tissue

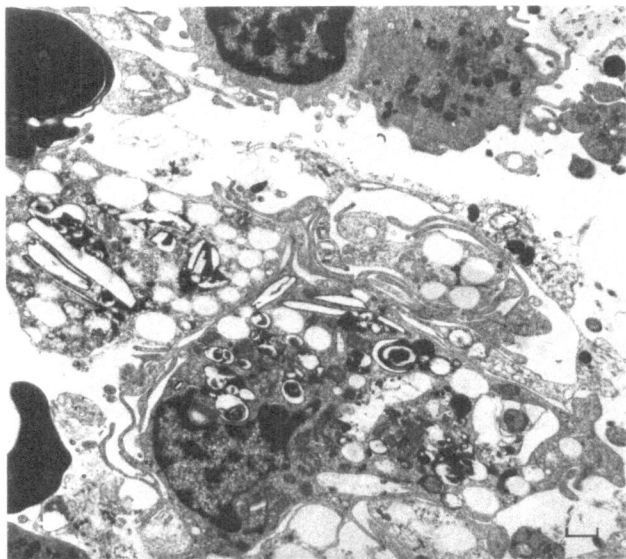

Fig. 6. *Tangier disease.* Bone marrow biopsy. Phagocytic reticulum cell contains lipid droplets composed mainly of cholesteryl esters in liquid state and solid crystalline material surrounded by an electron dense lysosomal matrix. When the fatty acids are split off by lysosomal acid lipase, they can be catabolized further; however, the cholesterol moiety of cholesteryl esters remains due to HDL deficiency within the cell and becomes stored in the form of crystalline residual bodies. *Scale bar, 1 µm*

about an accumulation of lipids (mainly cholesteryl-oleate) in smooth muscle cells of the rectal mucosa and of the peritoneum, in Schwann cells, in gingival fibroblasts, and in nevus cells [1, 18, 31, 33]. The mechanism leading to this peculiar distributory pattern is hard to rationalize in detail. The role that macrophages play in this disease can be defined as follows. Whereas the cells listed above mainly contain membraneless lipid droplets, and in places these deposits seem to originate from ergastoplasmic cisternae, in macrophages the major part of the lipid droplets are membrane bounded and can be localized within secondary lysosomes. In some of these lysosomes solid crystals of cholesterol surrounded by dense ceroid-like material precipitate (Fig. 6).

These findings suggest that Tangier macrophages hydrolyze cholesterylesters and that fatty acids are catabolically removed. Due to the deficiency of HDL, the surplus cholesterol cannot be shuttled to the excretory bile pathway of the liver and remains within residual bodies. This type of cholesterol (ester) storage is most prominent in macrophages with a high lipid turnover, as for instance due to cellular uptake and breakdown. So the macrophages of the bone marrow and the spleen are converted to foam cells as well as the histiocytes occurring in tonsils, in the lamina propria coli and in the inflamed mucosa of endocervical ektopia [1, 30].

The birefringence of stored cholesterol crystals accompanied by maltese cross-birefringent lipid droplets rich in cholesterylesters which is seen in histiocytic foam cells distributed all over reticuloendothelial tissues and which occurs at sites of chronic inflammation is highly suggestive for the diagnosis of Tangier disease.

2. The Recruitment of Macrophages in Gaucher's Disease

In adult patients suffering from Gaucher's disease the bone marrow becomes increasingly filled with Gaucher cells (Fig. 7). In time these cerebroside-storing tissue macrophages obliterate the hematopoietic tissue. This process is incompletely compensated for by a progressive widening of the bone marrow space leading to Erlenmayer-flask-like deformities of the femura and disseminated osteolytic lesions, well known from X-ray examination. In the spleen and in the liver, too, the advanced accumulation of Gaucher cells attains large tumor-like volumes, and we remember that Gaucher [10] had entitled his original thesis "De l'épithelioma primitif de la rate", taking the latter storage disease for an epithelial tumor.

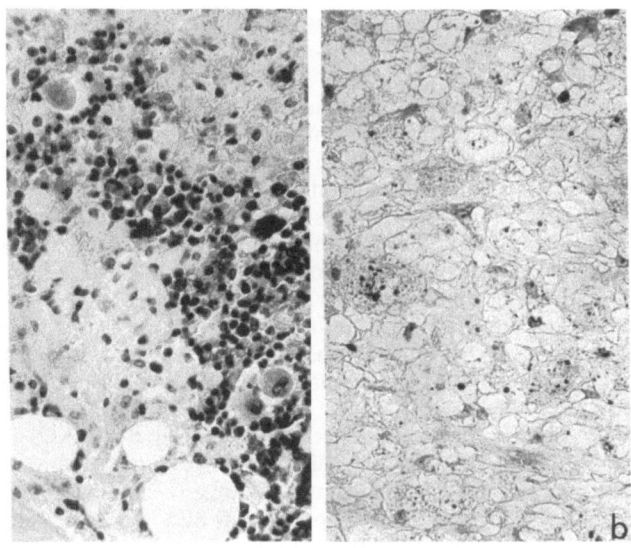

Fig. 7a, b. Progressive obliteration of the bone marrow due to accumulation of glucocerebroside-storing reticulum cells. **a** Biopsy obtained from a 35-year-old female and **b** from a 50-year-old male patient, both suffering from *Gaucher's disease*

Taking this obvious augmentation of storing macrophages into account, the question arises whether new cells are brought up by proliferative self-renewal of local resident macrophages or recruit themselves from blood monocytes.

We have studied blood monocytes from a child (juvenile type 3) and from three adults (adult type 1) suffering from Gaucher's disease. On light as well as on electron microscopical study the general morphology of monocytes appeared normal. Few monocytes seemed to be less mature. However, there were no storage phenomena and the Golgi area was somewhat enlarged in the more mature cells. But these findings are hard to objectify. Interestingly enough, these nearly normal-looking blood monocytes exhibited a clear-cut enzyme histochemical abnormality in that the acid phosphatase was mainly tartrate resistant [35]. We have detected monocytes with a much weaker activity of tartrate-resistant acid phosphatase in rare cases of chronic myelomonocytic leukemia and in patients treated with prophylactic doses of chloroquine against malaria [35]. In normal monocytes acid phosphatase is completely inhibitable by tartrate.

We do not know whether this monocytic tartrate-resistant acid phosphatase is identical to the tartrate-resistant isoenzyme known from hairy cell leukemia [22]. But if we apply this very histochemical test to the Gaucher cells it becomes readily evident that monocytes share the tartrate-resistant isoenzyme known to occur in the cerebroside-storing macrophages [44], notwithstanding that these cells may contain additional species of acid phosphatase inhibitable by tartrate. The unique and constant finding of a strong activity of tartrate-resistant acid phosphatase in blood monocytes of patients suffering from Gaucher's disease is in line with the assumption that Gaucher cells are derived from blood monocytes. In addition to the given histochemical evidence, it even can be taken from the electrophoretic studies [22] that the pathologic process of cerebroside storage induces peculiar isoenzymes of acid phosphatase (and probably of other hydrolases) in macrophages. Since in advanced disease, especially in the bone marrow, Gaucher cells undergo regressive transformation, these enzymes are shed and can be detected in the blood [7, 41, 42]. The same inductive stimuli seem to act on blood monocytes preparing themselves for transformation to macrophages. In any case, the detection of tartrate-resistant acid phosphatase in blood monocytes provides an easy histochemical screening test for Gaucher's disease.

3. Functional Heterogeneity of the Mononuclear Cell Macrophage System as Disclosed by Storage Diseases

In the previous literature there has been considerable debate as to whether the fat-storing Ito cells [16, 17] and the Kupffer cells of the liver merely represent different functional states of an unitary pool of convertible litoral cells [2, 24] or rather constitute a different cell type of their own [5, 15]. With this question in mind, we have studied by electron microscopy liver biopsies obtained from three cases of adult-type Gaucher's disease. In both specimens, masses of cerebroside-storing cells obliterating in part the sinusoids and crowding out hepatocytes have been found. The cytoplasm was engorged with giant secondary lysosomes exhibiting their typical tubular substructure. Since any normal-looking Kupffer cells were missing, these Gaucher cells seem to represent transformed Kupffer cells. Quite in contrast, the cytoplasmic structure of Ito cells remained completely unchanged (Fig. 3). This finding is consistent with the present day leading trend of opinion which considers Ito cells to be separate from the macrophage system.

With respect to this problem a converging line of evidence emanates from the histological analysis of a hitherto unreported type of acquired storage disease. In the course of repeated hemodialysis three adult patients suffering from chronic renal failure developed ascites, hepatomegaly, and increasing hematopoietic deficiency. Biopsies of the liver and of the iliac crest revealed an advanced infiltration of the bone marrow and of the liver by large histiocytic foam cells. The development of ascites could be ascribed to a widespread obliteration of hepatic sinusoids by the pathologic histiocytes. The same cells could be isolated from the peritoneal fluid. Similar to the lipid-storing macrophages of certain types of metabolic inborn errors, in this apparently acquired abnormality the histiocytes contain many empty appearing large lysosomes. Up to now, any attempts to identify the content of these vacuoles have failed. Because of the water-like transparence as seen in the electron microscope we have coined the term *hydrops lysosomalis* in order to designate this systemic macrophage disease. Contrary to the complete absence of any detectable lipids in these hydropic macrophages, Ito cells remained completely unaltered; they contained an usual amount of lipid droplets and did not engage in the pathological process of storage (Fig. 8).

Lipid storage in macrophages represents an event common to most types of hyperlipoproteinemia. However the differences in onset and regional distribution of those phenomena are characteristic for the various entities of hyperlipoproteinemia as typed by Fredrickson et al. [9] and Beaumont et al. [3]. Studying the dermal lesions of an individual suffering

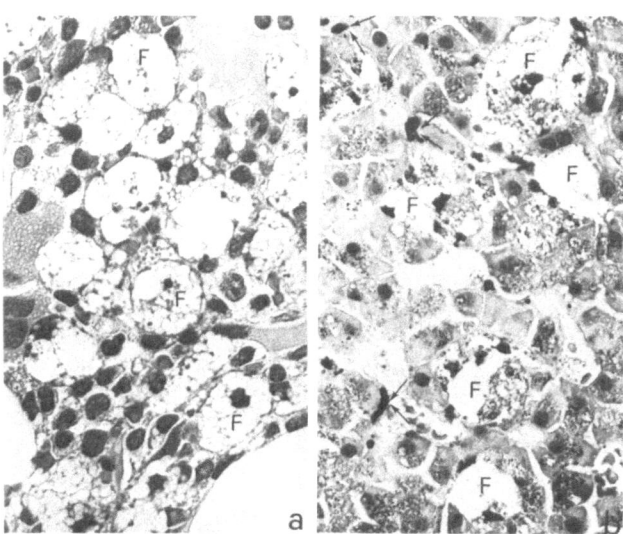

Fig. 8a, b. So-called *hydrops lysosomalis*, a hemodialysis-associated secondary storage disease. Massive infiltration of bone marrow **a** and of liver **b** with histiocytic foam cells. Note lipid-filled Ito cells (**b**, *arrows*) not participating in the storage process to be seen in phagocytic reticulum cells of the bone marrow and in the von Kupffer cells of the liver. Biopsy specimens **a** HE-stained, **b** osmicated and oil red O-stained

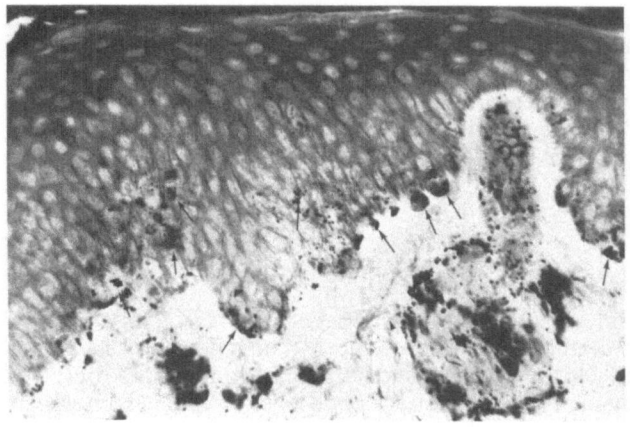

Fig. 9. Palmar dermal biopsy (oil red O-stained cryostate section) from a patient suffering from *familial type 3 hyperlipoproteinemia*. Besides large perivascular fat-storing histiocytes, small dendritic cells containing smaller lipid droplets occur in the basal epidermal strata (*arrows*). Compare with Fig. 10 for further identification of the latter cells

from the familiar type-3 hyperlipoproteinaemia and presenting xanthomata tuberosa et striata palmaria, we observed the bulk of lipid deposits to occur within histiocytes surrounding the small intrapapillary blood vessels and gathering focally within the dermal tissue. Unlike as in other types of dermal xanthomata, we came over small intraepidermal cells storing lipid droplets, too. These cells are disseminated mainly among the keratinocytes of the near-basal layer of the epidermis (Fig. 9). Because of their dendritic outshape these fat storing cells were assumed to represent Langerhans cells. However on electronmicroscopical observation, it became readily clear that the Langerhans cells remained quite normal looking. The intraepidermal lipid deposits were found strictly confined to the melanocytes (Fig. 10). It can be assumed that the intradermal xanthomata result from phagocytosis of lipoproteins leaking out of the blood vessels. Since in hyperlipoproteinemia type-3 there is an increase of very low density lipoprotein (VLDL) with an abnormal electrophoretic β-mobility and a smaller mean diameter of about 350 Å [26, 28] instead of a normal VLDL particle size of 400 Å, it seems likely that β-VLDL may enter the dermal and epidermal interstitial

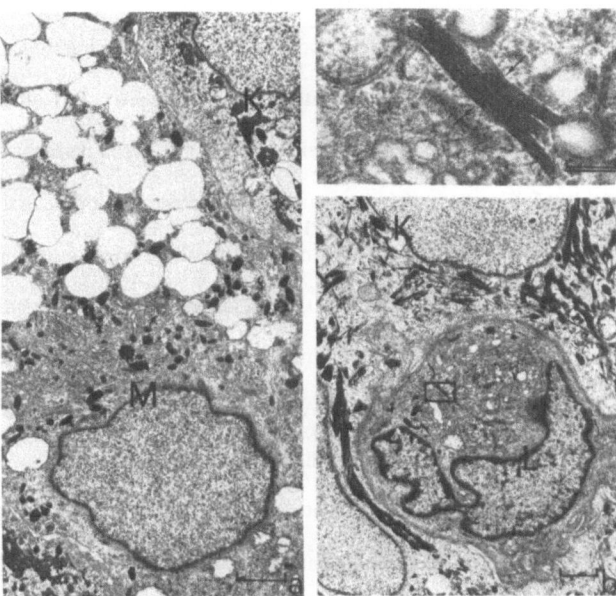

Fig. 10.a By electron microscopy the intraepidermal lipid-storing cells to be seen in *hyperlipoproteinemia type 3* (compare with Fig. 9) are identified as melanocytes (*M*) differing from kerotinocytes (*K*) by the absence of tonofilaments and their contents of electron-dense melanosomes. **b** Langerhans cells (*L*) containing typical Bierbeck granula (see higher enlargement of the inset area at top) do not store lipids. *Single scale bar, 1 μm; double scale bar, 100 μm*

spaces as well. It is interesting to note that under those pathological conditions only the intraepidermal melanocytes take up and store lipids. Langerhans cells residing within the same microenvironment refrain from any phagocytotic activity exerted on the surrounding lipoproteins. It has been argued that Langerhans cells originate from and belong to the monocyte–macrophage system [29, 36, 39]. In any case, the functional behavior as expressed in type-3 hyperlipoproteinemia points to an ultimate differentiation of Langerhans cells far away from a phagocytic cell.

4. Concluding Comments

The examples mentioned to delineate the role of macrophages in storage disease seem rather restricted and depend on the author's casual experiences. With regard to *cytopathology* it has been generally appreciated that the characteristic storage phenomena to be observed in macrophages provide a pertinent diagnostic tool. It should be stressed that comprehensive citation of all the accumulated data on the histophathological appearance of storage diseases lies beyond the scope of this paper. In this respect our knowledge has far advanced. On the other hand, inborn errors of metabolism bring about manifold exceptional states uncovering the functional heterogeneity of the monocyte-macrophage system. With regard to *cytobiology* a more systematic approach is needed to fully exploit the information that may be derived from a detailed study of the behavior these cells display in storage disease.

References

1. Assmann G, Schaefer HE (1980) Possible mechanisms of lipid storage in Tangier disease. In: Gotto AM, Smith LC, Allen B (eds) Atherosclerosis V. Springer, Berlin Heidelberg New York, p 666
2. Aterman K (1963) The structure of the liver sinusoids and the sinusoidal cells. In: Rouiller C (ed) The liver: morphology, biochemistry, physiology, vol I. Academic Press, New York, p 61
3. Beaumont JL, Carlson LA, Cooper GR, Fejfar Z, Fredrickson DS, Strasser T (1971) Classification of hyperlipoprotein–emias. Bull WHO 43:891
4. Brady RO, King FM (1973) Niemann-Pick's disease. In: Hers HG, von Hoof F (eds) Lipids and lipidosis. Academic Press, New York, p 439
5. Bronfenmajer S, Schaffner F, Popper H (1966) Fat-storing cells (lipocytes) in human liver. Arch Pathol 82:447
6. Diezel PB (1955) Histochemische Untersuchungen an den Globoidzellen der familiären diffusen Sklerose vom Typus Krabbe. Virchows Arch Pathol Anat 327:206
7. Fraccaro M, Magrini U, Scappaticci S, Zacchello F (1968) In vitro culture of spleen cells from a case of Gaucher's disease. Ann Hum Genet 32:209
8. Fredrickson DS (1964) The inheritance of high density lipoprotein deficiency (Tangier disease). J Clin Invest 43:228
9. Fredrickson DS, Levy RJ, Lees RS (1967) Fat transport in lipoproteins – an integrated approach to mechanisms and disorders. New Engl J Med 276:32, 94, 148, 215, 273
10. Gaucher PCE (1882) De l'épithélioma primitif de la rate; hypertrophie idiopathique de la rate sans leucémie. Medical dissertation, University of Paris
11. Golde DW, Schneider EL, Bainton DF, Pentchev PG, Brady RO (1975) Pathogenesis of one variant of sea-blue histiocytosis. Lab Invest 33:371
12. Hallervordem J (1948) Eine Speicherungshistiocytose des kindlichen Gehirns (Gauchersche Krankheit?). Verh Dtsch Ges Pathol 32:96
13. Henderson LO, Herbert PN, Fredrickson DS, Heinen RJ (1978) Abnormal concentration and anomalous distribution of apolipoprotein A-I in Tangier disease. Metabolism 27:165
14. Hoffmann HN, Fredrickson DS (1965) Tangier disease (familial high density lipoprotein deficiency): clinical and genetic features in two adults. Am J Med 39:582
15. Hübner G (1968) Zur Feinstruktur und Funktion der sog. Fettspeicherzellen der Leber. Anat Anz 121:495
16. Ito T (1951) Cytological studies on stellatae cells of Kupffer and fat storing cells in the capillary wall of the human liver. Acta Anat Jpn 24:42

17. Ito T, Nemoto M (1952) Über die Kupfferschen Sternzellen und die „Fettspeicherungszellen" ("fat storing cells") in der Blutkapillarenwand der menschlichen Leber. Okajimas Folia Anat Jpn 24:243

18. Katz SS, Small DM, Brook JB, Lees RS (1977) The storage lipids in Tangier disease. J Clin Invest 59:1045

19. Kurtz HJ, Fletcher TF (1970) The peripheral neuropathy of canine globoid-cell leukodystrophy (Krabbe-type). Acta Neuropathol (Berl) 16:226

20. Lazarus SS, Vethamany VG, Schneck L, Volk BW (1967) Fine structure and histochemistry of peripheral blood cells in Niemann-Pick disease. Lab Invest 17:155

21. Lee RE (1968) The fine structure of the cerebroside occurring in Gaucher's disease. Proc Natl Acad Sci USA 61:484

22. Li CY, Yam LT, Lam KW (1970) Studies of acid phosphatase isoenzymes in human leukocytes. Demonstration of isoenzyme cell specifity. J Histochem Cytochem 18:901

23. Lynn R, Terry RD (1964) Lipid histochemistry and electron microscopy in adult Niemann-Pick disease. Am J Med 37:987

24. Matsuo U (1959) Electron microscopic studies on the reticuloendothelial system in the normal rabbit's liver and the formation of the epithelioid tubercle. Kobe Igaku Kijo 15:265

25. Patrick AD, Lake BD (1968) Cystinosis: electron microscopic evidence of lysosomal storage of cystine in lymph node. J Clin Pathol 21:571

26. Patsch RJ, Sailer S, Braunsteiner H (1975) Lipoprotein of the density 1.006–1.020 in the plasma of patients with type III hyperlipoproteinemia in the postabsorptive state. Eur J Clin Invest 5:45

27. Rowden G (1980) Expression of Ia antigens on Langerhans cells in mice, guinea pigs, and man. J Invest Dermatol 75:22

28. Sata G, Havel RJ, Jones AL (1972) Characterization of subfractions of triglyceride-rich lipoproteins separated by gel chromatography from blood plasma of normolipemic and hyperlipemic humans. J Lipid Res 13:757

29. Scaravilli F, Jacobs JM (1981) Peripheral nerve grafts in hereditary leukodystrophic mutant mice (twitcher). Nature 290:56

30. Schaefer HE, Assmann G (1977) Die Manifestation der Tangier-Krankheit (sog. An-alpha-Lipoproteinämie) an der Portio vaginalis uteri. Verh Dtsch Ges Pathol 61:401

31. Schaefer HE, Assmann G (1979) Cholesteatosis naevi naevocellularis – ein bisher unbekanntes Phänomen bei der Tangier-Krankheit. Verh Dtsch Ges Pathol 63:708

32. Schaefer HE, Assmann G, Fischer R (1976) Morphologische und biochemische Untersuchungen bei Niemann-Pickscher Erkrankung. Verh Dtsch Ges Pathol 60:259

33. Schaefer HE, Assmann G, Gheorghiu T (1976) Licht- und elektronenmikroskopische Untersuchungen zur Tangier-Krankheit (sog. An-alpha-Lipoproteinämie). Verh Dtsch Ges Pathol 60:473

34. Schaefer HE, Hellriegel KP, Fischer R (1977) Vorkommen von tartrat-resistenter saurer Phosphatase in verschiedenen Zelltypen des lymphoretikulären und hämatopoetischen Zellsystems. Blut 34:393

35. Schaefer HE, Flentje M, Fischer R (1980) Tartrate resistant acid phosphatase (TSP) in monocytes – a hitherto unknown phenomenon. Fifth meeting, European Division, International Society of Hematology, Hamburg IV, A 42

36. Stingl G, Katz SI, Shevach EM, Rosenthal AS, Green I (1978) Analogous functions of macrophages and Langerhans cells in the initiation of the immune response. J Invest Dermatol 71:59

37. Suzuki K, Suzuki Y (1978) Galactosylceramide lipidosis: globoid cell leukodystrophy (Krabbe's disease). In: Stanbury JB, Wyngaarden JB, Fredrickson DS (eds) The metabolic basis of inherited diseases, 4th edn. Blakiston, New York, p 747

38. Sweeley CC (1978) Fabry's disease (alpha-galactosidase A deficiency. In: Stanbury JB, Wyngaarden JB, Fredrickson DS (eds) The metabolic basis of inherited diseases, 4th edn. Blakiston, New York, p 812

39. Thorbecke GJ, Silberberg-Sinakin I, Flotte TJ (1980) Langerhans cells as macrophages in skin and lymphoid organs. J Invest Dermatol 75:32

40. Tondeur M, Resibois A (1969) Fabry's disease in children: an electron microscopic study. Virchows Arch [Cell Pathol] 2:239

41. Tuchman LR, Suna H, Carr JJ (1956) Elevation of serum acid phosphatase in Gaucher's disease. J Mount Sinai Hosp 23:227

42. Tuchman LR, Goldstein G, Clyman M (1959) Studies on the nature of the increased serum acid phosphatase in Gaucher's disease. Am J Med 27:959

43. Vethanamy VG, Welch JP, Vethamany SK (1972) Type D Niemann-Pick disease (Nova Scotia variant): ultrastructure of blood, skin fibroblasts, and bone marrow. Arch Pathol 93:537
44. Woodruff CM (1979) Tartrate-resistant acid phosphatase-positive Gaucher cells. Am J Clin Pathol 71:361
45. Yunis EJ, Lee RE (1969) The ultrastructure of globoid (Krabbe) leukodystrophy. Lab Invest 21:415

Haematology and Blood Transfusion Vol 27
Disorders of the Monocyte Macrophage System
Edited by F. Schmalzl, D. Huhn, H.E. Schaefer
© Springer-Verlag Berlin Heidelberg New York 1981

Interaction of Lipoproteins with Macrophages

G. Assmann and G. Schmitz

The suggestion that the cholesterol accumulation in cardiovascular lesions is caused by an interaction of serum lipoproteins with arterial cells has been supported by numerous investigations. The lipid-loaden foam cells which accumulate cholesterol in atheroma are thought to arise from either smooth muscle cells [1–4] or monocytes–macrophages [1, 2, 5, 6]. At the present time, however, it has not been possible to demonstrate that smooth muscle cells can be overloaded with cholesterol by any of the naturally occurring or cholesterol-induced lipoproteins. These cells, like fibroblasts, lymphocytes and endothelial cells, can prevent excessive uptake and accumulation of lipoprotein-cholesterol due to the presence of specific cell surface receptors for plasma lipoproteins [7]. The binding of the lipoproteins to these receptors initiates a number of intracellular events, including internalization and degradation of the lipoproteins and the subsequent regulation of cellular cholesterol metabolism. The receptors interact with both low density lipoproteins (LDL), which contain the B-apoprotein, and certain high density lipoproteins (HDL), which contain the E-apoprotein, and are, therefore, designated as apo B, E receptors [8, 9].

In contrast to the apo B, E receptor cells, macrophages can be overloaded with cholesteryl esters in the presence of certain serum lipoproteins. Evidence has been provided that the lipoprotein-induced conversion of macrophages to foam cells is an important mechanism underlying xanthoma and atheroma formation. In addition to the occurrence of macrophage-derived foam cells in xanthoma and atheroma, several disorders of lipid metabolism are known where tissue macrophages convert to lipid-laden foam cells.

This short review describes various mechanisms of lipoprotein–cholesterol influx into macrophages and the potential role of HDL in removing cholesterol from these cells. Figure 1 gives a schematic illustration of these processes and lists several disorders related to lipoprotein metabolism in macrophages.

1. LDL and Macrophages

Among the cholesterol-carrying lipoproteins, LDL contains most of the circulating cholesterol, and LDL cholesterol is highly correlated with the total cholesterol level [10]. The relationship of coronary heart disease with LDL cholesterol is well documented; in particular, the strong association of increased LDL cholesterol concentrations and premature coronary heart disease in familial hypercholesterolemia suggests an important role of these lipoproteins in atherogenesis.

It has been previously speculated that the atherogenic mechanism of LDL is simply associated with increased filtration of this lipoprotein from blood plasma into the intima of blood vessels. According to several new developments, however, another theory appears to

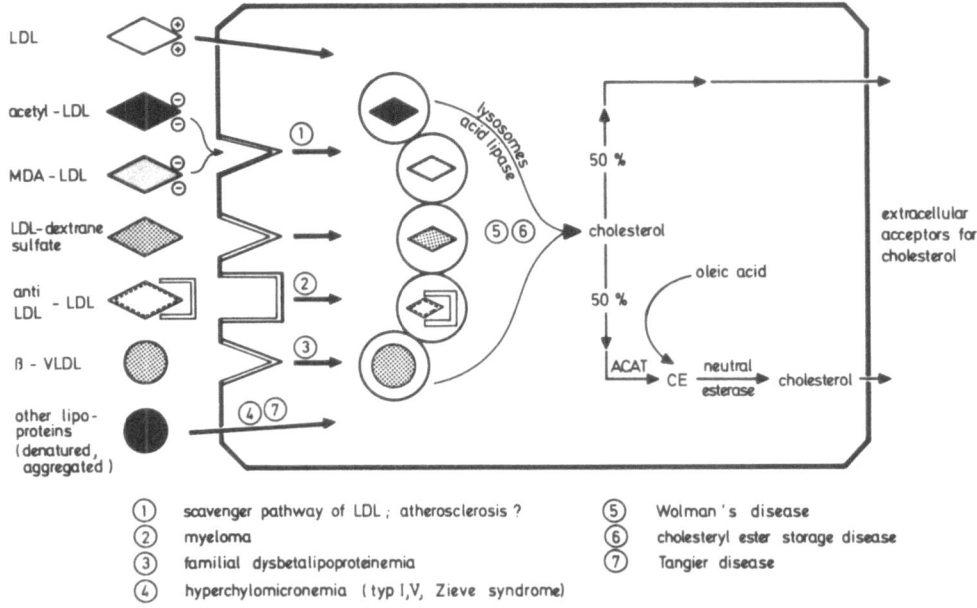

LDL	
acetyl - LDL	
MDA - LDL	
LDL-dextrane sulfate	
anti LDL - LDL	
ß - VLDL	
other lipo-proteins (denatured, aggregated)	

lysosomes acid lipase

50 %

cholesterol

50 %

oleic acid

ACAT

CE

neutral esterase

cholesterol

extracellular acceptors for cholesterol

① scavenger pathway of LDL; atherosclerosis ?
② myeloma
③ familial dysbetalipoproteinemia
④ hyperchylomicronemia (typ I,V, Zieve syndrome)

⑤ Wolman's disease
⑥ cholesteryl ester storage disease
⑦ Tangier disease

Fig. 1. Schematic illustration of lipoprotein–cholesterol influx and efflux pathways of macrophages and several disorders related to the lipoprotein metabolism in macrophages

provide a more adequate explanation. Goldstein and Brown [7] have shown that under physiologic conditions LDL is bound to specific surface receptors of peripheral nonhepatic and nonphagocytic cells (apo B, E receptor cells). After internalization of LDL, the endogenous cholesterol synthesis of the receptor cells is suppressed by cholesterol which is released from LDL by lysosomal acid lipase. Furthermore, there is a negative feedback between the uptake of LDL cholesterol and the number of cell surface receptors [11]. From these observations it has been concluded that LDL uptake normally is feedback regulated in peripheral apo B, E receptor cells. Disturbances in this mechanism, i.e., an insufficient number of cell surface receptors [12], lead to a reduced uptake of LDL by the specific apo B, E receptor mediated pathway, to high LDL plasma concentrations and to enhanced synthesis of endogenous cholesterol. Patients homozygous for familial hypercholesterolemia completely lack functioning apo B, E receptors [7] and show extreme elevations of LDL cholesterol in plasma. In persons heterozygous for this desease [7], only 50% of specific apo B, E receptors may be found and LDL cholesterol concentrations are elevated two- to threefold.

Besides the specific binding and uptake of LDL a nonspecific uptake of LDL also takes place [7]. Unlike the specific uptake, this nonspecific uptake is proportional to the LDL concentration in the interstitial fluid. During the catabolic degradation of LDL in man up to 30% of the LDL is degraded via the high-affinity binding–uptake process. The remaining 70% of the LDL might be degraded via a scavenger mechanism. The cells which are held to be responsible for this scavenger mechanism are mainly macrophages.

However, exposure of mouse peritoneal macrophages to native human LDL does not lead to cholesteryl ester accumulation within these cells [13, 14]. By contrast, exposure of macrophages (e.g., mouse peritoneal macrophages and guinea pig Kupffer cells) to chemically modified LDL (acetylated human LDL, malondialdehyd LDL, LDL conjugated with dextran sulfate) [13–18] induces a rapid and massive cholesteryl ester accumulation. The uptake of these chemically modified LDL occurs via specific cell surface receptors (Fig. 1). In contrast to the apo B, E receptors on fibroblasts and arterial smooth muscle cells, synthesis of the macrophage receptors is not depressed by the internalized LDL and therefore cholesteryl ester accumulation occurs via the hydrolysis–re-esterification of LDL cholesterol. Ultrastructural studies have revealed that most of the cholesteryl esters deposited in macrophages are present in lipid droplets without a limiting membrane [15, 19].

Goldstein et al. postulated that the macrophage receptors serve for specific LDL uptake by scavenger cells (phagocytic cells) [7]. However, the precise biologic mechanism by which LDL undergoes charge modification in vivo and a subsequent macrophage uptake is not yet fully understood. One mechanism of biological LDL modification may relate to platelet aggregation. Blood platelet aggregation results in the release of considerable amounts of malondialdehyd (1 mol malondialdehyd is produced per mol thromboxane A-2) [16]. Thus, in vivo modification of LDL might occur in the context of platelet aggregation at sites of arterial injury. Alternatively, interaction of LDL with glycosaminoglycanes may occur in vivo as soon as LDL has permeated the endothelial barrier [18]. At present, however, there is no experimental evidence that the naturally occurring sulfated glycosaminoglycans enhance LDL uptake and binding in mouse peritoneal macrophages, even though several of these, especially heparin and dermatane sulfate, form complexes with LDL. Nevertheless, the finding of widespread macrophage-cholesteryl ester storage in patients with apo B, E receptor deficiency (familial hypercholesterolemia) suggests an important role of these cells in the scavenger uptake of plasma LDL and xanthoma as well as atheroma formation.

To what extent the IgG or C 3 receptors of macrophages in certain pathologic conditions are involved in the uptake of LDL bound to immunoglobulins needs to be clarified [20–22]. Patients suffering from monoclonal gammopathy occasionally develop widespread planar xanthomatosis due to macrophage cholesteryl ester storage [23–25]. It is not unlikely, that immunoglobulins in this and other pathologic conditions behave as autoantibodies against lipoproteins and deliver lipoprotein-cholesterol via immunoglobulin receptors into macrophages.

2. The β-VLDL and Macrophages

A characteristic feature of profound hyperlipidemia in cholesterol-fed animals (dogs, rats, rabbits, swine and monkey) is the occurrence of abnormal cholesterol-rich lipoproteins with a density of very low density lipoproteins (VLDL, density < 1.006 g/ml) [26]. In contrast to normal triglyceride-carrying VLDL, which show pre-beta-mobility on electrophoresis, these abnormal cholesterol-rich lipoproteins have beta-mobility and are designated as β-VLDL. In addition to being cholesterol rich, the β-VLDL contain the B and E apoproteins as major protein constituents and interact with a specific, high affinity receptor on the surface of macrophages which differs from the apo B, E receptors of fibroblasts and arterial smooth muscle cells [27, 28]. Macrophages incubated with β-VLDL exhibit a 100-fold increase in cholesterol ester content and morphologically they resemble arterial foam cells.

Familial type III hyperlipoproteinemia (dysbetalipoproteinemia) is characterized by the occurrence of β-VLDL in plasma, severe xanthomatosis, and premature atherosclerosis [29]. To what extent the β-VLDL of these patients resemble β-VLDL of cholesterol-fed animals in the capacity to induce macrophage-cholesteryl ester storage needs to be investigated. However, it appears from this analogy that the understanding of the precise mechanism of the interaction of β-VLDL with macrophages may add important information regarding the pathogenesis of human atherosclerosis.

3. HDL and Macrophages

Biochemical, clinical, and epidemiological data have been accumulated to support a specific role of HDL in atherosclerosis [30]. It is generally assumed that HDL serves a carrier function in clearing cholesterol from the arterial wall and other tissues and that a low concentration of HDDL in plasma leads to an increased risk of coronary heart disease. Based upon the hypothesis of HDL as a critical molecule in cholesterol removal from cells, one should expect that patients with Tangier disease (absence of HDL from plasma) suffer from widespread cholesterol accumulation in peripheral body cells. Extensive investigations of the tissues of these patients have shown, however, that cholesteryl ester storage is limited to tissue histiocytes, intestinal smooth muscle cells, and Schwann cells, while apo B, E receptor cells, including arterial smooth muscle cells, are unaffected. The predominant change in the Tangier tissues is the conversion of large numbers of histiocytic cells in many organs to depots of cholesteryl esters. The major clinical manifestations of histiocytic lipid storage in Tangier disease include tonsillar hypertrophy, splenomegaly, enlargement of lymph nodes, and alterations in the cornea, intestinal mucosa, skin, and bone marrow[32]. Within the histiocytic foam cells the major part of the cytoplasmic lipid droplets are unbound by membranes and occur largely independent of lysosomes. The finding of extralysosomal cholesteryl ester storage in tissue histiocytes in Tangier disease is in contrast to the lysosomal cholesteryl ester storage in patients affected with acid lipase deficiency (Wolman's disease, cholesteryl ester storage disease) [34].

The distribution of lipid-storing histiocytes within different Tangier organs offers some explanation as to the source of the accumulated material[31]. Histiocytes occur in all tissues which are engaged in the breakdown of cells under physiological or pathological conditions, or both. For example, lipid-loaden macrophages in the bone marrow and in the spleen pulp are known to degrade erythropoietic cell residues or senescent granulocytes. A similar cell phagocytosis occurs in chronically inflamed areas. It is likely, therefore, that at least part of the histiocytic lipid content derives from phagocytosis of cell debris. Alternatively, tissues rich in histiocytes may also take up lipoprotein remnants. In Tangier plasma, grossly abnormal lipoproteins that could be considered as chylomicron surface remnants and possibly as targets for phagocytosis can be detected. Ultrastructural data are consistent with the concept that abnormal products of chylomicron catabolism are components of the lipid deposits in various cells, particularly histiocytes [34]. Thus, the foam cell infiltration of Tangier tonsils, spleen, rectal mucosa, the inflamed stroma of a cervical ectopia, and other chronically inflamed areas may have a common pathophysiological principle and primarily may originate on the basis of the obligate intake of cholesterol in the form of cell debris, membranes, and denatured lipoproteins [31].

It is tempting to speculate that the removal of the ingested cholesterol portion from histiocytes requires the extracellular presence of HDL, and that the depletion of HDL from the plasma of Tangier patients specifically causes cholesterol storage within these cells. This notion is particularly supported by experiments of Brown et al. who demonstrated a specific capacity of HDL in removing cholesterol from cholesteryl ester laden macrophages [15, 35].

The findings in Tangier disease imply that those cells which do not store lipid at least quantitatively differ from cholesteryl ester storage cells (e.g., tissue histiocytes) with respect to cholesterol uptake and removal mechanism. It appears that HDL may not be required for cholesterol removal as long as cells have the ability to regulate their uptake of cholesterol primarily through regulation of the apo B, E receptor sites and regulation of endogenous cholesterol synthesis. As long as these feedback regulatory mechanisms are operative, cholesterol input into the cell is controlled so that there may be no need for the cells to excrete cholesterol. By contrast, macrophages cannot themselves regulate cholesterol uptake and, in the presence of increased cholesterol influx or diminished cholesterol efflux (via HDL), become susceptible to cholesteryl ester storage. Verification of this concept has obvious implications for the understanding of cholesteryl ester storage in macrophages in various disorders including atherosclerosis.

References

1. Ross R, Glomset JA (1973) Atherosclerosis and the arterial smooth muscle cell. Proliferation of smooth muscle is a key event in the genesis of the lesions of atherosclerosis. Science 180:1332
2. Baumgartner H-R, Studer A (1978) Smooth muscle cell proliferation and migration after removal of

arterial endothelium in rabbits. In: Schettler G, Stange E, Wissler RW (eds) Atherosclerosis, is it reversible? Springer, Berlin Heidelberg New York

3. Gaton E, Wolman W (1977) The role of smooth muscle cells and hematogenous macrophages in atheroma. J Pathol 123:123

4. Doerr W (1978) Arteriosclerosis without end. Principles of pathogenesis and an attempt at a nosological classification. Virchows Arch [Pathol Anat] 380:81

5. Benditt EP (1976) Implications of the monoclonal character of human atherosclerotic plaques. Beitr Pathol 158:406

6. Adams CWM, Bayliss-High OB (1980) Mononuclear phagocytes in atherosclerosis. In: Gotto AM Jr, Smith LC, Allen B (eds) Atherosclerosis V. Springer, Berlin Heidelberg New York, p 130

7. Goldstein JL, Brown MS (1977) The low-density lipoprotein pathway and its relation to atherosclerosis. Annu Rev Biochem 46:897

8. Innerarity TL, Mahley RW (1978) Enhanced binding by cultured human fibroblasts of apo-E-containing lipoproteins as compared to low-density lipoproteins. Biochemistry 17:1440

9. Mahley RW (to be published) Cellular and molecular biology of lipoprotein metabolism in atherosclerosis. Diabetes

10. Kannel WB, Castelli WP, Gordon T (1979) Cholesterol in the prediction of atherosclerotic disease. New perspective based on the Framingham Study. Ann Intern Med 90:85

11. Brown MS, Goldstein JL (1979) Receptor mediated endocytosis: Insights from the lipoprotein receptor system. Proc Natl Acad Sci USA 76:3330

12. Brown MS, Donna SE, Goldstein JL (1975) Receptor-dependent hydrolysis of cholesterol ester contained in plasma low density lipoproteins. Proc Natl Acad Sci USA 72:2925

13. Goldstein JL, Ho YK, Basu SK, Brown MS (1979) Binding site on macrophages that mediates uptake and degradation of acetylated low density lipoprotein, producing massive cholesterol deposition. Proc Natl Acad Sci USA 76:333

14. Traber MG, Kayden HJ (1980) Low density receptor activity in human monocyte derived macrophages and its relation to atheromatous lesions. Proc Natl Acad Sci 77:5466

15. Brown MS, Goldstein JL, Krieger M, Ho YK, Anderson RGW (1979) Reversible accumulation of cholesteryl esters in macrophages incubated with acetylated lipoproteins. J Cell Biol 82:597

16. Fogelman AM, Shechter J, Saeger J, Hokom M, Child JS, Edwards PA (1980) Malondialdehyde alteration of low density lipoproteins leads to cholesteryl ester accumulation in human monocyte-macrophages. Proc Natl Acad Sci 77:2214

17. Shechter J, Fogelman AM, Haberland ME, Saeger J, Hokom M, Edwards PA (1981) The metabolism of native and malondialdehyde altered low density lipoproteins by human monocyte-macrophages. J Lipid Res 22:63

18. Basu SK, Brown MS, Ho YK, Goldstein JL (1979) Degradation of low density lipoprotein-dextran sulfate complexes associated with deposition of cholesteryl esters in mouse macrophages. J Biol Chem 254:7141

19. Goldstein JL, Anderson RGW, Buja LM, Basu SK, Brown MS (1977) Overloading human aortic smooth cells with low density lipoprotein-cholesteryl esters reproduces features of atherosclerosis in vitro. J Clin Invest 59:1196

20. Riesen W, Noseda G (1975) Antibodies against lipoproteins in man. Occurrence and biological significance. Klin Wochenschr 53:253

21. Beaumont JL, Beaumont V (1977) Autoimmune hyperlipidemia. Atherosclerosis 26:405

22. Feiwel M (1974) Xanthomatosis, lipoproteins, immunoglobulins and paraproteins. Dermatologica 149:308

23. Bazek A, Dupré A, Christol-Jalby B (1970) Xanthomatoses et dysglobulinémies. Bull Soc Fr Dermatol Syphiligr 77:654

24. Krain LS (1974) Cutaneous xanthomatosis and multiple myeloma: analysis and review. Cutis 14:423

25. Rivat M-H, Colomb D, Normand J, Cavailles M (1979) Xantomes et dysglobulinémies myélomateuses. A propos d'un cas associé également à une amyloidose systématisée, dite primitive, type Lubarsch-Pick. Revue de 42 cas de la littérature. Dermatol Venereol 106:755

26. Mahley RW (1978) Alterations in plasma lipoproteins induced by cholesterol feeding in animals including man. In: Dietschy JM, Gotto AM Jr, Ontko JA (eds) Disturbances in lipid and lipoprotein metabolism. American Physiological Society, Bethesda, p 181

27. Mahley RW, Innerarity TL, Brown MS, Ho YK, Goldstein JL (1980) Cholesteryl ester synthesis in macrophages: stimulation by β-very low density lipoproteins from cholesterol-fed animals of several species. J Lipid Res 21:970

28. Goldstein JL, Ho MS, Innerarity TL, Mahley RW (1980) Cholesteryl ester accumulation in macrophages resulting from receptor-mediated uptake and degradation of hypercholestereolemic canine β-very low density lipoproteins. J Biol Chem 255:1839

29. Fredrickson DS, Goldstein JL, Brown MS (1978) Familial hyperlipoproteinemia. In: Stanbury JB, Wyngaarden JB, Fredrickson DS (eds) The metabolic basis of inherited disease, 4th edn. McGraw-Hill, New York, p 604

30. The lipid research clinics program prevalence study (1980) Epidemiology of plasma high density lipoproteins cholesterol levels. Circulation [Suppl 4] 62

31. Assmann G (1979) Tangier disease and the possible role of high density lipoproteins in atherosclerosis. In: Gotto AM, Paoletti R (eds) Atherosclerosis reviews, vol 6. Raven, New York, p 1

32. Herbert PN, Gotto AM, Fredrickson DS (1978) Familial lipoprotein deficiency. In: Stanbury JB, Wyngaarden JB, Fredrickson DS (eds) Metabolic basis of inherited disease. McGraw-Hill, New York

33. Fredrickson DS, Ferrans VJ (1978) Acid cholesteryl ester hydrolase deficiency. In: Stanbury JB, Wyngaarden JB, Fredrickson DS (eds) Metabolic basis of inherited disease. McGraw-Hill, New York

34. Assmann G, Schaefer H-E (1978) High density lipoprotein deficiency and lipid deposition in Tangier disease. In: Carlson LA, et al. (eds) International conference on atherosclerosis. New York, pp 97-101

35. Ho YK, Brown MS, Goldstein JL (1980) Hydrolysis and excretion of cytoplasmic cholesteryl esters by macrophages: stimulation by high density lipoproteins and other agents. J Lipid Res 21:391

Haematology and Blood Transfusion Vol 27
Disorders of the Monocyte Macrophage System
Edited by F. Schmalzl, D. Huhn, H.E. Schaefer
© Springer-Verlag Berlin Heidelberg New York 1981

The Role of Macrophages in Atherosclerosis

H.-E. Schaefer

The development of arteriosclerosis results from diverse pathogenetic sources, and it is conceivable, therefore, that the course of morphogenesis as well as the ultimate picture of arteriosclerosic lesions appears more or less heterogeneous [8]. Nevertheless, the formation of atheroma (fibrofatty plaque, gruel plaque) represents a hallmark of what is called in a strict sense atherosclerosis. Most atherosclerosis-linked symptoms of clinical relevance can be considered as direct sequelae of fibrofatty plaques. For instance, an atheroma can induce coronary heart disease by its occlusive action, and secondary ulcerative lesions lead to thromboembolic complications.

Because of the key pathogenetic role atheroma plays, much interest has focused on its histogenesis and structure. It is well established that atheromatous lesions develop primarily in the tunica intima. A full-grown lesion consists of a necrotic center covered by a fibrous cap. This fibrous boundary as well as the other areas of the thickened intimal tissue are rich in elastic and in collagenous fibers. In most cases the necrotic center is surrounded by a belt of lipid-laden foam cells. Smaller groups of similar foam cells constitute the so-called fatty streaks, which represent early atherosclerosic lesions which are distinguished from atheromata by their smaller dimension and by the absence of intimal fibrosis and necrosis [16, 23].

According to his "unitarian cellular hypothesis for atherosclerosis" Wissler [32] has derived the origin of atheroma from focal proliferation of smooth muscle cells [14, 4], which are assumed to migrate from the tunica media to the intima [18, 29], where one part of the cells converts to "myocytogenous" foam cells, and another part produces fibers.

The finding that the major part of collagen isolated from atherosclerotic lesions belongs to Type I, instead of Type III constituting the normal product of the media smooth muscle cells [15], favors the opposite contention that cells known from other scarring processes, e.g. myofibroblasts [9, 28], synthesize atherosclerotic fibers rather than media-derived smooth muscle cells.

The local accumulation of lipid within the necrotic center of atheroma represents a further unexplained problem. On the whole, in atherosclerosis the lipid contents of the tunica intima are increased. In part lipids are diffusely distributed all over the intimal tissue. There is circumstantial evidence [24] that these lipids are derived from apolipoprotein B containing lipoproteins [12, 26] which have entered from the blood plasma and to some extent become bound to fibers [25, 27]. However, biochemical analysis has established clear-cut differences between the lipids piled up within the necrotic centers of atheromata on the one hand and the diffusely distributed intimal lipids on the other hand. While the latter resemble more or less low density lipoproteins (LDL) and very low density lipoproteins (VLDL), the necrotic center lipids distinguish themselves by a disproportionate concentration of phospholipids and unesterified cholesterol [13, 25] and by a tripartite composition of their physicochemical phases [24]. In most fibrofatty plaques, unesterified cholesterol monohydrate has precipitated, forming plate-like crystals [13] which have been described in detail as early as 1858 by Virchow [30].

The purpose of this paper is not to paraphrase those well known principles of atherosclerosis but to offer some morphological and histochemical evidence that the foam cells of fatty streaks and of atheromata represent true macrophages which are derived from monocytes rather than from smooth muscle cells. Furthermore, there are some observations indicating that the peculiar lipid deposits of the atheroma centers are produced by macrophages. Thus the action of macrophages seems to constitute an essential part in the morphogenesis of atherosclerosis.

1. Material and Methods

At autopsy specimens of the distal thoracic aorta have been collected from 12 individuals ranging from 40 to 80 years of age. This material comprised initial as well as advanced developmental stages of atherosclerotic lesions. Intensely calcified foci, ulcerative or thrombotic atheromata have been excluded.

All enzyme histochemical reactions were applied to cryostat sections obtained from unfixed tissue or from specimens fixed for 12 h at 4 °C in a freshly prepared solution of 0,1 M calcium acetate, 1% formaldehyde, and 0.5% glutaraldehyde. After fixation, the tissue was washed for a 12 h period at 4 °C in a solution of 1% gum arabic and of 30.1% sucrose which was changed at least twice.

Lysosomal hydrolytic enzymes have been demonstrated by the *hexazonium pararosaniline technique* [7] in order to obtain reaction products insoluble in lipid deposits, thus preventing diffusion artifacts. Hexazonium solution was prepared from 50 mg pararosaniline hydrochloride dissolved in 3 ml 1-N HCL, mixed subsequently with 0.5 ml 1-M NaNO$_2$. After a reaction time of about 5 min, this hexazonium solution is to be used immediately, or it can be stored at –70 °C for several months.

This medium for acid *phosphatase* consisted of 10 mg naphthol–AS–BI–phosphate dissolved in 0.5 ml dimethylsufoxide mixed subsequently with 100 ml 0.1-M acetate buffer, pH 5.6. Addition of 1 ml of hexazonium solution (v.s.) decreases the pH of the medium to an ultimate range of about 5.2–5.3. For the demonstration of the tartrate resistant isoenzyme, 150 mg L(+) tartaric acid dissolved in 1.8 ml 1-N NaOH was added to obtain an inhibitor concentration of 10^{-2} M.

Esterase were demontrated by a medium composed of 100 ml (1/15) M phosphatase-buffer, pH 6.7, titrated by addition of 1 ml hexazonium solution (v.s.) to pH 6.3. Alternatively, α-naphthylacetate (10 mg dissolved in 0.1 ml dimethylsufoxide), or α-naphthyl-butyrate (10 mg dissolved in 0.1 ml dimethylsufoxide) were added as substrates. Optionally, NaF was added to obtain an inhibitor concentration of $10^{-2}M$.

Unfixed cryostat section were stained for *alkaline phosphatase* by a solution of 10 mg naphthol-AS-BI-phosphate sodium salt dissolved in 0.5 ml dimethylsulfoxide, 50 mg fast red TR salt, and 100 ml 0,05 M tris-(hydroxy)-methyl-aminomethane.

After an incubation at room temperature for 3 h (acid phosphatase) or for 60 min (esterase and alkaline phosphatase) the tissue sections were rinsed with water, counterstained by Mayer's haemalaun and covered with Kaiser's glycerine gelatine.

Lipids were stained (1) by oil red 0 dissolved in 85% propylene glycol [6], 2. by a solution of Sudan black B in 70% ethanol containing horse radish peroxidase in order to prevent unspecific staining of endogenous peroxidase (e. g., granulocytes) [20], and 3. by the OBS method [19] which represents a combined treatment by OsO$_4$ prior to paraffin embedding, bleaching of polar lipids by ammonium peroxosulfate, and final staining of polar lipids by oil red O. Unstained tissue sections were studied in plane-polarized light.

2. Results

Foam cells occur in early atherosclerotic lesions (fatty streaks) and in full-grown gruel plaques as well, forming dense clusters adjacent to the necrotic centers. These foam cells react negative for alkaline phosphatase, but they constantly show a high enzyme activity of esterase and of acid phosphatase (Fig. 1). The reaction products appear diffusely distributed all over the cytoplasm, leaving unstained the contents of the vacuolar lipid inclusions. In places a breakdown of foam cells around the necrotic foci can be visualized histochemically by a spillage of esterase and acid phosphatase (Fig. 2). Obviously these lysosomal enzymes are

Fig. 1a, b. Adjacent cryostat sections of an atheroma a stained for lipids by oil red 0, and b stained for acid phosphatase. Dark staining lipids and cholesterol crystals are deposited within the necrotic center *(C)* as well as in the surrounding foam cells which are positive reacting for acid phosphatase

leaking from the necrotizing foam cells to the interstitial space where they become diffusely mixed with the lipids concentrated within the necrotic plaque centers. Media smooth muscle cells display a much weaker activity of esterase and of acid phosphatase. Unlike in foam cells, these enzymes are restricted to few lysosomal granules situated in the paranuclear region. Smooth muscle cells migrating from the media to the intima or cell types representing intermediate developmental stages between smooth muscle cell and intimal foam cells are not encountered.

However, in most sections stained for acid phosphatase or for esterase, a few smaller mononuclear cells are to be seen which are comparable to the lipid-storing foam cells with respect to their high activity of lysosomal enzymes. The cytoplasm of these mononuclear cells stains diffusely for acid phosphatase and for esterase. These macrophage-like cells are disseminated all over the intimal tissue, in places they have been found in the tunica media, and more frequently they occur in the adventitia surrounding the vasa vasorum.

Beside foam cells and a few smaller macrophage-like cells, there are fibroblast-like cells with slender spindle-shaped or embranched cytoplasm, which are best seen in sections conducted in parallel to the inner surface of the vessel wall. These cells constitute part of the fibrous cap tissue and otherwise they are loosely distributed over the intimal tissue. These cells exhibit a varying moderate activity of alkaline phosphatase diffusely distributed over

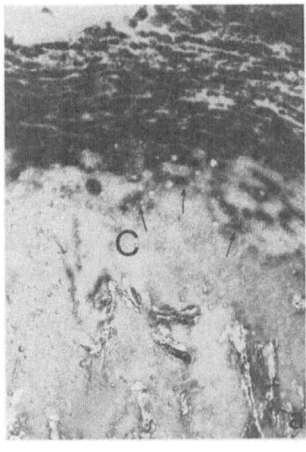

Fig. 2.a On the boundary between the necrotic atheroma center *(C)* and the acid phosphatase-positive belt of foam cells individual macrophages *(arrows)* undergo necrosis and spill their acid phosphatase. The necrotic centre *(C)* contains in part birefringent cholesterol cristals. b Unstained smear of the contents expressed from a gruel plaque containing many flatly spread rhomboidal crystals of cholesterol monohydrate

the whole cytoplasm, and in part the enzyme seems to be linked to the cell wall. These fibroblast-like cells display very low activities both of esterase and of acid phosphatase rather comparable to the behavior of media smooth muscle cells.

By the techniques applied, intra- and extracellular lipids have been traced. In most specimens numerous small lipid droplets of about chylomicron size are diffusely distributed over the more or less sclerotic intimal layers. These extracellular deposits are not birefringent, and they mainly exhibit the staining pattern of apolar lipids. In contrast, the lipids constituting the atheroma areas are in part polar ones. Furthermore, these necrotic centers include masses of birefringent cholesterol crystals (Fig. 2).

Intracellular lipids are mainly restricted to foam cells especially to cells with a high esterase and acid phosphatase activity. Lipid droplets have been scarcely found in the fibroblast-like cells (one case with a known history of diabetes mellitus). Often the perinuclear lysosomes of smooth muscle cells contain lipofuscin exhibiting a variable affinity for lipid stains. These lipid-like structures are readily distinguished from the various lipids found in the tunica intima by their autofluorescence in the ultraviolet light.

The lipids of foam cells display mainly a polar staining pattern, and they show a variable degree of birefringence. In some cells the optical qualities are isotropic ones. It is interesting to note that in particular the foam cells located next to the necrotic atheroma centers contain anisotropic lipids exhibiting a maltese cross or a crystalline type of birefringence.

3. Discussion

The findings reported here allude to the cytogenetic problem of the origin of the intimal foam cells as well as to the histogenetic problem of how an atheroma develops. In our view, both these questions are intimately connected.

The first point seems to be readily answered. The enzyme-histochemical pattern of the lipid-storing foam cells is compatible with that of macrophages. Our findings are corroborated by similar results indicating that atherosclerotic foam cells can be discriminated from smooth muscle cells by their higher activity of catalase [3], cytochrome oxidase [1], naphthol-AS-D-acetate-esterase [10], acid lipase [21, 33], and β-galactosidase [2]. It appears obvious that foam cells develop themselves from smaller mononuclear macrophages which are not yet lipid laden and which can be traced by staining the tissue for acid phosphatase or for esterase. There is electron microscopical evidence, too, that these macrophages are monocyte derived [11]. It may be assumed from the high frequency of macrophages surrounding the vasa vasorum that monocytes may leave the blood at these sites. Moreover, mononuclear cells have been detected *en passage* between endothelial cells of the aorta [17], and it has been demonstrated that macrophages constitute cells common to the intimal tissue of normal aorta [22]. However, there is some evidence that the cell clusters forming individual atherosclerotic lesions have at least in part monoclonal characteristics [5]. If there are monoclonal foam cell foci, we have to assume that those foam cell clusters represent the product of local proliferation of a few monocytic precursor cells. The finding of a high foam cell to lipid-free macrophage ratio is in line with this view.

The second point touches the role that foam cells play in the histogenesis of atheromatosis. At a first glance, the foam cells surrounding the necrotic plaque center could be interpreted as cells actively resorbing the pathologic lipid precipitates. So far, the presence of foam cells in an atheroma would indicate a sort of repair mechanism.

But if we analyze our findings, there emerge several facts incompatible with this assumption. Firstly, there are obvious degenerative changes occurring in foam cells located next to the atheroma centers. Some of these foam cells undergo complete necrosis resulting in visible spillage of lysosomal enzymes. It is rather improbable that those cells are capable of resorptive activity. Secondly, there is no explanation of how the local accumulation of lipids constituting the plaque centers is generated. As has been pointed out, the diffusely distributed lipids to be found in atherosclerotic intimae originate from plasma-derived lipoproteins. But histochemically as well as biochemically the lipid deposits of the atheroma centers differ from the diffusely distributed intimal lipids. If the atheroma center lipids, too, are derived from plasma lipoproteins, the original lipids must have been modified by specific metabolic processes.

There are several observations indicating that such a process may be brought about by the action of macrophages. As has been shown by the study in plane-polarized light, part of the foam cells contain optically isotropic lipids similar to the diffusely distributed interstitial lipid droplets. Another part of foam cells is filled with anisotropic lipids and with small crystals of unesterified cholesterol. In particular, those cells gather around atheroma centers. It seems likely, therefore, that macrophages start with taking up the directly lipoprotein-derived interstitial lipid droplets distributed all over the intimal tissue of aging vessels. The internalized lipids are split up by lysosomal hydrolysis and become catabolized to some extent. Degradation of lipoproteins by macrophages has been demonstrated by the intracellular disappearance of anti-apolipoprotein-B immunofluorescence. This function is readily performed by macrophages as opposed to the slow acting smooth muscle cells [31]. At least cholesterol remains undigestible and becomes stored intracellularly in as much as its transfer to high density lipoprotein (HDL) is inhibited as a consequence of the age-dependent general decrease of plasma HDL. Moreover the increasing density of intimal fibrosis inhibits the fluid perfusion of the wall of aging vessels. When the equilibrium between catabolic formation of cholesterol and efflux of cholesterol via HDL from macrophages to the liver is out of balance, the process of intracellular storage accelerates. Exhaustion of storing capacity ultimately leads to the breakdown of foam cells. The accumulating debris of necrotic cells rich in cholesterol gives rise to the development of a fibrofatty plaque by local stimulation of fiber-producing cells and by chemotaxis of additional macrophages.

The nucleus of atheroma results from incidental necrosis of macrophages transformed to foam cells when scavenging the arterial vessel wall from lipoprotein-derived extracellular lipids. Thus, an atheroma may be considered as a graveyard of lipid-laden macrophages.

References

1. Adams CWM, Bayliss OB (1976) Detection of macrophages in the atherosclerotic lesions with cytochrome oxidase. Br J Exp Pathol 57:30
2. Adams CWM, Bayliss-High OB (1980) Mononuclear phagocytes in atherosclerosis. In: Gotto AM, Smith LC, Allen B (eds) Atherosclerosis V. Springer, Berlin Heidelberg New York, p 130
3. Adams CWM, Bayliss OB, Turner DR (1975) Phagocytes, lipid-removal and regression of atherosclerosis. J Pathol 116:225
4. Baumgartner H-R, Studer A (1978) Smooth muscle cell proliferation and migration after removal of arterial endothelium in rabbits. In: Schettler G (ed) Atherosclerosis, is it reversible, Springer, Berlin Heidelberg New York
5. Benditt EP (1976) Implications of the monoclonal character of human atherosclerotic plaques. Beitr Pathol 158:405
6. Chiffele TL, Putt FA (1951) Propylene and ethylene glycol as solvents for Sudan IV and Sudan black B. Stain Technol 26:51
7. Davis BJ, Ornstein L (1959) High resolution enzyme localization with a new diazo reagent, "hexazonium pararosaniline". J Histochem Cytochem 7:297
8. Doerr W (1978) Arteriosclerosis without end. Principles of pathogenesis and an attempt at a nosological classification. Virchows Arch [Pathol Anat] 380:91
9. Feigl W (1975) Reaktionsformen der glatten Muskelzelle der menschlichen Arterienwand – ihre Bedeutung für die Atherosklerose. Wien Med Wochenschr [Suppl] 32:1
10. Gaton E, Wolman W (1977) The role of smooth muscle cells and hematogenous macrophages in atheroma. J Pathol 123:123
11. Gerrity RG, Naito HK, Richardson M, Schwartz CJ (1979) Dietary induced atherogenesis in swine. Morphology of the intima in prelesion stages. Am J Pathol 95:775
12. Hoff HF, Heideman CL, Gotto AM, Gaubatz JW (1977) Apolipoprotein B retention in the grossly normal and atherosclerotic human aorta. Circ Res 41:684
13. Katz SS, Shipley GG, Small DM (1976) Physical chemistry of the lipids of human atherosclerotic

lesions. Demonstration of a lesion intermediate between fatty streaks and advanced plaques. J Clin Invest 58:200

14. Knieriem HJ, Kao VCY, Wissler RW (1968) Demonstration of smooth muscle cells in bovine arteriosclerosis. J Atheroscler 8:125
15. McCullagh KG, Duance VC, Bishop KA (1980) The distribution of collagen types I, III, and V (AB) in normal and atherosclerotic human aorta. J Pathol 130:45
16. Movat HZ, Haust MD, More RH (1959) The morphologic elements in the early lesions of arteriosclerosis. Am J Pathol 35:93
17. Poole JCF, Florey HW (1958) Changes in the endothelium of the aorta and the behaviour of macrophages in experimental atheroma of rabbits. J Pathol Bacteriol 75:245
18. Ross R, Glomset JA (1973) Atherosclerosis and the arterial muscle cell. Proliferation of smooth muscle is a key event in the genesis of the lesions of atherosclerosis. Science 180:1332
20. Schaefer H-E, Fischer R (1972) Peroxydaseaktivität als Ursache stabiler Sudanophilie und als allgemeine Fehlermöglichkeit beim histochemischen Lipidnachweis. Acta Histochem [Suppl] 12:319
19. Schaefer H-E (to be published) Der Nachweis apolarer Lipide am paraffineingebetteten Gewebe unter besonderer Berücksichtigung der pathologischen Verfettung bei der Tangierkrankheit. Acta Histochem [Suppl]
21. Schaffner T, Elner VN, Bauer M, Wissler RW (1978) Acid lipase: A histochemical and biochemical study using Triton X-100-naphthyl palmitate micelles. J Histochem Cytochem 26:696
22. Schwartz CJ, Gerrity RG, Sprague EA, Hagens MR, Reed CT, Guerro DL (1980) Ultrastructure of the normal arterial endothelium and intima. In: Gotto AM, Smith LC, Allen B (eds) Atherosclerosis V. Springer, Berlin Heidelberg New York, p 112
23. Sinapius D (1975) Frühveränderungen der Atherosklerose beim alten Menschen. Verh. Dtsch Ges Pathol 59:354
24. Small DM (1980) Summary of concepts concerning the arterial wall and its atherosclerosis lesions In: Gotto AM, Smith LC, Allen B (eds) Atherosclerosis V. Springer, Berlin Heidelberg New York, p 520
25. Smith EB (1965) The influence of age and atherosclerosis on the chemistry of aortic intima. 1. The lipids. J Atheroscler Res 5:224
26. Smith EB, Slater RS (1972) Relationship between low-density lipoprotein in aortic intima and serum lipid levels. Lancet 1:463
27. Smith EB, Evans PH, Downham MD (1967) Lipid in the aortic intima. The correlation of morphological and chemical characteristics. J Atheroscler Res 7:171
28. Smith P, Heath D (1980) The ultrastructure of age-associated intimal fibrosis in pulmonary blood vessels. J Pathol 130:247
29. Titus JL, Weilbaecher DG (1980) Smooth muscle cells in atherosclerosis. In: Gotto AM, Smith LC, Allen B (eds) Atherosclerosis V. Springer, Berlin Heidelberg New York, p 126
30. Virchow R (1858) Genauere Geschichte der Fettmetamorphose. In: Die Cellularpathologie und ihre Begründung auf physiologische und pathologische Gewebelehre. Von Hirschwald, Berlin, p 309
31. Walton KW, Dunkerley DJ, Johnson AG, Khan MK, Morris CJ, Watts RB (1976) Investigations by immunofluorescence of arterial lesions in rabbits on two different lipid supplements and treated with pyridinol carbamate. Atherosclerosis 23:117
32. Wissler RW (1968) The arterial medical cell, smooth muscle cell or multifunctional mesenchyme? J Atheroscler Res 8:201
33. Wolman M, Gaton E (1976) Macrophages and smooth muscle cells in the pathogenesis of atherosclerosis. J Israel Med Assoc 99:450

Clinical Syndromes

Haematology and Blood Transfusion Vol 27
Disorders of the Monocyte Macrophage System
Edited by F. Schmalzl, D. Huhn, H.E. Schaefer
© Springer-Verlag Berlin Heidelberg New York 1981

Acute Monocytic Leukemias[1]

F. Schmalzl

1. Abstract

Normal monocytes and macrophages are characterized by peculiar ultrastructural and cytochemical features and, in addition, show characteristic membrane properties, a variety of special functional capacities, and important secretory activities. Almost all these cytological
features can also be detected in leukemic monocytic cells, and it is quite conceivable that
these peculiar features may influence or determine the clinical syndrome associated with
the leukemic accumulation of monocytic cells. The morphological identification of monocytic leukemias is a very intriguing diagnostic problem and some controversies still exist
concerning their cytological classification. For clinical as well as scientific purposes the diagnosis of monocytic leukemias should rely on the demonstration of specific monocytic features of the leukemic cells. Clinical findings frequently associated with acute monocytic leukemia include increased frequency of leukemic tissue infiltrations as well as increased tendency to hypokalemia and − especially in the "immature" variants − to disorders of hemostasis.

2. Introduction

The problem of monocytic leukemias − Schilling or Naegeli type or "leukemic reticuloendotheliosis" − was one of the most debated hematological controversies in the first half
of this century and still gives rise to discussions about the origin and the cytological features
of monocytic or "histiocytic" leukemias [28, 31, 56]. In the last 2 decades a lot of information
on the physiological properties and the reactive changes of blood monocytes and
macrophages has been accumulated [7, 18, 32, 37]. The recent knowledge provides the
criteria for the identification of leukemic or tumorous cell populations as monocytic and
helps us to understand the polymorphous manifestations of the neoplastic disorders affecting the monocyte–macrophage system (Table 1).

Several aspects of monocytic leukemias are discussed separately in this monograph;
therefore, I shall restrict my presentation to outline what we consider "monocytic leukemia"
and to indicate some peculiar cytological and clinical features which characterize this disorder as a separate entity among the acute leukemias.

Considering the physiological maturation of the monocyte–macrophage system, different forms of neoplastic disorders can be expected. The differences depend upon the degree
of the maturation capacity preserved in the accumulating leukemic or tumorous cell populations. In this presentation we are concerned with those leukemias which are characterized
by the accumulation of immature promonocyte-like cells or of cells which cytologically and
functionally reach the stage of mature blood monocytes (see Fig. 1). Neoplastic disorders
characterized by macrophage-like cells will be the topics of later presentations.

[1] The investigations have been supported by grants from the funds "Kampf dem Krebs"

Table 1. Cytological and functional characterization of leukemic monocytes (AMoL)

Monocytic character of leukemic population	Suggested	Strongly suggested	Proved
By:			
Morphology	X		
Ultrastructure		X	
Cytochemistry		X	
Lysosomal enzymes	X		
Lysozyme determinations	X		
Production of CSA	X		
Surface receptors (IgG, compl.)	X		
"Monocyte-specific" antibodies		X?	
Type of growth in agar culture	X?		
Transformation into macrophages		X	
Migration into inflammatory exudates		X	
Glas adherence	X		
Phagocytosis	X?		

Only by combination of several tests

3. Materials and Methods

Our presentation is based on the personal clinical experience of 42 cases of monocytic leukemia (24 "mature" and 18 "immature") and 12 cases of "myelomonocytic" leukemia (see following discussion). In addition, cytochemical examinations have been performed in over 30 cases of monocytic and "mye-lomonocytic" leukemias sent for examination from other hospitals.

The experimental techniques employed for the study of the leukemic cell populations are commonly known and precise instructions on the methodology can be found in the cited literature. In the personally studied cases each of the leukemic cell populations has been studied by several means, although the whole panel of the reported tests could not be carried out in all patients. However, the diagnoses always have been based on careful cytochemical examinations [51] and, in addition, on lysozyme determinations in urine [1, 42]. Skin window tests (in 32 out of 54 cases) [44, 48, 50, 52], electron microscopic studies (22/54) [26], IgG receptor (8/54) and complement receptor tests (6/54) [4, 9, 22, 25, 30], and in vitro cultivation in liquid media (6/54) [53] and in soft agar (2/54) [27] as well as the NBT test [19] have been performed less frequently. Only occasionally have cell homogenates of the leukemic cells been studied for esterase and for lysomal enzyme activities [33, 45].

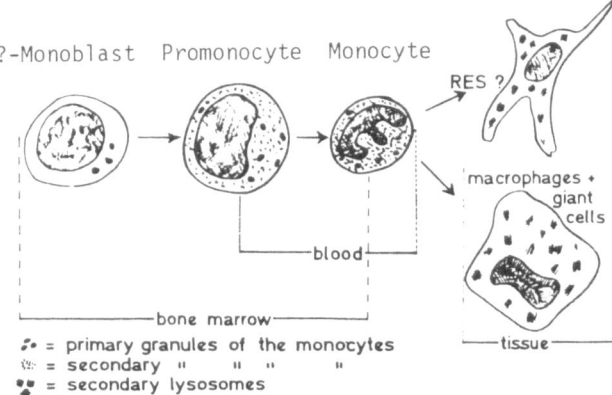

?-Monoblast Promonocyte Monocyte

RES ?

macrophages + giant cells

blood

bone marrow

tissue

$\cdot\cdot$ = primary granules of the monocytes
$\cdot\cdot$ = secondary " " " "
$\cdot\cdot$ = secondary lysosomes

Fig. 1. Monocyte–macrophage system: Sequence of maturation stages

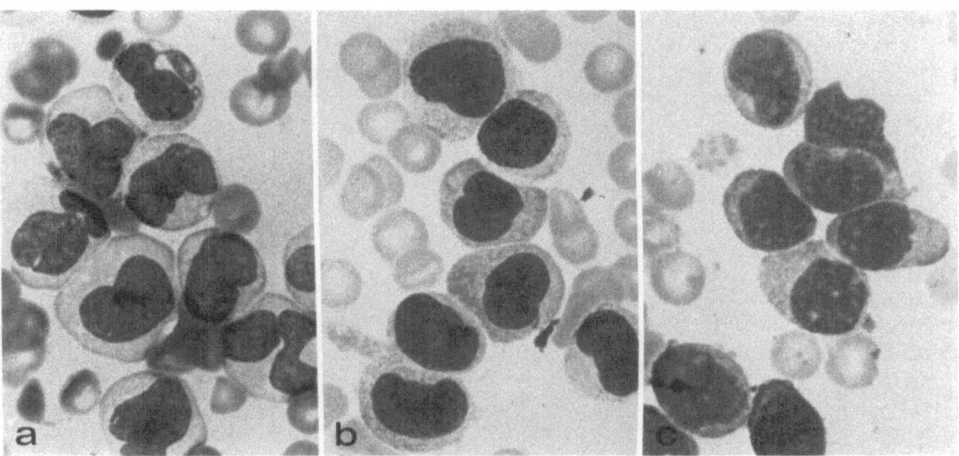

Fig. 2a-c. May-Grünwald-Giemsa stain of acute monocytic leukemia (AMoL). Leukemic monocytes may show varying degrees of differentiation. **a** and **b** Mature AMoL; **c** Immature AMoL (monoblastic or promonocytic). Orig. magnif. × 960

4. Results and Comments

Morphologic examination alone is of limited value in the diagnosis of monocytic leukemia, despite the fact that it was the most important diagnostic criterium until 20 years ago. The morphological heterogeneity of leukemic monocytes is familiar to all hematologists (Fig. 2). Nuclear and chromatin structure, cytoplasmic basophilia, and the amount of granules identifiable in panoptic stains are not strictly related to the maturation of the cells identifiable by cytochemical and ultrastructural examination. Marked dissociation of nuclear and cytoplasmic maturation is common [6, 54].

Electron microscopic examination discloses lobulated or folded nuclei presenting varying heterochromatization and prominent nucleoli. The cytoplasm contains moderate amounts of mono- and polyribosomes, a well developed ergastoplasm, a conspicuous Golgi apparatus, and small electron dense granules. Bundles of fibrils and microfilaments, usually in perinuclear localization, and pseudopod-like protrusions add to the typical shape of leukemic monocytes. However, the cell shape and the subcellular structure may be subjected to considerable variations within the same and among different patients depending upon

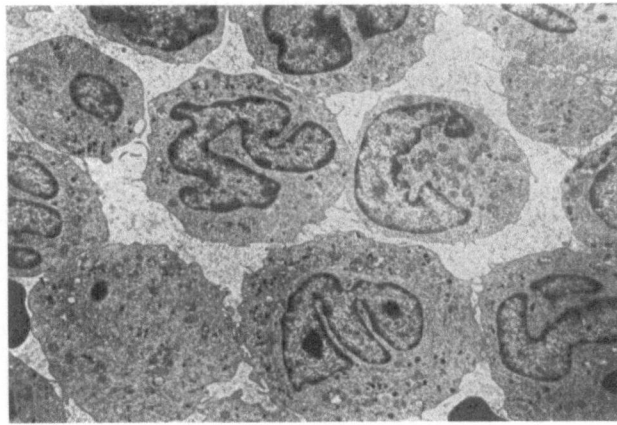

Fig. 3. Ultrastructure of leukemic monocytes (for explanation see text). (Photomicrograph by H.E. Schaefer, Cologne/FRG). Orig. magnif. × 1940

Table 2. Cytochemistry of leukemic monocytes (AMoL)

	Mature MoL	Immature MoL
Pyroninophilia	$- ++$	$++ - +++$
Nonspecific esterases		
(NaF-sensitive N-AS-ac.-est.)	$++ - +++$	$+ - ++$
(α-N-butyrate-esterase)		
(α-N-acetate esterase)		
Aminopeptidase	$++ - +++$	$0 - +$
Lysozyme (immunocytology)	$++ - +++$	$0 - ++$
Acid phosphatase	$++ - +++$	$+ - ++$
Peroxidase	$0 - + (++)$	$0 - +$
N-AS-D-chloroacetate esterase	$0 - +$	$0 - +$

the degree of cellular maturation [17, 20, 26, 35, 43, 63; Manoharan, this volume p. 205]. (See Fig. 3.)

The distinction of leukemic monocytic cell populations according to their degree of maturation is further supported if *cytochemical criteria* are considered for the cytologic characterization [16, 35, 48, 59]. We termed mature monocytic leukemia those cases in which the leukemic populations consisted mainly of cells featuring the cytochemical pattern of normal blood monocytes (see Table 2) [26, 48, 49]. Moderate or strong naphthylamidase activities were observed only in the most mature appearing monocytic cell populations. Leukemic cell populations showing only weak activities of sodium fluoride sensitive monocytic esterase and of acid phosphatase and presenting weak or absent naphthylamidase activity were labeled as immature (see Table 2) [48]. These findings point to the lower content of specific monocytic granules associated with the features of less cytological differentiation, but without relevant alterations of the primary granules [38].

Special importance has been attributed to the sodium fluoride sensitive naphthol-AS-D-acetate esterase as a marker for normal monocytes and promonocytes and in some instances for monocyte-derived macrophages [15, 50]. The amount of intracellular acid phosphatase is related to the production and presence of monocyte granules and macrophage lysosomes and can be regarded as a valuable tool for the estimation of the cellular maturation [7, 48, 49]. We were not able to find any correlation between "histiocyte"-like appearance, cytoplasmic granulation, and peroxidase and naphtolth-AS-D-chloroacetate esterase staining. In the nitroblue tetrazolium test (NBT [19] normal and leukemic monocytes show a positive staining with and without phagocytic stimulation.

Enzyme deletions are well known from other myelogenous leukemias and preleukemias and may occur also in monocytic leukemias [8, 11, 55]. These defects, for instance, a major defect of NaF-sensitive Naphthol-AS-D-acetate esterase, may cause considerable diagnostic troubles. However, according to our experience this is a rather rare event – 2 out of about 80 classified cases of acute monocytic and myelomonocytic leukemia.

Biochemical examinations of cell homogenates of monocytic leukemia disclosed the presence of considerable amounts of lysosomal or granule-bound enzymes, which also can be used for diagnostic purposes [27]. According to our own experience these results are in good accordance with the cytochemical findings.

Futher information on the enzyme patterns of leukemic monocytes was added in 1969 when the immunocytological localization of *lysozyme* (muramidase) was studied together with Asamer [2]. The strong immunocytological labeling for lysozyme protein in mature leukemic monocytes correlated well with the demonstration of high concentrations of this enzyme within the same cells. Secretion by intact and liberation from disintegrating leukemic monocytes induce increased levels of lysozyme in serum and, beyond a kidney threshold, induce strongly increased urinary excretion of this enzyme [24, 42]. In cases of monocytic leukemia presenting less differentiated cell populations both the immunocytological

Table 3. Immunocytological demonstration and quantitative determination of lysozyme in leukemic monocytes, serum, and urine (AMoL)

	Immunocytologic demonstration	Intracellular content $\mu/10^6$ cells	Serum level µg/ml	Urinary excretion µg/ml
Normal monocytes	$++ - +++$	2–6	5–10	< 4
Leukemia:				
Mature monocytic	$++ - ++++$	4–30	> 30	100–800
Immature	$0 - ++$	< 2	5–17	4– 40

labeling as well as the urinary excretion were lower than in mature monocytic leukemia [48, 49]. The urinary excretion approximated the upper normal limits (see Table 3).

Several other *biologically active proteins* are synthesized by leukemic monocytes and may be secreted. Golde et al. [21] reported that the excretion of *colony-stimulating factor* is strongly increased in monocytic leukemia. In addition the secretion of colony-stimulating activity can be assessed by culturing leukemic monocytes in feeder layers of agar cultures [63]. Thromboplastin-like, procoagulant, or plasminogen activities and some other enzymes or factors derived from leukemic monocytes may interfere with coagulation [37, 39, 62]. Worwood et al. reported that leukemic monocytes contained very high concentrations of *ferritin* [64]. Increased levels of *vitamin B 12-binding proteins* have also been reported in leukemic monocytes [10, 65]. Normal monocytes are involved in the production of complement components; however, the complement metabolism has not been extensively investigated in monocytic leukemia [Schorlemmer, this volume p. 59].

Functional properties typical for the normal monocytes can also be demonstrated in leukemic monocytic cell populations. Testing the growth of leukemic human monocytes in liquid cultures we could confirm previous reports of Nowell [40] and Fischer and Gropp [14] who observed that leukemic monocytes were capable *of transforming into macrophages* in vitro [53]. This transformation is associated with a strong increase of lysosomal enzymes within the cells [53]. Leukemic myeloblasts showed some degree of maturation, producing peroxidase and naphthol-AS-D-chloroacetate esterase positive granules, but macrophages were detected only occasionally [53]. Transformation of leukemic monocytes into macrophages as demonstrated also in vivo in skin window exudates [52]. The transformation into macrophages was somewhat impaired in the case of immature monocytic leukemia. When *cultured in agar* only very few colonies are produced; frequently the production of clusters is even more impaired than in other types of acute leukemia, and leukemic cells remain in a monocellular distribution; commonly a large proportion of these cells transform into enzyme-rich macrophages [63].

Sundström and Nilsson [60] reported on a cell line derived from a histiocytic lymphoma which exhibited strong NaF-sensitive nonspecific esterase activities and retained the capacity to produce lysozyme.

Skin window tests, performed according to Rebuck and Crowley [44], yielded a striking but still unexplained result. In normal persons as well as in patients with a large variety of diseases, even with considerable monocytosis, the neutrophil granulocytes constitute 90%–95% of the exudate cells during the first hours of the skin window experiment. During these first hours monocytes range between 1% and 3%;, later on monocytes rise constantly up to more than 50% at 7 to 10 h. In monocytic leukemia, both mature and immature monocytes constitute a considerable proportion of the exudate cells even in the first hours, occasionally over 50%. These findings, which have been confirmed by several authors [41, 47], may be related to unexplained chemotactic factors. The cytochemical pattern of the exudate monocytes is exactly the same as that of the leukemic monocytes in the peripheral blood, which means that no difference in the migratory capacity exists between mature and immature cell types and apparently no selection occurs among the cells migrating in the

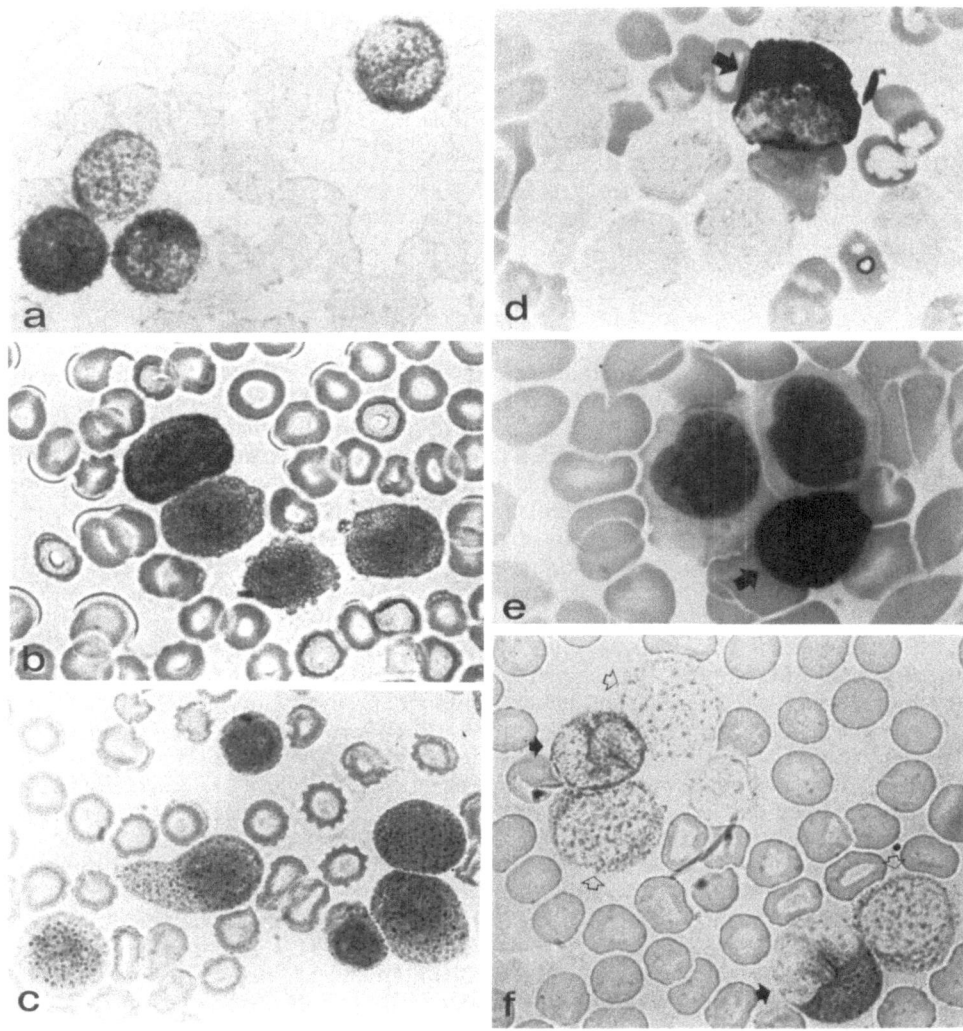

Fig. 4a–f. Cytochemistry of mature leukemic monocytes. The cells contain strong activities of naphthol-AS-D-acetate esterase **a**, and of naphthylamidase (substrate: alanyl-β-naphthylamid) **b**; moderate staining for naphthylamidase in a less mature leukemic population **c**. Leukemic monocytes show only very weak or uncertain activity neutral protease (substrate: naphthol-AS-D-chloroacetate) **d** and of peroxidase **e**, whereas promyelocytes are strongly stained for both enzyme activities (↑). **f** Acute myelomonocytic leukemia; double staining for naphthol-AS-D-acetate esterase (monocytic cells stained blue in the original, ↑) and for neutral protease (promyelocytic cells contain red azo dye, ↑). Orig. magnif. × 945

irritated areas. Leukemic monocytes show also marked *adhesivity to glass* or other surfaces, and this peculiar feature has also been used to characterize immature leukemic cell populations as monocytic [34, 47, 63].

Evidence for the *phagocytic capacity* of the leukemic cells can be appreciated in skin window exudates. As a simple screening the phagocytosis of latex particles can be evaluated in the NBT test [19]. Furthermore, examining the presence of Fc receptors by adding IgG-coated erythrocytes considerable phagocytic capacity may occasionally be observed.

Several authors used phagocytosis tests as indicators for the monocyte nature of immature leukemic cells [34, 47, 63]; however, the phagocytic capacity seems to be subjected to considerable variability. The degree of maturation seems to be one of the governing factors [34, 48]. Signs of in vivo phagocytosis can rarely be appreciated in post mortem examinations.

The *motility* of leukemic monocytes was studied extensively by Lichtmann and Weed [34]. In our hands the almost routinely performed skin window tests permitted a rough evaluation of cell motility in comparison to that of the polymorphonuclear neutrophils [48, 52].

Using the method of Huber and Fudenberg [25] we tested the presence of *IgG receptors* on the surface of cytochemically and ultrastructurally characterized leukemic monocytes. EA rosette formation and consequent phagocytosis of IgG-coated erythrocytes occurred in both mature and immature leukemic monocytes. Strikingly the receptors for IgG and complement are present even in very immature-appearing monocytic cells, and this typically differs from leukemic promyelocytes [4, 9, 22]. By contrast Koziner et al. [30] failed to show Fc receptors in several cases of immature monocytic leukemia. Figure 4 shows the cytological and cytochemical characterization of leukemic promonocytic cells presenting weak activities of acid phosphatase and NaF-sensitive naphthol-AS-D-acetate esterase as well as a very poor ultrastructural evidence of maturation; however, these cells exhibited moderately avid Fc receptors.

Surface immunoglobulins have been consistently demonstrated on leukemic monocytes of the peripheral blood by Gordon and Hubbard [22] and Koziner et al. [30]; in more immature cells of the bone marrow, however, no surface IgG was detected.

Monoclonal *monocyte-macrophage-specific antibodies* have been produced which may be helpful for the characterization of leukemic monocytes [3, 58]. *Cell electrophoresis* has been employed by Lichtmann and Weed [34]; the authors found that leukemic monocytes showed a similar behavior in the electric field as did normal monocytes.

Diagnosis of Acute Monocytic Leukemia

Some authors attributed relative little importance to the exact cytological classification of the acute leukemias. With this in mind the term "acute myelomonocytic leukemia" now is frequently used to indicate poorly differentiated myelogenous leukemias. The old term gained new actuality due to the development of a "unifying concept" of the myelogenous leukemias. Its use was further supported by the difficulties encountered by several authors in the cytological classification of the most undifferentiated myelogenous leukemias. However, little accordance usually existed several years ago among the authors which used this term.

The diagnosis of monocytic leukemia should rely on the identification of monocyte-specific cytological features in the leukemic cells and at least in cases with atypical leukemic cell populations several diagnostic criteria should be studied simultaneously. Single criteria on, for instance, ultrastructural, morphological, and surface receptors, may be not conclusive enough and, in some cases, even cytochemical examination may be insufficient. Combination of cytochemical and/or ultrastructural examination with functional tests seems to us the most suitable method for assessing this difficult diagnosis.

It is quite clear that for general clinical purposes cytochemical or ultrastructural examinations or lysozyme determinations in addition to the inconsistent morphological classification may be conclusive enough. However, for scientific purposes a panel of different investigative techniques has to be employed in order to exactly identify the type and the degree of differentiation of the leukemic cells.

The FAB classification considers these cytological features and distinguishes two kinds of monocytic leukemias: the "myelomonocytic type" – ("M 4") characterized by the presence of monocytic and granulocytic differentiation at the same time, and the "monocytic type" ("M 5") with almost exclusively monocytic or monoblastic differentiation of the leukemic cells.

Recently most authors agree on this characterization [13, 23]. The importance of the distinction between "pure monocytic" and "myelomonocytic" leukemia is stressed in regard to

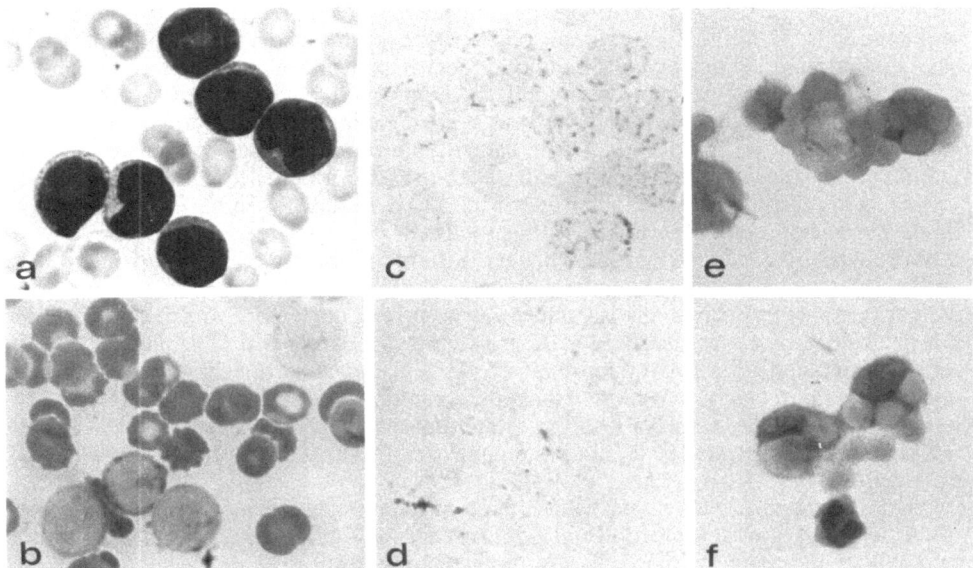

Fig. 5a–f. Immature monocytic leukemia **a** May-Grünwald-Giemsa; **b** weak staining for acid phosphatase and **c** for naphthol-AS-D-acetate esterase. The esterase activity is clearly inhibited by sodium fluoride (1.5 mg/ml) **d. e** and **f** Rosette formation and ingestion of IgG-coated erythrocytes. Orig. magnif. × 880

the type of stem cell affected by the malignant transformation. However, clinical hematologists are aware of the frequently changing cytological patterns during the course of myelogenous leukemias; according to our experience the admixture of other myeloid elements to monocytic leukemias may vary considerably during their course, especially in relapses. The cytological classification of acute leukemias is a challenge to clinical pathologists, but the cytological diagnosis gains its real clinical value if indications for the clinical course, possible complications, or therapy can be deduced.

Until now little attention has been payed to the *clinical course* and the *clinical pathology of monocytic leukemia* and it is generally accepted that they are quite similar to those of the other acute myeloid leukemias – except promyelocytic leukemia [12, 28, 46, 49, 50, 57, 61]. Reports indicating peculiar clinical or pathological features are rare [12, 23, 31, 50, 57, 61]. However, it is quite conceivable that the specific cytological properties of leukemic monocytes may induce peculiar clinical features. Some authors brought the attention towards the increased frequency of skin infiltrations [50], a topic which will be covered in a following paper by Dr. Burg [this volume p. 221]. Gum infiltrations have also been reported to be more frequent in monocytic than in other kinds of acute leukemia [57]. Perivascular infiltration, infiltration of the myocardium, nerves, and kidney, increased frequency of meningeal and cerebral infiltrations, and the migration of leukemic monocytes into inflammatory exudates, for instance into pneumonic lesions, are strictly related to the physiological capacity of the monocytes to leave the vascular system and to accumulate at inflammatory sites. Exaggerated phagocytic activity of the neoplastic cells are prominent features of disorders termed as "histiocytosis" or "reticulosis".

The course of monocytic leukemia may be complicated by the action of metabolic products secreted by intact monocytes. Rapid destruction of these cells due to aggressive therapy may induce sudden increases in the serum levels of such substances, like lysozyme, or factors interfering with coagulation. Induction of kidney tubular lesions has been related to excessive lysozyme excretion; loss of potassium and sodium may be the consequence [36, 49]. According to our experience severe complications of this kind are not frequent, but they

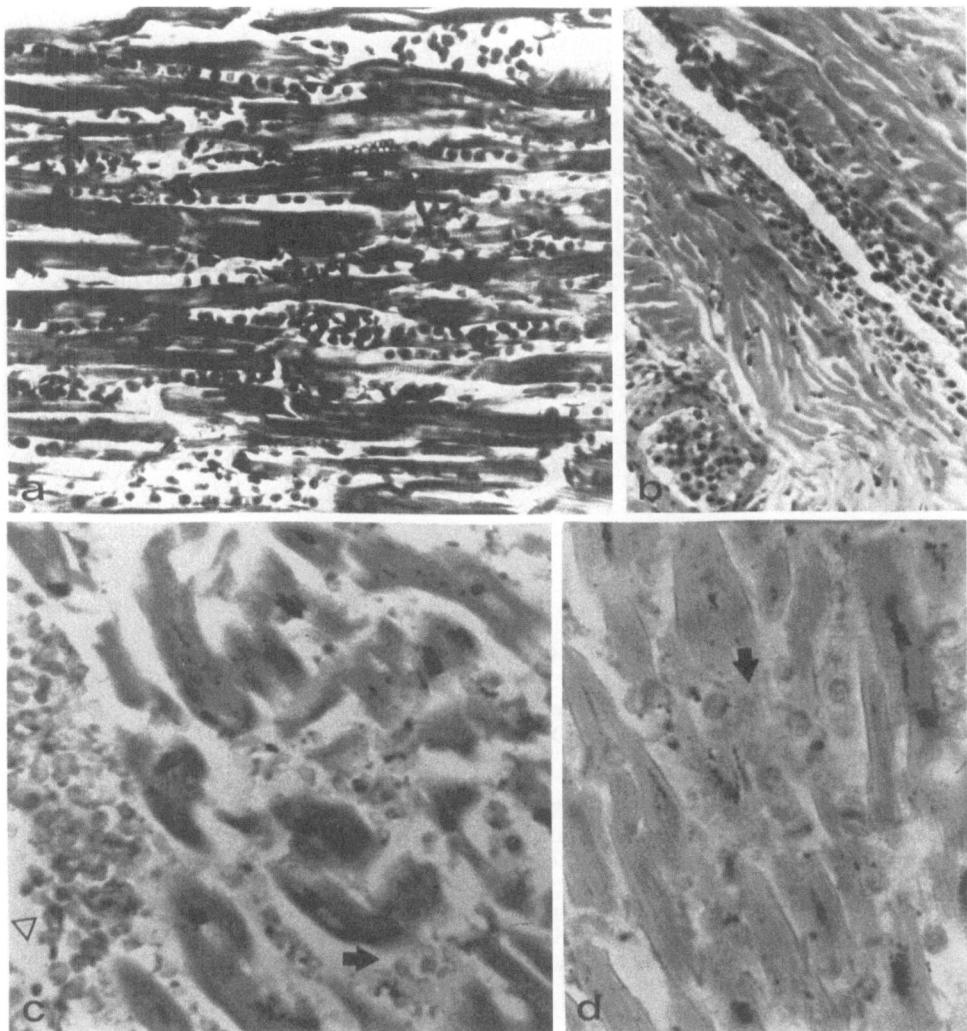

Fig. 6a-d. Acute monocytic leukemia; leukemic infiltration of myocardium **a** and dura mater **b** (hematoxylin-eosin). **c** and **d** Myocardium: naphthol-AS-D-acetate esterase reactions; leukemic monocytes in vascular (△) and perivascular sites (↑). Orig. magnif. × 99 (**a, b**), × 158 (**c**), and × 237 (**d**)

may develop suddenly and are especially dangerous if cardiac infiltrations are present. Hemorrhage is the most important and characteristic complication of hypergranular or mature promyelocytic leukemia. However, in our study hemorrhage is also a frequent cause of death in immature monocytic leukemias, whereas it is a rather rare event in mature monocytic leukemia. In our study the frequency of lethal hemorrhagic disorders is 2 out of 22 in mature and 12 out of 16 in immature monocytic leukemia [49]. The pathogenesis of this disorder is not satisfactorily understood. In promyelocytic leukemia the high concentrations of neutral proteases have been considered as the cause of destruction of coagulation factors or of disseminated intravascular coagulation [62]. However, immature monocytic leukemias show only weak or moderate amounts of granules and contain relatively low activities of lysosomal enzymes. It has been reported that macrophages and leukemic monocytes produce a tissue − a thromboplastin-like factor − and perhaps also a plasminogen activator; tentatively, a major role in the initiation of the coagulation disorder could be attributed to these factors [49].

4. Conclusion

Concluding I would like to emphasize that leukemic monocytes are characterized by almost the same qualitative cytological features and the same functional properties as normal promonocytes, monocytes, and macrophages; quantitative differences in the expression of these properties may depend upon the varying degree of maturation of the leukemic cell populations. The diagnosis of monocytic leukemia should be based on the identification of these properties, which may also determine the presentation of the leukemic disorder and may be of great importance for the clinical course. Depending upon the degree of maturation of the leukemic cells, mature and immature monocytic leukemias can be distinguished, and depending upon the admixture of other myeloid elements, myelomonocytic leukemia can be distinguished from fairly "pure" monocytic leukemia. We believe that the discussed clinical features favor the distinction of monocytic leukemias as separate clinical entities among the acute leukemias.

References

1. Asamer H, Schmalzl F, Braunsteiner H (1971) Die diagnostische und prognostische Bedeutung der Muramidase-(Lysozym-)Bestimmung in Leukozytenlysaten, Serum und Harn von Leukämiepatienten. Klin Wochenschr 49:587–593
2. Asamer H, Schmalzl F, Braunsteiner H (1971) Immunocytological demonstration of lysozyme (muramidase) in human leukaemic cells. Br J Haematol 20:571–574
3. Baker MA, Falk RE, Falk J, Greaves MF (1976) Detection of monocyte-specific antigen on human acute leukemia cells. Br J Haematol 32:13–19
4. Barrett S, Garatty E, Garatty G (1979) Cell surface heterogeneity of human blood neutrophils and monocytes. Br J Haematol 43:575–588
5. Bennet JM, Catovsky D, Daniel MT, Flandrin G, Galton DAG, Gralnick HR, Sultan C (1976) Proposals for the classification of acute leukaemias. French American British (FAB) Cooperative Group. Br J Haematol 33:451–458
6. Bessis M (1973) Living blood cells and their ultrastructure. Springer, Berlin Heidelberg New York
7. Braunsteiner H, Schmalzl F (1970) Cytochemistry of monocytes and macrophages. In: van Furth R (ed) Mononuclear phagocytes. Blackwell, Oxford, pp 62–80
8. Breton-Gorius J (1979) Abnormalities of granulocytes and megakaryocytes in preleukemic syndromes. In: Schmalzl F, Hellriegel K-P (eds) Springer, Berlin Heidelberg New York
9. Burns GF, Cawley JC (1979) Membrane receptors of human leukaemic myeloid cells: sequential expression of the γ-Fc-receptor. Br J Haematol 42:499–505
10. Carmel R, Hollander D (1978) Extreme elevation of transcobalamin II levels in multiple myeloma and other disorders. Blood 51:1057–1063
11. Catovsky D, Galton DAG, Robinson J (1972) Myeloperoxidase-deficient neutrophils in acute myeloid leukemia. Scand J Haematol 9:142–147
12. Cline MJ, Golde DW (1973) A review and reevaluation of the histiocytic disorders. Am J Med 55: 49–60
13. Cline MJ, Golde JW, Billing RJ, Groopman JE, Zieghelboim J, Gale RP (1979) Acute leukemia: Biology and treatment. Ann Intern Med 91:758–773
14. Fischer R, Gropp A (1964) Cytologische und cytochemische Untersuchungen an normalen und leukämischen in vitro gezüchteten Blutzellen. Klin Wochenschr 42:111–114
15. Fischer R, Schmalzl F (1964) Über die Hemmbarkeit der Esteraseaktivität in Blutmonozyten durch Natriumfluorid. Klin Wochenschr 42:751
16. Flandrin G, Daniel MT (1973) Practical value of cytochemical studies for the classification of acute leukemias. In: Mathé G, Pouillard P, Schwarzenberg L (eds) Nomenclature, methodology and results of clinical trials in acute leukemia. Springer, Berlin Heidelberg New York, pp 43–56
17. Freeman AJ, Journey LJ (1971) Ultrastructural studies on monocytic leukemia. Br J Haematol 20:225–231

18. Furth R van, Raeburn JA, van Zwet TL (1979) Characteristics of human mononuclear phagocytes. Blood 54:485–500
19. Gifford RH, Malawista SE (1970) A simple rapid micromethod for detecting chronic granulomatous disease of childhood. J Lab Clin Med 75:511–519
20. Glick AD, Horn RG (1974) Identification of promonocytes and monocytoid precursors in acute leukaemia of adults: Ultrastructural and cytochemical observations. Br J Haematol 26:395–403
21. Golde DW, Rothman B, Cline MJ (1974) Production of colony-stimulating factor by malignant leukocytes. Blood 43:749–756
22. Gordon DS, Hubbard M (1978) Surface membrane characteristics and cytochemistry of the abnormal cells in adult acute leukemia. Blood 51:681–692
23. Gralnick HR, Galton DAG, Catovsky D, Sultan C, Bennett JM (1977) Classification of acute leukemia. Ann Intern Med 87:740–753
24. Hansen NE (1974) Lysozyme activity in leukemia. Ser Haematol 7:70–87
25. Huber H, Fudenberg HH (1968) Receptor sites of human monocytes for IgG. Int Arch Allergy Appl Immunol 34:18
26. Huhn D, Schmalzl F, Demmler K (1971) Monozytenleukämie. Licht- und elektronenmikroskopische Morphologie und Zytochemie. Dtsch Med Wochenschr 96:1594–1605
27. Hultberg B, Sjögren U (1980) Diagnostic significance of lysosomal enzymes in different types of leukemias. Acta Med Scand 207:105–110
28. Kass L, Schnitzler B (1973) Monocytes, monocytosis, and monocytic leukemia. Thomas, Springfield
29. Konwalinka G, Glaser P, Odavic R, Bogusch E, Schmalzl F, Braunsteiner H (1980) A new approach to the morphological and cytochemical evaluation of bone marrow CFUc in agar cultures. Exp Hematol 8:434–440
30. Koziner B, McKenzie S, Straus D, Clarkson B, Good RA, Siegal FB (1977) Cell marker analysis in acute monocytic leukemias. Blood 49:895–901
31. Leder L-D (1967) Der Blutmonozyt. Springer, Berlin Heidelberg New York
32. Leder L-D (1967) The origin of blood monocytes and macrophages. Blut 16:86–98
33. Li CY, Lam KW, Yam LT (1973) Esterases in human leukocytes. J Histochem Cytochem 21:1–12
34. Lichtmann MA, Weed RI (1972) Peripheral characteristics of leukocytes in monocytic leukemia: Possible relationship to clinical manifestation. Blood 40:52–61
35. McKenna RW, Bloomfield CD, Dick F, Nesbit ME, Brunning RD (1975) Acute monoblastic leukemia: Diagnosis and treatment of ten cases. Blood 46:481–494
36. Muggia FM, Heinemann HO, Farhangi M, Ossermann EF (1969) Lysozymuria and renal tubular dysfunction in monocytic and myelomonocytic leukemia. Am J Med 47:351–366
37. Nathan CF, Murray HW, Cohn ZA (1980) The macrophage as an effector cell. N Engl J Med 303:622–626
38. Nichols BA, Bainton DF (1973) Differentiation of human monocytes in bone marrow and blood: sequential formation of two granule populations. Lab Invest 29:24–40
39. Niemetz J, Muhlfelder T (1980) Leukocytes and the initiation of intravascular coagulation. Proc. 18th Congr. Intern. Soc. Haemat. & 16th Congr. Intern. Soc. Blood Transf., Montreal
40. Nowell PC (1960) Differentiation of human leukemic leukocytes in tissue cultures. Exp Cell Res 19:267–273
41. Ohta H, Matsuda Y (1973) Ready release of intracellular muramidase (lysozyme) from mononuclear cells in the skin window exudates. Acta Haematol (Basel) 49:159–165
42. Osserman EF, Lawlor DP (1966) Serum and urinary lysozyme (muramidase) in monocytic and myelomonocytic leukemia. J Exp Med 124:921–930
43. Polliack A, McKenzie S, Gee T, Lampen N, De Harven E, Clarkson B (1975) A scanning electron microscopic study of 34 cases of acute granulocytic, myelomonocytic, monocytic, monoblastic and histiocytic leukemia. Am J Med 59:308–315
44. Rebuck JW, Crowley RH (1955) A method of studying leukocytic functions in vivo. Ann NY Acad Sci 59:757–763
45. Rindler, Hörtnagl H, Schmalzl F, Braunsteiner H (1973) Hydrolysis of a chymotrypsin substrate and of naphthol AS-D-chloro-acetate by human leukocyte granules. Blut 26:239
46. Rundles RW (1972) Monocytic leukemia. In: Williams WJ, Beutler E, Erslev AJ, Rundles RW (eds) Hematology, McGraw-Hill, New York
47. Schiffer CA, Sanel FT, Stechmiller BK, Wiernik PH (1975) Functional and morphological characteristics of the leukemic cells of a patient with acute monocytic leukemia: Correlation with clinical features. Blood 46:17–26

155

48. Schmalzl F (1971) Unreifzellige Monozytenleukämie. Blut 22:157–174
49. Schmalzl F, Abbrederis K (1979) Prognostic value of the cytological classification of monocytic leukemias. In: Mandelli F (ed) Therapy of acute leukemias. Lombardo, Rome, pp 108–112
50. Schmalzl F, Braunsteiner H (1968) Zur Zytochemie der Monozytenleukämie. Klin Wochenschr 46:1185–1195
51. Schmalzl F, Braunsteiner H (1971) The application of cytochemical methods to the study of acute leukemia. Acta Haematol (Basel) 45:209–217
52. Schmalzl F, Huber H, Asamer H, Abbrederis K, Braunsteiner H (1969) Cytochemical and immunhistological investigations on the source and the functional changes of mononuclear cells in skin window exudates. Blood 34:129–140
53. Schmalzl F, Pastner D, Abbrederis I, Braunsteiner H (1969) In vitro cultivation of leukemic monocytes. Acta Haematol (Basel) 41:225–233
54. Schmalzl F, Huhn D, Asamer H, Abbrederis K, Braunsteiner H (1972) Atypical (monomyelocytic) myelogenous leukemia. Cytochemical, electron microscopy and biochemical investigations. Acta Haematol (Basel) 48:72–88
55. Schmalzl F, Huhn D, Asamer H, Rindler R, Schmalzl F (1973) Cytochemistry and ultrastructure of pathologic granulation in myelogenous leukemia. Blut 27:243–260
56. Schnitzler B, Kass L (1973) Leukemic phase of reticulum cell sarcoma (histiocytic lymphoma). Cancer 31/57:547–559
57. Shaw MT, Nordquist RE (1978) Pure monocytic or histomonocytic leukemia. Am J Hematol 4:97–103
58. Stuart AE, Young GA, Grant PF (1976) Identification of human mononuclear cells by anti-monocyte serum. Br J Haematol 34:457–464
59. Sultan G, Imbert M, Richard MF, Sebaoun G, Marquet M, Brun B, Forgues L (1977) Pure acute monocytic leukemia. A study of 12 cases. Am J Clin Pathol 68:752–757
60. Sundström C, Nilsson K (1976) Establishment and characterization of a human histiocytic lymphoma cell line (U-937). Int J Cancer 17:565–577
61. Tobelem G, Jacquillat C, Chastang C, Auclerc MF, Lechevallier T, Weil M, Daniel MT, Flandrin G, Harrousseau JL, Schaison G, Boiron M, Bernard J (1980) Acute monoblastic leukemia: A clinical and biologic study of 74 cases. Blood 55:71–76
62. Trobisch H, Egbring R (1976) Hämostatische Defekte bei akuten myeloischen Leukämien des Erwachsenen. Dtsch Med Wochenschr 101:1430–1433
63. Weetman RJ, Chilcote RR, Rierden WJ, Baehner RL (1978) In vitro characteristics of childhood leukemic monoblasts. Exp Hematol 6:9–17
64. Worwood M, Summers M, Miller F, Jacobs A, Whittaker JA (1974) Ferritin in blood cells from normal subjects and patients with leukemia. Br J Haematol 28:27–35
65. Zittoun Z, Zittoun R, Marquet J, Sultan C (1975) The three transcobalamins in myeloproliferative disorders and acute leukemia. Br J Haematol 31:287–298

Haematology and Blood Transfusion Vol 27
Disorders of the Monocyte Macrophage System
Edited by F. Schmalzl, D. Huhn, H.E. Schaefer
© Springer-Verlag Berlin Heidelberg New York 1981

Subacute and Chronic Myelomonocytic Leukemia

R. Zittoun

Subacute and chronic myelomonocytic leukemia have been previously described under different names. Recent works have underlined the most common features of these syndromes [3, 8, 10, 16, 28, 29]. Since numerous other papers refer to them, giving additional etiological and biological data, this talk will focus on certain specific topics. I shall try to clarify the problem of evolution and prognosis, to discuss treatment, and to point out what we have learned from this type of leukemia.

1. Characterization of Subacute and Chronic Myelomonocytic Leukemia

Subacute (SMML) and chronic (CMML) myelomonocytic leukemia are diseases with a paucity of symptoms and a long time lapse between the first symptoms and diagnosis; many cases are diagnosed incidentally from routine blood examination. Clinical findings are limited to pallor and a moderate splenomegaly in about one-third of the patients. Hepatomegaly, lymphadenopathy, and bruises are found in some cases, but gingival hypertrophy is generally not observed. Infections represent the major complication and were noticed in half of our cases.

Diagnosis is based on blood monocytosis higher than 10^9/liter, generally $5-10 \times 10^9$/liter, with immature and atypical circulating monocytes. Bone marrow examination shows a moderate increase in the percentage of monocytes and promonocytes to 10%–30%. Myeloid involvement is characterized by a slight neutrocytosis with some immature circulating forms, a frequently low neutrophilic alkaline phosphatase score, and a prominent granulopoietic proliferation in the marrow. All three major myeloid lines are involved, as pointed out by Saarni and Linman [19]. The myeloid abnormalities observed are similar to those described in the myelodysplastic syndromes [6]: dyserythropoiesis with increased sideroblasts, abnormal megakaryocytes, and a partial or complete lack of neutrophilic granules. The most common abnormality is an excess of myeloblasts in the bone marrow from 5% to 30%. Chromosomal abnormalities have been described in some patients, with occasional aneuploidy affecting the C, F, or Y chromosomes, but the Ph1 chromosome has never been observed [10, 16, 29]. However, cytogenetic studies using banding techniques are still missing.

Once inflammatory and malignant causes of monocytosis have been excluded [13], each case of myelomonocytic leukemia raises specific nosological and diagnostic difficulties. First, SMML can be differentiated from *acute myelomonocytic leukemia*. This acute form, known as the M4 type of acute myeloid leukemia according to the French-American-British classification [1], is characterized by frequent skin, gum, and lymph node involvement. Myeloblasts in the bone marrow are ≧30%, with Auer rods occasionally seen.

Subacute myelomonocytic leukemia is not clearly separated from *subacute myeloid leukemia* as defined by Cohen et al., who found no prognostic value for the monocytosis

Table 1. Relative frequency of chronic and subacute myelomonocytic leukemia, Hotel-Dieu, 1970–1980, compared to other types of leukemia

SMML and CMML	38 (2.8%)
SMML and oligoblastic L	35 (2.6%)
Chronic granulocytic L	238 (17.7%)
Chronic lymphocytic L	309 (23 %)
Hairy cell L	27 (2 %)
Acute L:	
– Lymphoblastic	147 (10.9%)
– Myeloblastic	310 (23 %)
– Myelomonocytic	91 (6.8%)
– Monoblastic	33 (2.4%)
– Erythroleukemia	24 (1.8%)
– Promyelocytic	33 (2.4%)
– Unclassified	60 (4.4%)

observed in 50% of their cases of acute and subacute myeloid leukemia [4]. I feel that the distinction between SMML and the subacute myeloid leukemia must be kept just as that between the M2 and M4 type of acute myeloid leukemia, since one cannot yet rule out that later on it might be of therapeutic and prognostic importance.

Secondly, the distinction between CMML and *chronic monocytic leukemia* is mainly semantic, as pure monocytic leukemia without involvement of the other myeloid cell lines seldom occurs. CMML must also be distinguished from *refractory anemia* with slight monocytosis: in CMML, anemia is absent in 10% to 40% of cases, whereas blood and bone marrow monocytosis is the major feature.

2. Incidence of SMML and CMML

The incidence of SMML and CMML remains to be determined. We have tried to get an indirect idea of their incidence by comparing the number of SMML and CMML registered in our department at Hotel-Dieu with that of the major types of leukemias seen in adults. Cases of refractory anemia developing secondarily into SMML were excluded. Table 1 shows that SMML and CMML are less frequent than the common types of leukemia such as chronic granulocytic, chronic lymphocytic, or acute myeloblastic leukemia. Their frequency is similar to that of acute promyelocytic leukemia or of hairy cell leukemia. Among the leukemias involving the monocytic cells, SMML and CMML are less frequent than acute myelomonocytic leukemia and as frequent as acute monoblastic leukemia. This frequency must not be underestimated, since during the last 8 years 136 cases have been described in six large series. The major risk seems actually to be the misdiagnosis of one of these syndromes.

The question may be raised of a higher frequency in some etiological conditions. Cohen et al. pointed out a history of possible myelotoxic exposure in 14 out of 31 patients with subacute myeloid leukemia – half of these were of the myelomonocytic type [4]. This rate seems higher than the one usually observed for the common types of acute myelogenous leukemia. There is also a high frequency of the myelomonocytic type among the so-called secondary leukemias, mainly in patients treated for multiple myeloma [11], for Hodgkin disease, and for different malignant diseases treated with cytostatic drugs [25]. It should be noted that preleukemic states may also occur in patients exposed to alkylating drugs [24].

3. Subacute Versus Chronic Myelo monocytic Leukemia

Many reports in the literature inaccurately use the terms subacute or chronic myelomonocytic leukemia – as we did when we published our first series [29]. Clinical observation

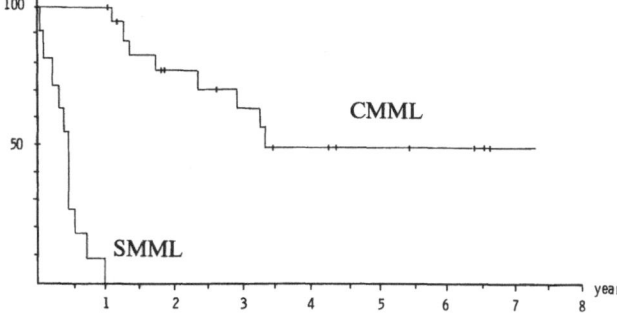

Fig. 1. Subacute and chronic myelomonocytic leukemia

shows that some patients have a rapid evolution and die within months after diagnosis either from infection or hemorrhage or after the transformation into a true overt picture of acute leukemia. On the other hand, other patients have a very steady course and live for many years, even without any specific treatment.

The international nomenclature adopted as early as 1950 the terms chronic, subacute and acute leukemia [23]; at that time efficient drugs were still not used in acute leukemia and the natural course could be observed to a greater extent than today. The theoretical importance of keeping such denominations is obvious, since the decision for submitting patients to highly toxic chemotherapy depends mainly on their spontaneous life expectancy. Cohen et al. [4] characterized subacute myeloid leukemia by abnormalities in all cell lines and the presence of some maturation of the myeloid line beyond the promyelocyte stage. The survival of their patients varied from less than 1 to more than 68 months. The authors found a significantly shorter survival in patients presenting at the time of diagnosis with anemia, hepatosplenomegaly, thrombocytopenia or increased white blood cell count or in patients between 50 and 70 years of age [4].

We have tried to ascertain some differences in the presenting signs of our cases of SMML and CMML. We have defined CMML as patients surviving more than 1 year without a transformation into acute leukemia during the 1st year. Patients with 30% or more blast cells at the first bone marrow examination were considered as acute leukemias and excluded. SMML, therefore, was characterized by less than 30% blast cells, a progressive course leading to overt acute leukemia and/or death within 1 year after the diagnosis. Due to the number of patients lost from the follow-up, the final study includes 20 patients with CMML and 11 patients with SMML. The subacute form consequently seems less frequent than the chronic one. The actuarial survival curves exhibit a marked prognostic difference between SMML and CMML (Fig. 1): the median survival of SMML is 6 months and all patients died within 12 months, whereas CMML is characterized by a far longer survival, with a median survival of 40 months and 7 of 20 patients surviving from 3.5 up to 7 years.

Table 2 shows some of the differences that we found between the two syndromes: the high M/F sex ratio seems restricted to CMML. SMML was characterized by a higher frequency of splenomegaly, a lower hemoglobin level, and a higher white blood cell count involving both the granulocytic and monocytic series. Platelet counts varied greatly in both series. The mean percentage of blast cells in the bone marrow is 15.5 in SMML and 7.8 in CMML. Finally a transformation into acute leukemia was observed in 6 of 20 patients with CMML and 5 of 11 patients with SMML. As already emphasized, all the blastic transformations were of the monoblastic or myelomonocytic type. The acute leukemic phase resulted from a blastic crisis after a steady chronic phase in CMML. On the other hand, delimitation of subacute and acute phases in SMML was often only a matter of semantics, since the course in these cases was characterized by a progressive proliferation and increase of blast cells, passing at some point the limit of 30% arbitrarily chosen to define acute leukemia. SMML appears, therefore, as a peculiar cytological type of an acute leukemia, whereas CMML is a chronic disease which, potentially, can evolve into acute leukemia.

159

Table 2. Subacute versus chronic myelomonocytic leukemia: main distinctive features (at presentation, mean values)

	CMML 20 cases	SMML 11 cases	P
Sex ratio (M/F)	2.33	0.83	
Age: median and (range)	68 (25–38)	68 (47–83)	
Duration prediagnosis (mth)	0–60	0–6	
Splenomegaly	4[a]	6	
Bruising	5	3	
Hemoglobin (g/l)	106 ± 27	85 ± 29	< 0.05
WBC ($\times 10^{-9}$/l)	12.4	61.4	NS
PMN ($\times 10^{-9}$/l)	5.9	18.7	NS
Monocytes ($\times 10^{-9}$/l)	3.35	25.7	NS
Platelets ($\times 10^{-9}$/l)	91	103	NS
Circulating blasts ($\times 10^{-9}$/l)	0.01	2.76	< 0.02
Bone marrow blasts (%)	7.8	15.5	< 0.01
Bone marrow monocytes (%)	11.6	15.2	NS
Blastic evolution			
– number (and %)	6 (30)	5 (45)	
– after (months)	13–33	1–6	

[a] Splenomegaly appeared later in three other patients

Some of our data, namely, the hemoglobin level with its attached prognostic value, are close to the conclusions of Cohen et al. [4].

On the other hand, Dresch et al. [5] failed to find any differences in the clinical and hematological data between long- and short-term surviving patients except for a higher but not significant percentage of bone marrow blasts in patients surviving less than 1 year. However, their patients were not all studied from the time of diagnosis, since 7 out of their 13 short survival patients were CMML at a late phase of their disease; their series included in fact 6 SMML and 17 CMML.

Since some statistical differences can be found between SMML and CMML, the question remains of a possible identification in individual patients at the time of diagnosis. We have tried to diagnose the two categories by reviewing the slides from the patients at diagnosis (SMML and CMML being randomly mixed). Eight cytological criteria were selected in blood (monocytes >30%, immature cells >5%, blasts + promyelocytes >5%, and morphological abnormalities of the three myeloid cell lines) and bone marrow (blasts >20%, young or abnormal monocytes >10%, erythroblasts <10% or >35% or abnormal, and polymorphonuclears <10%). SMML was characterized by more than four criteria in 9 out of 11 patients, whereas 12 of 13 CMML had ≦ 4 criteria (P < 0.001).

However, these clinical and hematological findings are insufficient to define a clear prognosis, consequently, we have tried to consider other biological aspects, hoping to learn more about the prognosis and the physiopathology of these diseases.

4. Serum and Urinary Lysozyme

Lysozyme, a cationic protein with a hydrolase activity, is produced by the granulocytic and the monocytic macrophagic series; however, while it is excreted by the living monocyte, it is only released by the dying granulocyte [9]. Relatively high levels of serum lysozyme are observed in myeloproliferative disorders as well as in myelodysplastic syndromes with ineffective granulopoiesis. High levels of serum lysozyme are also commonly found in acute monocytic and myelomonocytic leukemia. In these acute diseases, large amounts of lyso-

160

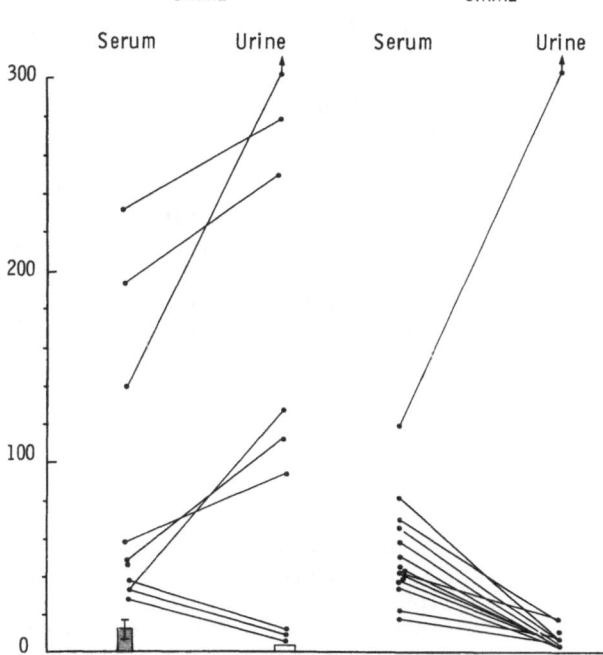

Fig. 2. Serum and urine lysozyme in subacute and chronic myelo-monocytic leukemia

zyme are also excreted in urine probably due to high serum levels exceeding the tubular reabsorption capacity.

Figure 2 shows that only 7 of 23 patients had serum and urinary levels of lysozyme as high as the ones usually observed in acute myelomonocytic leukemia: 6 of these 7 patients had SMML and died within 6 months. The patient with CMML and high urinary lysozyme levels had 16 months later a blastic crisis and died. These very high levels were probably due to a marked proliferation of the monocytic leukemic cells. One of these two patients had no peripheral hyperleukocytosis at the time of diagnosis; thus, the high lysozyme levels observed in this case demonstrate that this abnormality is a more precise indicator of a large medullary and extramedullary leukemic infiltration than the white blood cell count.

On the other hand, CMML is characterized by moderately increased serum levels of lysozyme with in most cases normal urinary levels. This can be correlated to the fact that in this disease extramedullary involvement is uncommon. A moderate splenomegaly was found in only 4 out of 20 patients, and no other organomegaly was found. Two patients were splenectomized — one for splenic rupture, the other for hypersplenism. In these two cases, histological examination confirmed a myelomonocytic infiltration, which, however, did not destroy the normal splenic structure.

5. Serum Transcobalamins

The assay of serum transcobalamins is of interest in the myeloproliferative disorders [26]. It is well known that the increase in serum vitamin B 12 generally observed in chronic granulocytic leukemia (CGL) is explained by a high vitamine B 12 binding capacity (UBBC). Three different transcobalamins (TC) are currently separated by various methods. TC I and TC III originate in the granulocytes; consequently, CGL is characterized by an enormous increase of TC I. In the other myeloproliferative diseases TC I is only moderately increased. Moreover a high ratio of TC III/TC I is observed in polycythemia vera, and TC III seems to originate from the more mature forms of the granulocyte series.

The origin of TC II — largely unsaturated but of functional importance, since it is involved in delivery of vitamine B 12 to cells — is still hypothetical. The liver was initially considered as the major source but we were impressed by the marked increase of TC II in AML, and we have shown that normal and leukemic blast cells produce TC II. A shift occurs at the promyelocyte–myelocyte stage to the production of TC I, then TC III [27]. More recently, however, Rachmilewitz et al. [18] have shown that the monocyte–macrophage system might be one of the major sources of TC II, thus explaining the increased levels of TC II in many inflammatory diseases. We studied the TC in several cases of SMML and CMML.

A marked increase of TC II was observed in three out of the six CMML cases studied, with a moderate increase of TC I in two of these. On the other hand, two cases of SMML were characterized by high or very high levels of TC I with high levels of TC II in one of these. More extensive data are necessary before drawing any conclusion, but for the moment our findings are hardly in agreement with the hypothesis of Rachmilewitz et al. of a monocytic origin of TC II, since three cases of SMML and CMML have normal levels of TC II. In any case, very high levels of TC should be considered as evidence for a massive leukemic infiltration, as is seen in SMML.

6. Bone Marrow Cultures

In CMML a normal number of CFU-GM after bone marrow culture on semisolid medium was first observed by Sultan et al. [22]. This normal or even increased number of colonies was in contrast to the decreased number generally observed in myelodysplastic syndromes such as refractory anemia with excess of blasts. The data of Sultan were later confirmed by Milner et al. [17] with, however, marked variations from patient to patient and a decreased number of CFU-GM in some cases. More recently Dresch et al. [5] have extensively studied the results of blood and bone marrow culture in SMML and CMML, comparing the findings in short (< 1 year)- and long-term survival patients. These authors observed markedly overlapping results, with the number of colonies higher than normal, normal, or lower than normal. However, the cluster–colony ratio was generally increased, especially in the short survival group. A lower proportion of CFU-GM in S phase after hydroxyurea suicide was also observed in two patients dying within 1 year; this was related to a relatively low labeling index of myeloblasts and promyelocytes in the short survival group.

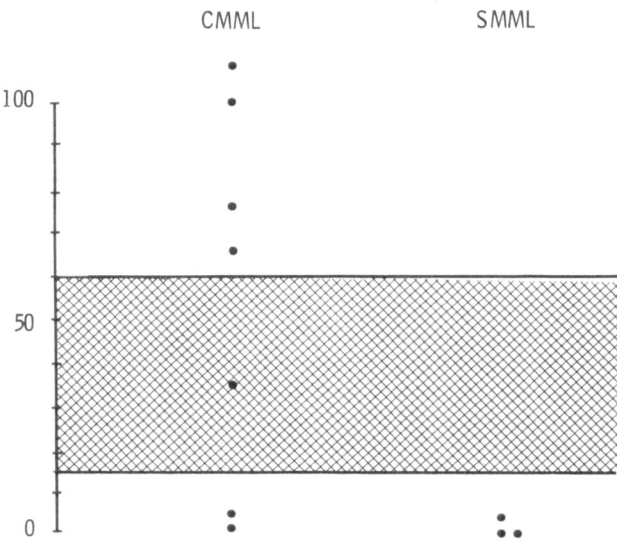

Fig. 3. Bone marrow CFU-GM in subacute and chronic myelomonocytic leukemia

Our own data, albeit in a small number of patients, are similar (Fig. 3): out of seven cases of CMML, a normal or high number of colonies was observed in five cases, but two patients had a decreased number or no colony.

Three cases of SMML were also studied using this method and had decreased number of colonies. A normal or high CFU-GM number seems consequently of good prognosis value.

The recent work from Dresch and his co-workers also sheds some light on the mechanism of CMML: using velocity sedimentation CFU-GM sedimented in a single peak between 4 and 5 mm/h, a pattern different from that of normal (6–8 mm/h), CGL (6–7 mm/h), and refractory anemia with an excess of blasts (5–6 mm/h). These findings as well as an autostimulation by the CSA originating from the leukemic monocytes themselves and a proliferative advantage of leukemic colony-forming cells involving the production of a leukemic inhibiting activity (LIA) [3] cannot yet explain in a simple way the high yield of colonies in CMML compared to the poor growth in acute and subacute forms. According to Dresch, however, a rapidly proliferating abnormal clone and an increase in LIA could explain the transition to subacute forms and the transformation into acute leukemia.

7. Are SMML and CMML Really Leukemias?

This question can only be raised for CMML, as SMML is obviously a disease very close to the M4 type of acute leukemia with proliferation of the two cell lines, peripheral immature and blast cells, increased blast cells in the bone marrow, and, finally, a progressive and lethal course.

On the other hand, CMML is frequently characterized by a moderate chronic monocytosis with a stable course. In the most benign cases one wonders whether the monocytosis is not secondary to an underlying disease rather than being primitive and leukemic. However, in our 20 cases the only associated diseases were current ones, not known to induce secondary monocytosis: diabetes, chronic renal insufficiency, malnutrition, head and neck cancer, chronic bronchitis, etc. Liver cirrhosis, which can be accompanied by a monocytosis, was suspected in three patients, but one of these evolved into an acute leukemia and another one had frank osteolytic lesions probably related to the myelomonocytic proliferation.

Nevertheless, the leukemic nature of a chronic monocytosis with reduced extramedullary involvement is questionable, especially in the cases in which abnormalities of erythrocytic and platelet lines are reduced or even absent [2, 5]. The course of such cases is similar to the non-evolutive types of chronic lymphocytic leukemia or to the benign monoclonal gammapathies. Their neoplastic nature is even less well defined, since their monoclonal character has not yet been demonstrated. We have to await such demonstration which might come from the study of cases occurring in patients heterozygous for G6PD deficiency. Such methods have already been applied to show the monoclonal and leukemic origin of the monocytic macrophagic cells in CGL [7].

The observation of visceral myelomonocytic infiltration in cases of CMML which could be autopsied [10, 16] and the occurrence of a blastic crisis in about 30% of cases remain at this time the two major arguments in favor of the leukemic nature of CMML. If the above hypothesis is accepted, the main theoretical interest of this disease remains the capacity to differentiate. Many recent works have shown that a number of factors have the property to induce differentiation of experimental leukemias along the granulocytic and monocytic lines [20]; SMML and CMML could represent a natural example in man of such differentiation capacity, and the search for inductive factors could be an important field for investigation in leukemia. The blastic crisis could be considered as the consequence of a secondary failure of these factors rather than a secondary cellular event. An argument against this theory was demonstrated by Geary et al. [8] of a cell modification in a case of blastic crisis of CMML, with the parallel appearance of a chromosomal abnormality (3; 16).

8. Treatment of SMML and CMML

A marked hypersensitivity to cytotoxic drugs has been noted in CMML and SMML, primarily after treatment by 6 mercaptopurine (6-MP) [2, 28]. We have observed a severe and frequently prolonged bone marrow aplasia in 10 patients out of 15 who received 6-MP after treatment with daily doses as low as 50 mg. It will be interesting to find out whether this hypersensitivity is specific for 6-MP. In this case, one could hypothesize a selective metabolic defect in CMML along the pathways involved in the biotransformation of 6-MP.

If this hypothesis is ruled out, a disturbed regulation of the balance between self-replication and differentiation of the myeloid stem cells could be postulated. In other words, the retained capacity for differentiation in CMML could be allowed only because of a limited capacity for self-replication.

Among the other drugs recently utilized for the treatment of CMML, VP 16-213, a podophyllotoxin derivative, must be pointed out. This drug was first shown to be active on the monocytic component of acute myelomonocytic leukemia [14] and in cases of acute monoblastic leukemia [15]. More recently Labedzki and Illiger [12] obtained a complete remission in one case of CMML by giving 100 mg IV VP 16-213 during 5 consecutive days, and then per os as a maintenance treatment. Our own experience with VP 16-213 as monochemotherapy is limited to a few cases of SMML; the results were poor and limited to transient decreases of the white blood cell count. More experience is needed with this drug using various schedules.

In fact the first important question is whether these diseases should be treated or not. Our data as well as the recent work of Dresch et al. show clearly that a distinction is possible between CMML and SMML. SMML, characterized by a short life expectancy, should be considered in patients with a severe anemia, a marked hyperleucocytosis, a high percentage of bone marrow blast cells, a decreased labeling index of blast plus promyelocytes, and an abnormal growth pattern in culture on agar. These patients should be treated as acute leukemias, and perhaps by the same protocols with chemotherapy, including anthracycline drugs, cytosine-arabinoside, and VP 16-213. Cohen et al. [4] have claimed however that half of their patients with subacute myeloid leukemia received chemotherapy with no demonstrated effect on survival in a noncontrolled, nonrandomized study.

On the other hand, chemotherapy should be avoided in CMML, except in cases with progressive myeloproliferation where courses of VP 16-213 should be tried.

References

1. Bennet JM, Catovski D, Daniel MT, Flandrin G, Galton DAS, Gralnick H, Sultan C (1976) Proposals for the classification of the acute leukaemias. French American British (FAB) cooperative group. Br J Haematol 33:451–458
2. Briere J, Dresch C, Briere JF, Faille A (1977) Leucémies myelomonocytaires subaigues ou chroniques de l'adulte. Etude clinique et données cinétiques. A propos de 46 observations. Actual Hematol (Paris) 53–80
3. Broxmeyer HE, Grossbard E, Jacobsen N, Moore MAS (1978) Evidence for a proliferative advantage of human leukemia colony-forming cells in vitro. J Natl Cancer Inst 60:513
4. Cohen JR, Creger WP, Greenberg PL, Schrier SL (1979) Subacute myeloid leukemia. A clinical review. Am J Med 66:959–966
5. Dresch C, Faille A, Poirier O, Balitrand N, Najean Y (1979) Bone marrow cell kinetics and culture in chronic and subacute myelomonocytic leukemia. Physiopathological interpretation and prognostic importance. Leuk Res 4:129–142
6. Dreyfus B, Rochant M, Sultan C (1969) Anémies réfractaires: enzymopathies acquises des cellules souches hématopoiétiques. Nouv Rev Fr Hematol 9:65–86
7. Fialkow PJ, Jacobson RJ, Papayannopoulou T (1977) Chronic myelocytic leukemia: clonal origin in a stem cell common to the granulocyte, erythrocyte, platelet, and monocyte-macrophage. Am J Med 63:125–130
8. Geary CG, Catovsky D, Wiltshaw E, Milner GR, Scholes MC, van Noorden S, Wadsworth LD,

Muldal S, MacIver JE, Galton DAG (1975) Chronik myelomonocytic leukaemia. Br J Haematol 30:289–302

9. Hansen NE (1974) Plasma lysozyme. A measure of neutrophil turnover. An analytical review. Ser Haematol 7:1–87
10. Hurdle ADF, Garson OM, Buist DGP (1972) Clinical and cytogenetic studies in chronic myelomonocytic leukaemia. Br J Haematol 22:773–782
11. Kyle RA, Pierre RV, Bayrd ED (1970) Multiple myeloma and acute myelomonocytic leukemia. Report of four cases possibly related to Melphalan. N Engl J Med 283:1121–1125
12. Labedzki L, Illiger HJ (1979) Erfolgreiche Behandlung der chronischen myelomonozytären Leukämie mit VP 16-213. Blut 38: 421–424
13. Maldonado JE, Hanlon DG (1965) Monocytosis: a current appraisal. Mayo Clin Proc 40:248–259
14. Mathe G, Schwarzenberg L, Pouillart P, Oldman R, Weiner R, Jasmin C, Rosenfeld C, Hayat M, Misset JL, Musset M, Schneider M, Amiel JL, de Vassal F (1974) Two epipodophyllotoxin derivatives, VM 26 and VP 16-213 in the treatment of leukemias, hemato-sarcomas and lymphomas. Cancer 34:985–992
15. McKenna RW, Bloomfield CD, Dick F, Nesbit ME, Brinning RD (1975) Acute monoblastic leukemia: diagnosis and treatment of ten cases. Blood 46:481–494
16. Miescher PA, Farcquet JJ (1974) Chronic myelomonocytic leukaemia in adults. Semin Hematol 11:129–139
17. Milner GR, Testa NG, Geary CG, Dexter TM, Muldai J, MacIver JE, Lajtha LG (1977) Bone marrow culture studies in refractory cytopenia and "Smouldering Leukaemia". Br J Haematol 35:251–261
18. Rachmilewitz B, Rachmilewitz M, Chaouat M, Schlesinger M (1977) The synthesis of transcobalamin II, a vitamin B 12 transport protein by stimulated mouse peritoneal macrophages. Biomedicine 27:213–214
19. Saarni MI, Linman JW (1971) Myelomonocytic leukemia: disorderly proliferation of all marrow cells. Cancer 27:1221–1230
20. Sachs L (1978) The differentiation of myeloid leukaemia cells: new possibilities for therapy. Br J Haematol 40:509–517
21. Sexauer J, Kass L, Schnitzer B (1974) Subacute myelomonocytic leukemia. Clinical, morphologic and ultrastructural studies of 10 cases. Am J Med 57:853–861
22. Sultan C, Marquet M, Joffroy Y (1974) Etude de certaines dysmyélopoieses acquises idiopathiques et secondaires par culture de moelle in vitro. Ann Med Interne (Paris) 125:599–602
23. Third, Fourth and Fifth reports (1950) of the Committee for clarification of the Nomenclature of Cells and Diseases of the Blood and Blood-Forming Organs. Am J Clin Pathol 20:562–579
24. Tulliez M, Ricard MF, Jan F, Sultan C (1974) Preleukaemic abnormal myelopoiesis induced by chlorambucil: a case study. Scand J Haematol 13:179–183
25. Vismans JJ, Briet E, Meijer K, Ottolander GJ (1980) Azathioprine and subacute myelomonocytic leukemia. Acta Med Scand 207:315–319
26. Zittoun J, Zittoun R, Marquet J, Sultan C (1975) The three transcobalamins in myeloproliferative disorders and acute leukaemia. Br J Haematol 31:287–298
27. Zittoun J, Marquet J, Zittoun R (1975) The intracellular content of the three transcobalamins at various stages of normal and leukaemic myeloid cell development. Br J Haematol 31:299–310
28. Zittoun R (1976) Subacute and chronic myelomonocytic leukaemia: a distinct hamatological entity. Br J Haematol 32:1–7
29. Zittoun R, Bernadou A, Bilski-Pasquier G, Bousser J (1972) Les leucémies myélo-monocytaires subaigues. Etude de 27 cas et revue de la littérature. Sem Hop Paris 48:1943–1956

Haematology and Blood Transfusion Vol 27
Disorders of the Monocyte Macrophage System
Edited by F. Schmalzl, D. Huhn, H.E. Schaefer
© Springer-Verlag Berlin Heidelberg New York 1981

Lysosomal Granules in Leukemic Monocytes and Their Relation to Maturation Stages

D. Catovsky and M. O'Brien

1. Summary

Material from 11 patients with acute monocytic leukaemia (M5), including two with mono-blastic transformation of chronic granulocytic leukaemia, were studied by light and trans-mission electron microscopy (TEM) cytochemical techniques. Conventional TEM examination revealed typical ribosome–lamella complexes in two cases. Acid phosphatase (AP) and myeloperoxidase (MPO) at TEM level helped to demonstrate three types of ly-sosomal granules: 1) primary, small (0.05 – 0.2 µm) were AP positive and MPO negative; 2) secondary, medium size (0.2 – 0.4 µm) were MPO and AP positive; 3) tertiary, of variable size (0.05 – 0.6 µm) were both AP and MPO negative. Primary granules were the only ones seen in monoblasts, secondary granules appeared during the promonocyte stage and ter-tiary granules were seen in more mature monocytic cells. These studies have helped us in the characterisation of leukaemia cells and to understand the sequential development of ly-sosomal enzymes during monocytic differentiation.

2. Introduction

Previous ultrastructural studies of monocyte precursor cells have described the appearance of MPO-containing granules as an early maturation event [4]. The promonocyte has often been regarded as the first identifiable cell of the monocyte series in normal bone marrow. These observations are not adequate to explain the frequent finding in poorly differentiated monocytic leukaemia of a negative MPO reaction whilst the cytochemical reactions for non-specific esterase (NASA and ANAE) and acid phosphatase (AP) are characteristically positive [1, 3, 7].

A study aimed at characterizing the granules present in monoblasts and promonocytes by means of the AP and MPO reactions at the ultrastructural level was carried out by our group in six cases of M5 leukaemia [5]. That study showed that the first lysosomal granule to appear in monoblasts is smaller than previously recognised and can be defined cytochemi-cally as AP (+) and MPO (−) [5]. The present report extends those earlier observations and describes the various types of lysosomal granules which can be observed during monocyte differentiation.

3. Material and Methods

Bone marrow and/or peripheral blood of 11 patients suffering from monocytic leukaemia (M5) were studied. The 11 patients included two cases with monoblastic transformation of chronic granulocytic leukaemia. The cases were divided as follows: seven poorly differentiated (M5PD) and four with dif-

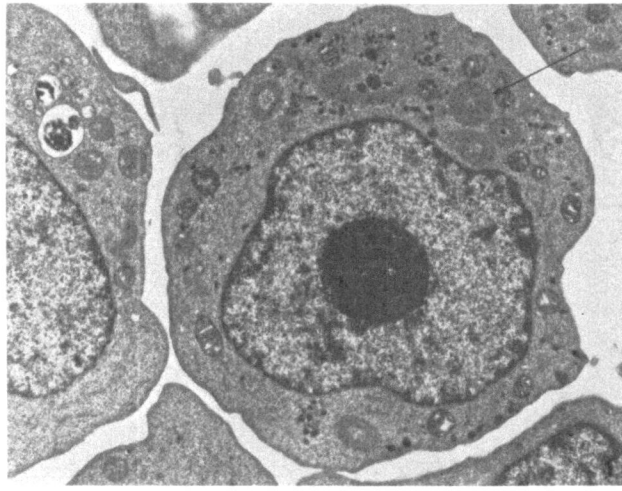

Fig. 1. TEM morphology of leukemic monocytes (lead citrate and uranyl acetate stain). Shown is a monoblast with numerous ribosome–lamella complexes *(arrow)* and small lysosomal granules in the cytoplasm. The latter were AP positive (illustrated in Figs. 3 and 4) and MPO negative (not shown). × 7000

ferentiation (M5D) according to the FAB classification [1]. The light microscopy techniques employed included morphological evaluation of Romanovsky-stained films (May-Grünwald Giemsa stain) and the cytochemical reactions for MPO, Sudan Black B, AP, naphthol-AS acetate esterase (NASA) and alpha naphthyl acetate esterase (ANAE) [3, 5]. Both esterases were tested with and without inhibition with NaF. Serum lysozyme was evaluated in seven cases along with the cytobacterial test [3] in peripheral blood and bone marrow samples.

Transmission electron microscopy (TEM) was used to examine morphology and for the evaluation and localization of MPO and AP activity according to techniques published elsewhere [5]. All the material processed for MPO and AP was viewed unstained.

4. Results

4.1 Light Microscopy

The cells of the seven cases of M5PD leukaemia were predominantly immature monoblasts and were positive in variable degree with the cytochemical reactions of AP, NASA and

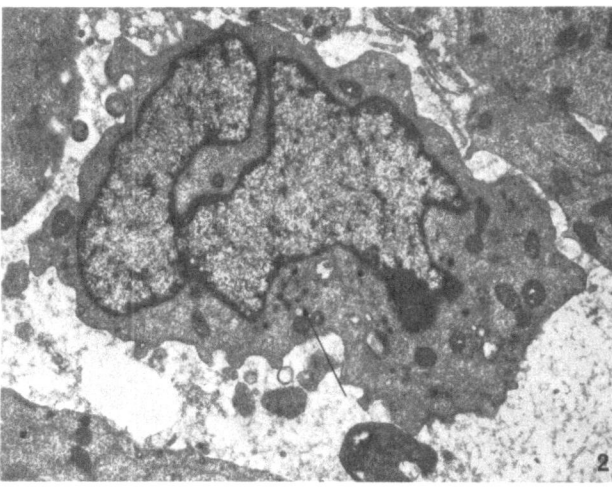

Fig. 2. Shown here is a promonocyte with a more irregular nucleus and more peripheral chromatin condensation than the cell in Fig 1. The cytoplasm contains small and medium size granules which are membrane bound and show the characteristic clear halo of promonocyte granules *(arrow)*. × 7000

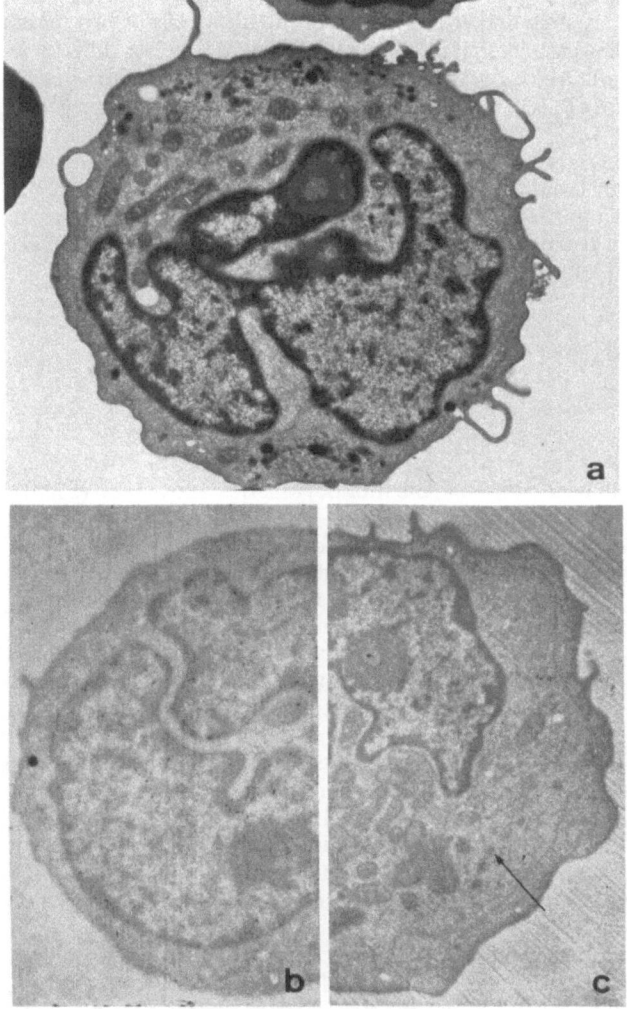

Fig. 3a–c. TEM from a case of well differentiated monocytic leukaemia. **a** (× 7500, lead and uranyl acetate stain) shows the typical cell morphology, with the cytoplasm containing numerous small and medium size granules (tertiary). These granules were AP negative, except for one granule **b** × 7500 and MPO negative **c** × 9800; the MPO unreactive granules are visible in this unstained section **c**, *arrow*

Table 1. Monocytic series – stages of maturation[a]

Granules (enzymes)	Monoblast I	Monoblast II	Promonocyte	Monocyte
Primary[b] (AP+)	+	++	+	±/+
Secondary[c] (MPO+, AP±)	–	–	+/++	±/+
Tertiary[d] (MPO–, AP–)	–	–	±	++

[a] AP, acid phosphatase; MPO, myeloperoxidase
[b] Small size (0.05–0.2 µm)
[c] Medium size (0.2–0.4 µm)
[d] Small & medium size (0.05–0.6 µm)

ANAE, the latter two inhibited by NaF. MPO was negative in all; the serum lysozyme levels were only moderately raised in one case and 20% and 50% of the blast cells of two others showed lysozyme activity with the cytobacterial method. One of the M5PD was difficult to classify on Romanovsky-stained films because the cells were undifferentiated with a deeply basophilic cytoplasm; ANAE and AP were positive, but NASA was only positive in 10% of them.

The cells of the four cases of M5D were predominantly promonocytes and monocytes. MPO activity was demonstrated in a minority (2%-5%) of leukaemic cells in all of them. AP was positive but less marked than in M5PD, and one case was negative; both NASA and ANAE were strongly positive in the four cases and the reaction was inhibited by NaF. Serum lysozyme was moderately to markedly raised in the two cases tested.

4.2 Transmission Electron Microscopy (TEM)

4.2.1 Morphology

Three stages of differentiation can be identified in the monocyte series: monoblast (Fig. 1), promonocyte (Fig. 2) and monocyte (leukaemic) (Fig. 3a). The former predominated in M5PD leukaemias and the latter two in the differentiated form, M5D.

Monoblasts have a very immature nucleus which is round or sometimes kidney shaped, a small nucleolus and little or no chromatin condensation. Bundles of microfibrils in the perinuclear region were seen in four cases. The cell membrane was regular with few projections. Numerous small membrane-bound granules (primary, Table 1) were scattered through the cytoplasm, often in the periphery [5]. The average number per cell was 12 but ranged from 1 to 145. In the case of M5PD which showed a deep basophilic cytoplasm in the Romanovsky-stained films, 80% of the monoblasts had the cytoplasm packed with ribosome lamella complexes (RLC) (Fig. 1). In another case with basophilic cytoplasm, 75% of the cells had 1–3 RLC which were smaller in size than in the previous case.

Promonocytes (Fig. 2) had an irregular nucleus, which was often kidney-shaped, a variable amount of peripheral chromatin condensation and small clumps of heterochromatin associated with a small nucleolus. The cytoplasmic granules were more prominent than in monoblasts and were generally larger (secondary, Table 1). Cytoplasmic projections were also conspicuous.

Leukaemic monocytes had a characteristic irregular nucleus and more condensed nuclear chromatin. The cytoplasm of these cells had more villous projections, microfibrils, vesicles of pinocytosis and long strands of endoplasmic reticulum. Granules were more numerous and of variable size (mainly tertiary, Table 1, see the following discussion).

4.2.2 Cytochemistry

The AP reaction was positive in the primary granules of monoblasts (Table 1; Fig. 4a and 4b) and, in some cells, in the Golgi membranes and endoplasmic reticulum as well. Often, more granules were observed with the AP reaction than suspected on the morphology preparations. The RLC of the two M5PD cases were AP negative. The monoblasts in two cases had a large number of small granules (100–145), the majority being AP positive. These are designated as Monoblasts II in Table 1 as they suggest a more advanced stage of maturation than the ones with few small granules (Monoblast I, Table 1). In promonocytes the AP reaction was positive in the medium size granules (secondary) and in some residual small ones (primary) (Fig. 4c). The number of smaller granules was considerably less than in monoblasts. In one of the cases with a predominance of promonocytes, no AP reaction was demonstrated by either TEM or light microscopy. In the monocytes, the number of AP-positive granules was less than in the less mature cells described previously (Fig. 3b).

170

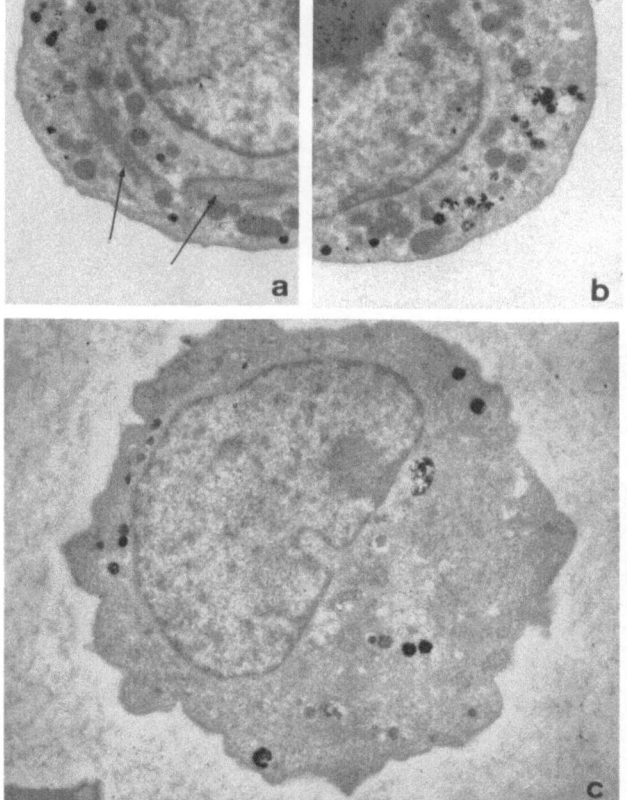

Fig. 4a–c. Acid phosphatase *(AP)* reaction, unstained sections. **a** and **b** (× 9000) show numerous small and medium size AP-positive granules (primary) in two monoblasts from the same case as Fig 1. It can be seen that two ribosome lamella complexes are unreactive (**a**, *arrow*). **c** shows the AP-positive reaction in a promonocyte. Note that the granules are slightly larger and less numerous. × 7250

The MPO reaction was negative (Fig. 5a) in all the cells with primary granules (AP positive, Table 1), but a positive reaction (Fig. 5b) was seen in granules of promonocytes (secondary, Table 1). In the more differentiated cells (monocytes) numerous granules visible by morphology (Fig. 3a) were seen to be MPO(−) (Fig. 3c) and AP(−) (Fig. 3b) and were classified as tertiary (Table 1).

5. Discussion

By combining observations with two cytochemical reactions at TEM level (AP and MPO) we were able to define three types of granules seen during monocytic maturation. These are summarised in Table 1. Primary granules are the smallest in size, are seen mainly in monoblasts (Figs. 1, 4a and 4b), are characteristically AP(t) and gradually decrease in number with maturation. Secondary granules are seen from the promonocyte stage onwards (Fig. 2), are of medium size and are MPO (+) Fig. 5b) and also have some AP activity (Fig. 4c), although less consistently than in the primary granules. They persist, but in smaller numbers, with maturation. In the more mature cells some of the earlier granules persist and there is a third generation of granules of variable size (Fig. 3a) which are both MPO (–) (Fig. 3c) and AP (–) (Fig. 3b).

Nichols and Bainton [4] have demonstrated the existence of what we have re-defined here as secondary and tertiary granules in normal and leukaemic cells. As discussed

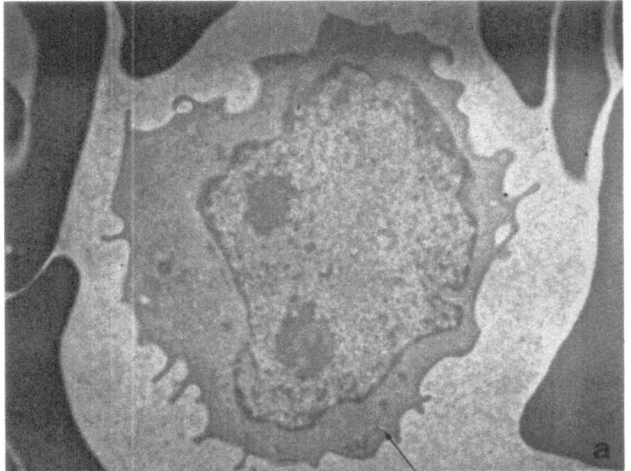

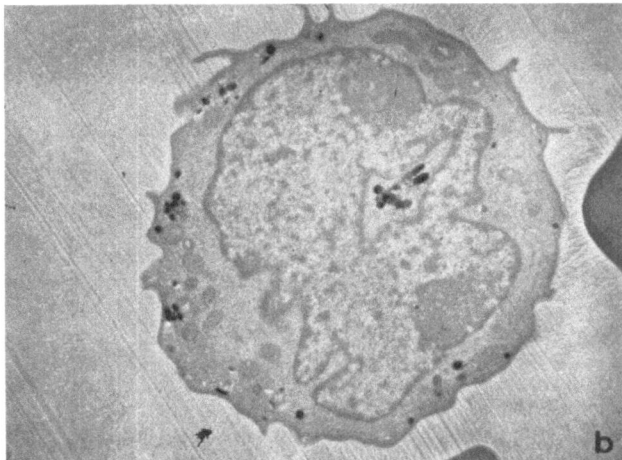

Figs. 5a, b. Myeloperoxidase (MPO) reaction, unstained sections (× 6600). **a** shows a less mature cell from the same patient with MPO negative granules *(arrow)*. In this case the majority of cells lacked MPO activity and the granules were classified as primary. **b** shows MPO-positive granules (secondary) in a promonocyte

elsewhere [5], the primary (AP-positive) granules have not been previously described. This is possibly due to two reasons: the fact that those cells are presumably very rare in normal bone marrow and the non-utilisation, in previous studies, of the AP reaction which appears to be a specific marker for those granules. When the techniques for the demonstration of non-specific esterase at the TEM level becomes more widely used, it might be possible to examine the question of the localisation of those enzymes in relation to the three types of lysosomal granules described here. The strong reaction seen with NASA and ANAE in M5PD cells suggests that they might be present, as AP, in the primary granules. Of interest, too, would be to find a suitable enzyme marker for the tertiary granules which appear to be AP and MPO negative.

The demonstration in the majority of cells of two cases of M5PD of typical RLC was of special interest. This structure, characteristic of hairy cell leukaemia (HCL) [6], has been reported in two other cases of monoblastic leukaemia by Brunning and Parkin [2] as well as in cells of lymphoid lineage. Considering the cases of monoblastic leukaemia, studied by us and by Brunning and Parkin, it would appear that in one-third (4 out of 13) of M5PD leukaemia cases we can expect to find RLC in a large proportion of blast cells. Although less common than in HCL [6], the incidence of cases with RLC is high enough to consider in future studies its possible diagnostic significance in this form of leukaemia.

6. Conclusion

Morphological and cytochemical studies by TEM have helped us to describe the sequential development of lysosomal granules during monocytic maturation. As this study was carried out exclusively on leukaemic monocytes, it is not proved that such a development necessarily reflects events during normal monocytopoiesis. Nevertheless, they have provided new interesting tools for the further characterisation and classification of monocytic leukaemia.

References

1. Bennett JM, Catovsky D, Daniel MT, Flandrin G, Galton DAG, Gralnick HR, Sultan C (1976) Proposal for the classification of the acute leukaemias. French-American-British (FAB) Cooperative Group. Br J Haematol 33:451
2. Brunning RD, Parkin J (1975) Ribosome-lamella complexes in neoplastic hematopoietic cells. Am J Pathol 79:565
3. Catovsky D, O'Brien M, Cherchi M (1978) Cytochemistry: an aid to the diagnosis and classification of the acute leukaemias. Recent Res Cancer Res 64:108
4. Nichols BA, Bainton DF (1973) Differentiation of human monocytes in bone marrow and blood. Sequential formation of two granule populations. Lab Invest 29:27
5. O'Brien M, Catovsky D, Costello C (1980) Ultrastructural cytochemistry of leukaemic cells: characterization of the early small granules of monoblasts. Br J Haematol 45:201
6. Rosner MC, Golomb HM (1980) Ribosome-lamella complex in hairy cell leukemia: ultrastructure and distribution. Lab Invest 42:236
7. Schmalzl F, Huhn D, Asamer H, Braunsteiner H (1975) Hairy cell leukemia ('leukemic reticuloendotheliosis'), reticulosarcoma, and monocytic leukemia. Acta Haematol (Basel) 53:257

Haematology and Blood Transfusion Vol 27
Disorders of the Monocyte Macrophage System
Edited by F. Schmalzl, D. Huhn, H.E. Schaefer
© Springer-Verlag Berlin Heidelberg New York 1981

Characteristics and Functions of Monocytes and Promonocytes in Monocytic Leukemia

R. van Furth, T.L. van Zwet, M.T. van den Barselaar and P.C.J. Leijh

Earlier studies have dealt with the morphological, cytochemical, and functional characteristics of human peripheral blood monocytes, human promonocytes, and monocytes of the bone marrow [5, 8], but the characteristics of mononuclear phagocytes in patients with monocytic leukemia are not well known. In the acute forms of leukemia the circulating cells are generally considered to be less mature than the monocytes in normal individuals as is indicated by the frequently used term "monoblast". The question remains, however, whether this view is correct and whether monocytes in the peripheral blood of patients with monocytic leukemia do divide more frequently than in normal individuals. The present contribution summarizes the findings in 27 patients with acute monocytic leukemia (AMoL) and chronic monocytic leukemia (CMoL); these findins will be published in detail elsewhere [3, 6].

1. Materials and Methods

Bone marrow was collected by sternal puncture and cell suspensions were prepared elsewhere [5]. Monocytes were obtained from heparinized venous blood, and monocyte suspensions were prepared by density centrifugation on a Ficoll-Isopaque gradient [2, 5].

1.1 Cytochemistry

Esterase activity was investigated according to Ornstein with α-naphthyl butyrate as substrate [1, 5, 9]. Peroxidase activity was determined according to Kaplow [5, 7].

1.2 Membrane Receptors

Fc receptors on promonocytes and monocytes were detected with IgG-coated sheep red cells; the IgG was isolated from mouse antiserum. C receptors were demonstrated with sheep red blood cell coated with IgM from rabbit antiserum and complement from mouse serum [5].

1.3 Functional Tests

The phagocytic activity of the mononuclear phagocytes originating from the bone marrow and peripheral blood was studied in 6-h cultures of cell monolayers on coverslips, with

Staphylococcus epidermidis serving as particle and 10% AB serum as opsonin [5]. The pinocytic activity was studied in similar preparations, with dextran sulphate (mol. wt. 500 000) as indicator substance [5]. The phagocytic activity of peripheral blood monocytes was also studied in cell suspensions under slow rotation with opsonized Staphylococcus aureus as particle according to the method described in detail elsewhere [4, 8]; the intracellular killing of bacteria by peripheral blood monocytes was studied in cell suspensions, also with S. aureus [4, 8].

1.4 ^{3}H-Thymidine Labeling

Bone marrow and blood monocyte cultures were incubated for 6 h in the presence of 0.1 µCi/ml ^{3}H-thymidine as described elsewhere [5]. Labeling of the nuclei of the promonocytes and monocytes was assessed in autoradiographs of these preparations [5].

2. Results

2.1. Peripheral Blood Monocytes

In this group of patients the percentage of monocytes in the peripheral blood ranged from 22 to 91, and the total number of peripheral blood monocytes from 6.0×10^9 to 1.2×10^{11} cell per liter.

The esterase activity of the monocytes lay within the normal range (95.7%–100%) in only 10 of the 27 patients; the others had lower values, and in 6 of these fewer than 10% of the monocytes were esterase positive.

The peroxidase activity of the monocytes was decreased in 75% of the patients, in about one-third of whom only 10% or less of the monocytes were peroxidase positive.

In about 85% of the patients, 80% or more of the monocytes had Fc receptors and complement receptors. The phagocytic activity of monocytes attached to a glass surface was moderately decreased in 6 out of 27 patients, whereas pinocytic activity varied widely: in a number of patients the monocytes were less active and in others more active than normal monocytes.

In the phagocytosis assay done in cell suspensions the rate of uptake of bacteria varied. In the majority of the patients the rate of phagocytosis was increased, but in 4 out of 13 patients it was lower than that shown by normal monocytes and in two patients the rate of ingestion was almost nil; in these two patients, however, the peripheral blood monocytes were morphologically very immature. The rate of intracellular killing by monocytes from patients with AMoL and CMoL also lay in the normal range, except in the two above-mentioned patients, whose monocytes did not ingest bacteria.

2.2 Bone Marrow Promonocytes and Monocytes

The esterase activity of promonocytes and monocytes varied, and in about a fourth of the patients the percentage of esterase-positive cells was lower than in normal bone marrow. In more than two-thirds of the patients the percentages of peroxidase-positive promonocytes and monocytes were lower than those occurring in the bone marrow of healthy individuals. Fc and C receptors were found in the majority of the promonocytes and monocytes, and only one-third of the patients showed values lying slightly below normal.

Phagocytosis of bacteria by promonocytes and monocytes was decreased in 35% of the cases. The pinocytic activity of these cells varied widely, as was also found for peripheral blood monocytes of these patients.

2.3 ^{3}H-Thymidine Labeling of Promonocytes and Monocytes

The labeling index of the promonocytes was high and ranged between 77%–98% except in one case (31.4%). However, the labeling index of the bone marrow monocytes was low: in four cases, 0%–0.5%; in six cases, between 1.0% and 2.0%, and in two cases, 3% and 3.6%. The labeling index of the peripheral blood monocytes showed similar values: in 14 of the 20 cases studied in this respect the index was between 0 and 0.2%, in four cases, between 1.0% and 2.0%, and in two cases, 4.0% and 5.5%.

3. Discussion

The present study has shown that the characteristics of promonocytes and monocytes of some patients with acute or chronic monocytic leukemia differ from those found in normal individuals. With respect to certain features of blood monocytes, i.e., peroxidase activity, esterase activity, Fcγ receptors, C3b receptors, phagocytosis, and intracellular killing, statistical analysis showed correlation only between the presence of Fcγ receptors and the ingestion of S. aureus (correlation coefficient: 0.95), between the phagocytosis and intracellular killing of S. aureus (correlation coefficient: 0.99), and between Fcγ receptor and intracellular killing of S. aureus (correlation coefficient: 0.90). In the majority of the patients the endocytic functions are in general not disturbed; phagocytosis and intracellular killing were severely impaired in only two patients who had very immature circulating monocytes.

The findings of the present study probably do not diverge greatly from those reported for another published series [10], but this is not certain, because in that report mention is only made of whether the cells were positive or negative for a given characteristic, whereas we determined the percentage of cells that were positive for each characteristic.

The labeling studies did not show an increased percentage of promonocytes in cell cycle, and there were even some cases in which the labeling index was lower than that found in normal individuals [5]. The labeling index of the monocytes of the bone marrow and blood was, with a few exceptions, also very low. From this very low labeling index of the circulating monocytes in acute and chronic monocytic leukemia it must be concluded that in fact no dividing cells circulate in the blood; the great majority are end cells.

Immature cells are called blast cells by hematologists on the basis of morphological characteristics of the nucleus and the cytoplasm. In this sense the term monoblast is also applied to cells in the peripheral blood of patients with monocytic leukemia. Cell biologists and others reserve the term blast cell for cells that divide – which in our opinion is correct – and use not only morphological but also other characteristics (i.e., DNA synthesis of cells, inhibitory effect of drugs on mitosis, and behavior shown by time-lapse cinematography) for the classification of a cell as a dividing cell. From the present study in which the morphological aspects and the proliferative activity of monocytes were investigated separately, it is evident that there is no correlation between the immature structure of the cells and the incorporation of ^{3}H-thymidine. It is therefore preferable to avoid using the term monoblast to denote circulating cells in monocytic leukemia, except when there is solid evidence that a large percentage of these circulating monocytes are dividing cells.

4. Summary

The characteristics of mononuclear phagocytes (promonocytes and bone-marrow and blood monocytes) of 27 patients with acute or chronic monocytic leukemia were studied. The percentages of promonocytes with peroxidase and esterase activity and of cells with Fcγ and C3b receptors showed some divergence from those found in normal individuals. The phagocytic activity of monocytes and their ability to kill ingested Staphylococcus aureus were not decreased in this group of patients. The ^{3}H-thymidine-labeling index of the promonocytes did not differ greatly from normal values, which indicates that the percentage

177

of dividing promonocytes had not increased. The very low labeling index of the circulating monocytes means that these cells do not divide and therefore cannot be blast cells or monoblasts.

References

1. Ansley H, Ornstein L (1970) Enzyme histochemistry and differential white cell counts on the Technicon Hemalog. J Adv Automated Anal 1:5
2. Bøyum A (1968) Separation of leucocytes from blood and bone marrow. Scand J Clin Lab Invest [suppl] 27:29
3. Furth van R, Zwet van TL (to be published) Cytochemical, functional and proliferative characteristics of monocytes and promonocytes from patients with monocytic leukemia
4. Furth van R, Zwet van TL, Leijh PCJ (1978) In vitro determination of phagocytosis and intracellular killing by polymorphonuclear and mononuclear phagocytes. In: Weir DM (ed) The handbook of experimental immunology, 3rd edn. Blackwell, Oxford, p 32.1.
5. Furth van R, Raeburn JA, Zwet van TL (1979) Characteristics of human mononuclear phagocytes. Blood 54:485
6. Furth van R, Leijh PCJ, Zwet van TL, Barselaar van der MT (to be published) Phagocytosis and intracellular killing by peripheral blood monocytic leukaemia.
7. Kaplow LS (1965) Simplified myeloperoxidase stain using benzidine dihydrochloride. Blood 26:215
8. Leijh PCJ, Barselaar van den MT, Furth van R (1981) Kinetics of phagocytosis and intracellular killing of Staphylococcus aureus and Escherichia coli by human monocytes. Scand J Immunol 13:159
9. Ornstein L, Ansley H, Saunders A (1976) Improving manual differential white cell counts with cytochemistry. Blood Cells 2:557
10. Straus DJ, Mertelsmann R, Koziner B, Mackenzie S, Harven de E, Arlin ZA, Kempin S, Broxmeyer H, Moore MAS, Phil D, Menedez-Botet CJ, Gee TS, Clarkson BD (1980) The acute monocytic leukemias: Multidisciplinary studies in 45 patients. Medicine (Baltimore) 59:409

Haematology and Blood Transfusion Vol 27
Disorders of the Monocyte Macrophage System
Edited by F. Schmalzl, D. Huhn, H.E. Schaefer
© Springer-Verlag Berlin Heidelberg New York 1981

Chronic Monocytic Leukemias

G. Meuret

It is impossible to derive clear definitions of chronic monocytic leukemias from the publications available today [1, 3–6, 8, 17, 19, 20]. The descriptions are not confined to pure forms of chronic monocytic leukemias. They do comprise, however, distinct types of myelomonocytic leukemias whose symptomatology is dominated by hyperproliferation of monocytopoiesis. Pure chronic monocytic leukemias are very rare. To our knowledge detailed and complete analyses of such cases are still lacking. Therefore, it seems justified to concentrate on findings obtained by systematic examination of two representative patients [10, 14].

One of these patients suffered from a disorder which corresponded to the general concept of pure chronic monocytic leukemia, which is characterized by the production of excessive amounts of monocytes in bone marrow. The investigations of the monocytopoietic cells failed to detect morphological or functional defects. The symptomatology of the second patient changed during the course of the disease. During the initial stage lymphadenopathy and splenomegaly predominated due to the local accumulation of atypical monocytopoietic cells. In the later stage atypical cells similar to those observed in peripheral lymphatic organs emerged in bone marrow and blood. Thus, it seemed probable that this disorder was caused by malignant transformation of mononuclear phagocytes (MNP) in lymphatic tissue.

1. Pure Chronic Monocytic Leukemia Probably Arising from Bone Marrow Without Cellular Anomalies

A 77-year-old patient was admitted to the hospital because of repeated infections. He demonstrated moderately enlarged lymphnodes and slight hepatomegaly. Histologic evaluation of biopsies detected increased amounts of histiocytes in these organs. Lysozyme level in serum was increased. The hematologic status of the patient was characterized by moderate anemia due to hemolysis and inadequate erythropoietic compensation, by moderate intermittent thrombopenia, increased amounts of plasma cells in bone marrow, severe monocytopoietic hyperproliferation, and monocytosis (Table 1). The serum protein level was increased to 9 g/100 ml due to monoclonal gammapathia of IgG type (Fig. 1).

1.1 MNP in Bone Marrow and Blood

Bone marrow smears were hypercellular due to tightly packed MNPs whose relative number exceeded the normal average by a factor of 15 (Table 1). The frequency distribution of the different types of monocytopoietic cells showed a shift in favor of the immature forms as follows: MNP with round–oval nuclei (Type I & II), MNP with slightly folded nuclei (Type

Table 1. Myelogram a) and blood monocytes b) of a patient with pure chronic monocytic leukemia

a)

Myelogram	Patient (%)	Normal mean (%) [13]
Erythroblasts	14.3	28.0
Granulopoietic cells	33.7	50.6
Pooled MNP	42.1	2.9
Type I & II MNP	21.8	1.0
Type III MNP	13.8	1.5
Type IV MNP	6.5	0.4
Lymphocytes	3.8	8.0
Plasma cells	5.7	1.4

b)

Blood monocytes	Patient (cells per μl blood)	Normal mean [11]
Total	8400	370
With round oval nuclei	1850	30
With slightly folded nuclei	2100	120
With distinctly folded nuclei	4450	220

III), and MNP with distinctly folded nuclei (Type IV) constituted 52%, 33% and 15%, respectively (normal = 34%: 52%: 14%[12, 13]). MNPs in the patient's bone marrow were morphologically identical to those of healthy individuals. The chromosomal pattern also was normal.

Blood monocyte counts scattered at around 8000/μl, thus exceeding the normal level by a factor of 20. Similar to MNP in bone marrow, blood monocytes demonstrated a "shift to the left" with an increased fraction of monocytes with round–oval nuclei: The percentages of monocytes with round–oval nuclei, slightly folded nuclei, and distinctly folded nuclei were 22%, 25%, and 53%, respectively (normal = 5% : 32% : 59% [11]).

Blood monocytes demonstrated normal morphology in both light and electron micro-

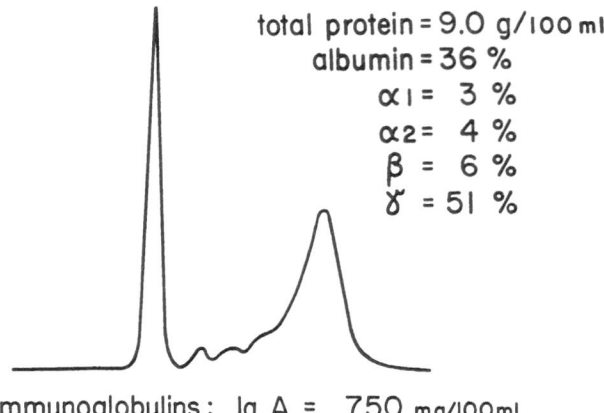

total protein = 9.0 g/100 ml
albumin = 36 %
α1 = 3 %
α2 = 4 %
β = 6 %
γ = 51 %

Immunoglobulins : Ig A = 750 mg/100ml
Ig M = 270 mg/100ml
Ig G = 5 225 mg/100ml

Fig. 1. Electrophoretic pattern of the first patient showing an IgG peak

Table 2. Activity indexes of cytochemical reactions in blood leukocytes in a patient with pure monocytic leukemia. Data in parentheses represent means of normal individuals [13]

Cytochemical reaction	Monocytes: nucleus morphology				
	round oval	slightly folded	distinctly folded	Neutrophils	Lymphocytes
NaF-sensitive naphthol-	50	72	151	0	0
AS-D-acetate esterase	(73)	(134)	(141)	(0)	(0)
Peroxidase	63	4	0	253	0
	(114)	(68)	(58)	–	–
Sudan black B	108	48	12	259	0
PAS	40	7	4	296	36
Naphthol-AS-D-chloro-	171	80	11	153	0
acetate esterase	(128)	(53)	(26)	–	–
Acid phosphatase	3	75	84	0	39
Neutrophil alkaline phosphatase	0	0	0	132	0

scopy and normal phagocytosis. Transformation into macrophages was undisturbed. Normal cell maturation was observed in vivo by the emergence of morphologically normal macrophages in pleural exudate during pneumonia. In culture, cell maturation was investigated by time-lapse microcinematography. Immature monocytes could be identified by round–oval nuclei and sluggish movements of nuclei and cytoplasma. With increasing differentiation time motility of the cells increased, and the round nuclei gradually transformed into folded ones.

The pattern of cytochemical reactions did not show relevant deviations from normal findings (Table 2). There was an age-dependent change in the reaction intensities similar to the one observed in monocytes of healthy individuals. Younger monocytes were characterized by round–oval nuclei, high reaction intensities of naphthol-AS-D-chloroacetate esterase, peroxidase, and low activity of NaF-sensitive naphthol-AS-D-acetateesterase. Mature monocytes showed distinctly folded nuclei, high activity of NaF-sensitive naphthol-AS-D-acetate esterase associated with low reaction intensity of naphthol-AS-D-chloroacetate esterase and peroxydase.

1.2 Cell Kinetics of MNP

Cell kinetic studies using ^{3}H-thymidine (^{3}H-TDR) pulse labeling confirmed the rise in immature monocyte influx entering the blood. This was reflected by the following tho phenomena (Fig. 2): 1) ^{3}H-TDR labeling indexes (^{3}H-TDR L.I.) of circulating monocytes were increased to 8% immediately (i.e., 10 min) after injection of the marker as compared to only 0.4% in normal individuals [11], and 2) during the initial 30 h of observation the ^{3}H-TDR L.I. of the immature monocyte population with round–oval nuclei increased more steeply than that of the mature forms with folded nuclei. During the later phase of observation the ^{3}H-TDR L.I. of immature monocytes formed a plateau, whereas those of mature monocytes continued to increase. These findings were interpreted by assuming intravascular maturation causing transformation of labeled monocytes with round–oval nuclei into those with folded nuclei during the observation period. This assumption was confirmed by simultaneous determination of mean silver grain counts per monocyte in autoradiographs. During the early phase of the examination grain counts of the two cell types markedly differed. Afterwards they gradually approached the same level (Fig. 3).

Direct information on intravascular behaviour of monocytes was obtained by analyzing in vitro labeled monocytes following autotransfusion. In this study ^{3}H-diisopropylfluorophosphate (^{3}H-DFP) was used as cell marker. Labeling indexes of immature monocytes with

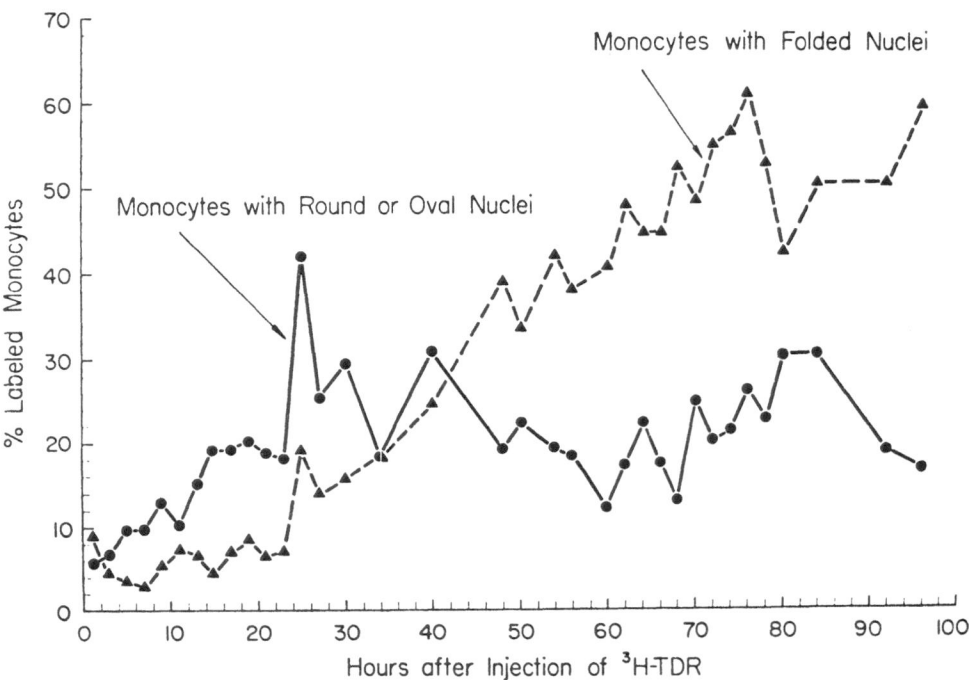

Fig. 2. Time course of ^{3}H-thymidine labeling indexes (^{3}H-TDR L.I.) of blood monocytes following ^{3}H-TDR pulse labeling

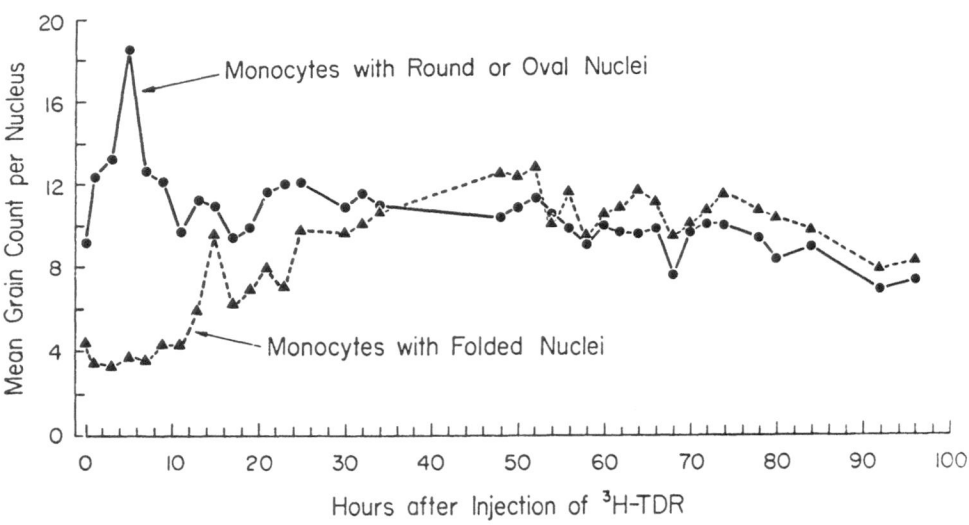

Fig. 3. Time course of mean silver grain counts per monocyte in autoradiographs determined during the ^{3}H-TDR pulse labeling study illustrated in Fig. 1

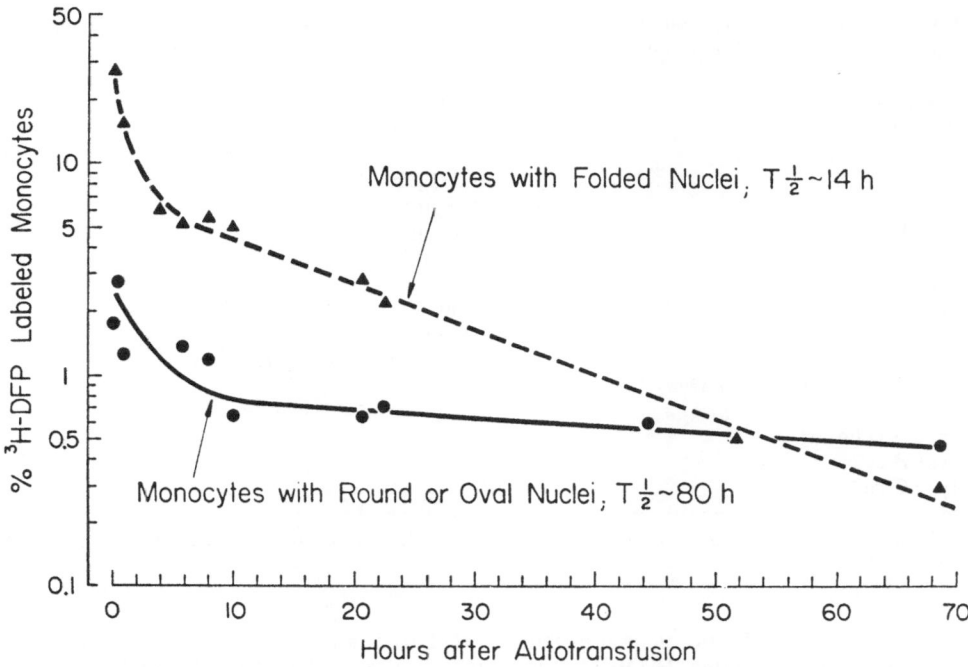

Fig. 4. Disappearance of monocytes labeled in vitro by ^{3}H-di-isoprophylfluorophosphate (^{3}H-DFP) after autotransfusion

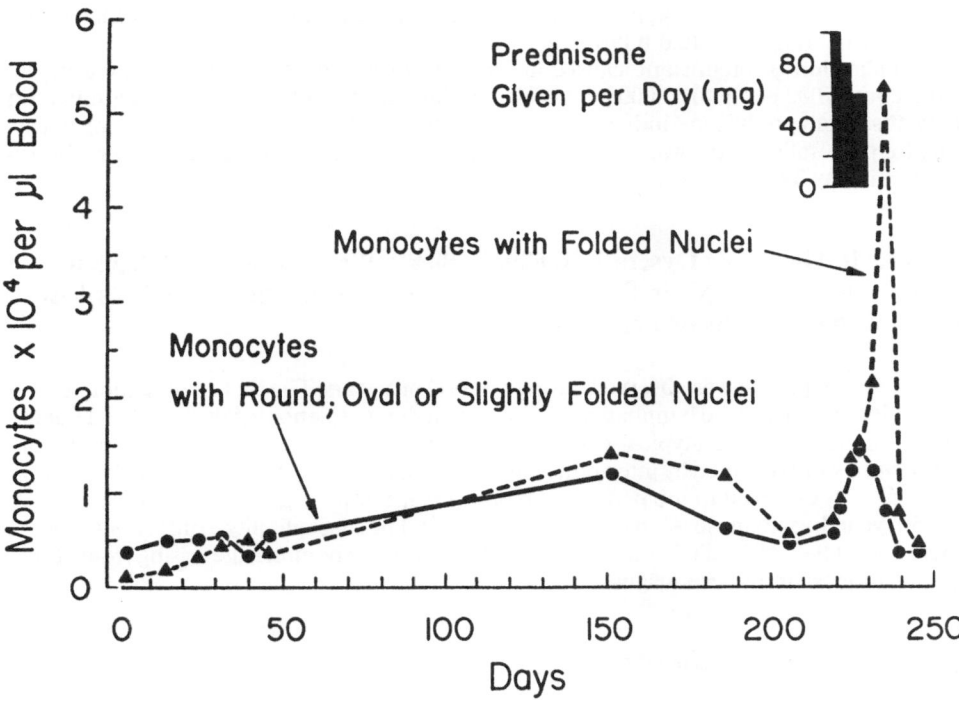

Fig. 5. Time course of monocyte blood counts during the observation period. Application of prednisone caused a marked increase in circulating monocytes

Table 3. Atypical cells in blood of the second patient whose disease seemed to arise from malignant transformation of MNP in peripheral lymphatic organs

Frequency	200–700/μl
Diameter	12– 26 μm
Area of nuclei:	
Range	40–330 μm^2
Type A cells	< 100 μm^2
Type B cells	100–150 μm^2
Type C cells	> 150 μm^2
DNA content[a]	
Type A cells	diploid
Type B cells	intereuploid
Type C cells	tetraploid or hypertetraploid
^{3}H-TDR labeling index (in vitro)	
Pooled cells	41%
Type A cells	15%
Type B cells	40%
Type C cells	80%

[a] Cytophotometric measurements of Feulgen-stained nuclei. Results refer to the majority of each cell type

round–oval nuclei did not decrease significantly between the 10th and 70th h after auto-transfusion (Fig. 4), while ^{3}H-DFP L.I. of mature monocytes with folded nuclei decreased with a half-time of about 14 h. This observation indicates that immature monocytes are hardly capable of leaving the vessels. They mature to monocytes with folded nuclei, acquiring emigration potential. The intravascular maturation may contribute to a slight prolongation of the half-disappearance time of monocytes with folded nuclei which lasted 14 h (normal range 4.5–10.0 h [9]).

Application of prednisone caused an increase of monocytes with segmented nuclei from about 7000 per μl to 55 000 per μl (Fig. 5). This intravascular monocyte accumulation may arise from prednisone-induced reduction of the cells' migration potential being necessary for penetration of the walls of the vessels. A similar effect of hydrocortisone was observed in mice by Thomson and van Furth [21].

2. Chronic Monocytic Dyscrasia Which Probably Was Caused by Malignant Transformation of MNP in Peripheral Lymphatic Organs and Which Was Associated with the Production of Atypical Cells

A 57-year-old patient was observed during a period of 5 years. The first examination revealed moderate generalized lymphadenopathy and moderate splenomegaly. Biopsies detected large amounts of large atypical cells in the peripheral lymphatic organs. Several partial remissions were obtained by intermittent chemotherapy. During the 4th year of observation atypical cells emerged in the peripheral blood. Their numbers gradually increased during the 5th year. Lysozyme levels in serum were normal. The susceptibility to infections was not increased. The patient died at the end of the 5th year from hemorrhage arising from atopic gastric mucosa in the esophagus.

2.1 Characteristics of Atypical Cells

The atypical cells were analyzed during the 5th year of observation. At this time the hematogram was characterized by moderate anemia, moderate thrombopenia and normal counts

Table 4. Activity indexes of cytochemical reactions to the blood leukocytes of the second patient

Reaction	Atypical cells	Monocytes	Neutrophils
Peroxydase	292	–	285
Naphthol-AS-D-chloroacetate esterase	287	63	251
Naphthol-AS-acetate esterase	243	251	150
PAS	52	96	484

of neutrophils, monocytes and lymphocytes. In bone marrow smears the relative number of atypical cells amounted to about 7%. During the last month of the patient's life blood counts of atypical cells rapidly increased from 200 to 700 per µl. The atypical cells showed extreme variations in size. Their diameters ranged from 12 to 26 µm (Table 3). The smallest cells (Type A, Table 3) demonstrated morphological characteristics of lymphoid cells. The majority of these small cells had diploid DNA content; their in vitro ^{3}H-TDR L.I. was 15%. Cells of intermediate size (Type B) demonstrated intereuploid DNA content and a ^{3}H-TDR L.I. of 40%. DNA content of the largest cells was either tetraploid or hypertetraploid. Their ^{3}H-TDR L.I. reached 80%. Most atypical cells demonstrated round- or oval-shaped nuclei, loose meshwork of chromatin with nucleoli, and grayish cytoplasma which sometimes contained azurophilic granules and vacuoles.

Cytochemical reactions of atypical cells were similar to those of relatively immature monocytes. Compared with normal mature monocytes, the atypical cells demonstrated higher reaction intensities of naphthol-AS-D-chloroacetate esterase and relatively low reaction intensities of non-specific esterase (naphthol-AS-acetate esterase) (Table 4).

2.2 Kinetics of Atypical Cells

The results of a ^{3}H-TDR pulse labeling study is illustrated in Fig. 6. The 1-h postinjection ^{3}H-TDR L.I. indicated that 44% of the atypical cells in circulation were in DNA synthesis phase. The time course of ^{3}H-TDR L.I. was characterized by two consecutive waves of labeled cell influx into the blood. During the first influx wave, which lasted about 30 h, the cells' mean grain counts fell from approximately 50 to 25 grains per cell. Thus, the first influx wave is indicative of one mitotic cell cycle of the precursor cells. The cell cycle time of 30 h is similar to that of proliferating monocytopoietic precursors in normal individuals [7, 12].

Blood monocytes as well as the atypical cells of the patient accumulated ^{3}H-diisopropyl-fluorphosphate similar to monocytes of healthy subjects [9]. This offered the opportunity to label both cell types in vitro and to study their intravascular fate following autotransfusion. Both labeled monocytes and atypical cell disappeared from the circulating blood in an exponential fashion. The half-disappearance time of both cell types approximated 13 h and thus moderately exceeded the range of normal monocytes (Table 5) [9].

3. Discussion

Two different types of chronic leukemic disorders of monocytopoiesis were analyzed. The first involved a patient who suffered from a rather pure chronic and idiopathic hyperproliferation of monocytopoiesis without detectable cell anomalites. The symptomatology of patients with this disease may comprise fever of unknown origin, increased susceptibility to infections, moderate splenomegaly, and lymphadenopathy due to macrophage accumulation within these organs. It has been speculated that fever may arise from release of pyrogenic products by monocytes and macrophages [2]. Final transformation into acute leukemia occurred in some of these cases [19]. Simultaneous hyperproliferation of plasma cells and monoclonal gammopathia, observed in two patients of ours and reported in the literature

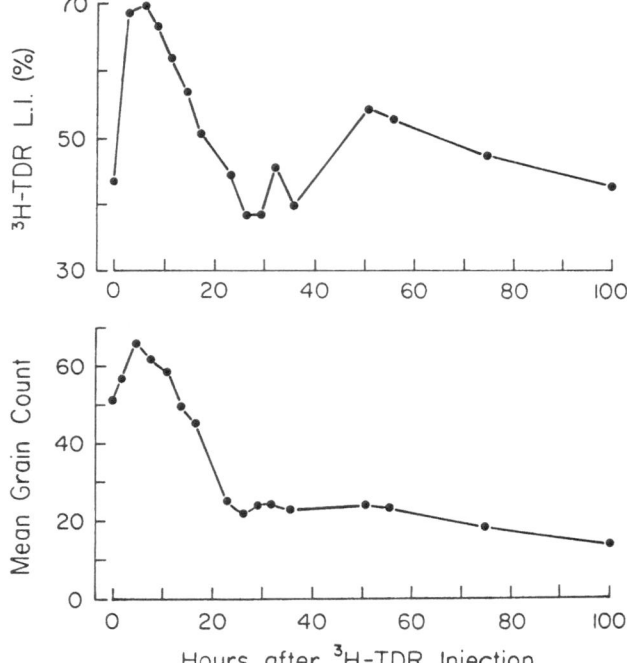

Fig. 6. Time course of ^{3}H-TDR L. I. and mean grain counts of atypical monocytopoietic cells following ^{3}H-TDR pulse labeling

[18], seemed to confirm the assertion of Osserman [16] of an interrelationship between monocytic and plasmacatic dyscracias. The absence of any detectable cell alteration sometimes renders the decision difficult whether the disease presents a leukemia, a preleukemic state, or a benign monocytosis. Cell kinetics of monocytopoietic cells and of monocytes were in accord with that of severe reactive monocytosis [11, 15]. Both states are characterized by a rise in monocyte production and by an increased influx of immature monocytopoietic cells into the blood. The immature cells underwent differentiation to mature monocytes within circulation before leaving the vessels.

The symptomatology of the second patient was different. During the initial 3 years of observation he demonstrated pronounced lymphadenopathy and splenomegaly due to the accumulation of atypical cells. During the 4th year of follow-up increasing amounts of atypical cells emerged in blood and bone marrow. The atypical cells in bone marrow, blood, and peripheral lymphatic organs were morphologically identical. According to their pattern of cytochemical reactions and their cell kinetic characteristics the atypical cells were identified as descendants of monocytopoiesis. In contrast to the leukemic cells of the first case, the leukemic cells of the latter were unable to produce measurable amounts of lysozyme, were not capable of differentiation into morphologically normal macrophages, demonstrated

Table 5. Kinetics of atypical cells and monocytes in the second patient's blood

	Atypical cells	Monocytes	Monocytes in normal individuals (mean) [9]
Count per μl blood	465	240	260
Total intravascular pool (cells × 10[6]/kg)	53	29	81
Half-disappearance time (hours)	12.8	13.0	8.4
Turnover rate (cells/kg × hour)	2.9	1.5	7

enhanced proliferation activity, and gave rise to a remarkably high fraction of cells with hypertetraploid DNA content.

During the first years lymphadenopathy due to accumulation of the atypical cells was the predominant feature of the disease. Therefore, it is probable that it arose from malignant transformation of MNP in the peripheral lymphatic organs. During the later stages the disease expanded into bone marrow and became leukemic. Then, proliferation of atypical monocytopoietic cells in bone marrow coexisted with normal monocytopoiesis. Transitional cell forms between the two cell lines were not observed. Therefore, this disease may be considered as a chronic leukemic monocytic disorder of the peripheral lymphatic tissue. The suggestion that the leukemic cells originated in bone marrow in one disorder and in peripheral lymphatic tissue in the other supports the hypothesis that stem cells of the MNP system, at least under pathologic conditions, may not reside exclusively in bone marrow.

References

1. Beattie JM, Seal RME, Crowther KV (1951) Chronic monocytic leukemia. Q J Med 20:131
2. Bodel P, Atkins E (1967) Release of endogenous pyrogen by human monocytes. N Engl J Med 276:1002
3. Broun OG Jr (1969) Chronic erythromonocytic leukemia. Am J Med 47:785
4. Kass L, Schnitzer B (1973) Monocytes, monocytosis and monocytic leukemia. Thomas, Springfield
5. Klein UE, Ude P (1975) Monozytenleukämien mit ungewöhnlichem Erkrankungsablauf. Med Klin 70:613
6. Labedzki L, Grips KH (1974) Chronische Monozytenleukämie. Dtsch Med Wochenschr 99:690
7. Meuret G (1974) Human monocytopoiesis. Exp Hematol 2:238
8. Meuret G (1978) Maligne Erkrankungen des Monozyten-Makrophagen-Systems. In: Queisser W (ed) Das Knochenmark, Thieme, Stuttgart, p 618
9. Meuret G, Hoffmann G (1973) Monocyte kinetic studies in normal and disease states. Br J Haematol 24:275
10. Meuret G, Südhoff A (1972) Zytochemische und zellkinetische Untersuchungen bei einer chronischen Erkrankung des Monozytensystems. Blut 24:226
11. Meuret G, Djawari D, Berlet R, Hoffmann G (1971) Kinetics, cytochemistry and DNA-synthesis of blood monocytes in man. In: Di Luzio (ed) The reticuloendothelial system and immune phenomena. Plenum, New York, p 33
12. Meuret G, Bammert J, Hoffmann G (1974) Kinetics of human monocytopoiesis. Blood 44:801
13. Meuret G, Batara E, Fürste HO (1975) Monocytopoiesis in normal man: pool size, proliferation activity and DNA-synthesis time of promonocytes. Acta Haematol (Basel) 54:261
14. Meuret G, Bundschu-Lay A, Senn HJ, Huhn D (1974) Funtional characteristics of chronic monocytic leukemia. Acta Haematol (Basel) 52:95
15. Meuret G, Detel U, Kilz HP, Senn HJ, van Lessen H (1975) Human monocytopoiesis in acute and chronic inflammation. Acta Haematol (Basel) 54:328
16. Ossermann EF (1969) Clinical and biochemical studies of plasmacytic and monocytic dyscrasias and their interrelationships. Trans Stud Coll Physicians Phila 36:135
17. Pearson HA, Diamond LK (1958) Chronic monocytic leukemia in childhood. J Pediatr 53:259
18. Poulik MD, Berman L, Prasad AS (1969) "Myeloma protein" in a patient with monocytic leukemia. Blood 33:746
19. Pretlow TG (1969) Chronic monocytic dyscrasia culminating in acute leukemia. Am J Med 46:130
20. Sinn CM, Dick RW (1956) Monocytic leukemia. Am J Med 20:588
21. Thompson J, van Furth R (1970) The effect of glucocorticoids on the kinetics of mononuclear phagocytes. J Exp Med 131:429

Haematology and Blood Transfusion Vol 27
Disorders of the Monocyte Macrophage System
Edited by F. Schmalzl, D. Huhn, H.E. Schaefer
© Springer-Verlag Berlin Heidelberg New York 1981

Therapy of Acute Monoblastic Leukemia

M. Weil, C. Jacquillat and G. Tobelem

By using ultrustructural and cytochemical techniques, monocytic precursors and granulocytic precursors can now be identified, and it is possible to isolate "pure acute monoblastic leukemia" [1]. In this study based on a retrospective study of 88 cases, we report the distinctive clinical and biologic features of acute monoblastic leukemia (AMol), which represents about 3.5% of all acute leukemias [2]. Described as a very acute leukemia in 1969 [3], AMol always has a very poor prognosis, despite progress in chemotherapy.

1. Materials and Methods

Eighty-eight cases of AMol have been studied. All were patients admitted to Hospital Saint-Louis between 1970 and 1978 and previously untreated. The diagnosis was based on examination of blood and bone marrow films according to the description of pure acute monocytic leukemia of the Fab classification, split into poorly differentiated (type M5, I) and well differentiated (type M5, II) groups [1], and confirmed by cytochemical methods: peroxidase- and monocyte-specific Naphthol AS-D acetate esterase activity as previously described [1, 4, 5]. Cytogenetic examinations were performed in eight patients.

Before treatment was started, blood and urine lysozyme, serum potassium, blood urea nitrogen, and creatinine were measured routinely. Hemostasis was investigated before treatment by means of the prothrombin time, activated partial thromboplastin time, and determination of factor V, fibrinogen, and fibrinogen degradation products.

Eighty-four patients were treated: prednisone, 6-mercaptopurine, and hydroxyurea were used in three cases; anthracyclines, in 81 cases; daunorubicin, in 18 cases; rubidazone, in 55 cases; the combination of rubidazone, cytosine arabinoside (Ara-C) and VP 16, in four cases; and the combination of rubidazone, Ara-C, and vincristine in four cases. After 1976, when hematologic remission was obtained, central nervous system (CNS) prophylaxis was performed in 41 patients. In the 55 patients treated with rubidazone, monthly reinductions of AMol protocol were given as described in Fig. 1. Statistical methods and significance levels are based on the log rank test [6, 7].

2. Results

2.1 Clinical Findings

The two sexes were involved with the same frequency: 42 females and 42 males. Two peaks were observed, one under 10 years and the second after 40 years of age. Of the 25 children under 10 years, 19 were less than 1 year old. The presence of tumor masses and extramedullary infiltration was frequent, especially of the skin and gums. The frequency of these different localizations was the same for both sexes and all ages.

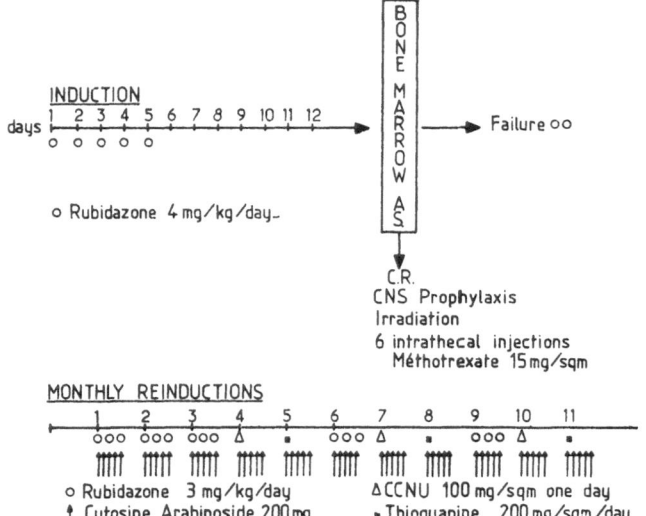

Fig. 1. Induction with rubidazone and reinduction using drug combinations in 55 patients

2.2 Hematologic Findings

The white blood cell count was, in most cases, over 10 000/mm^3, the median value for all the patients being 60 000/cu mm^3. Blast cells were usually, but not invariably, present in the peripheral blood. Hemoglobin level and platelet count were variable. In 67 patients, subclassification (M5, I and M5, II) could be used. Fifty-eight patients were M5, I, and 9 were M5, II.

2.3 Biologic Findings

2.3.1 Blood and Urine Lysozyme

Blood lysozyme was evaluated before treatment in 39 patients and urine lysozyme in 32 patients. Blood lysozyme was increased in 32 patients (85%) and over 100 µg/ml in 20 patients (normal values under 10 µg/ml). Urine lysozyme was increased in 19 patients (60%) and over 30 µg/ml in 18 cases (normal values under 5 µg/ml). Increase of blood and urine lysozyme was significantly correlated with hyperleukocytosis ($P < 0.0001$ for blood lysozyme and $P < 0.05$ for urine lysozyme).

2.3.2 Renal Failure

The findings of blood urea nitrogen over 0.70 g/liter and blood creatinine over 15 mg/liter before treatment has been considered as indicative of renal failure. Blood urea nitrogen and creatinine was measured before treatment in 69 patients, 28 (40.5%) of whom had renal failure by these criteria. The presence of renal failure was significantly correlated with hyperleukocytosis ($P < 0.05$). Renal failure and increased blood or urine lysozyme were also significantly correlated (>0.05 for blood lysozyme and $P< 0.03$ for urine lysozyme).

2.3.3 Kalemia

Sixty-nine patients had serum potassium measured before treatment. Fourteen cases (20.2%) had hypokalemia under 3 mEq/liter, four of them under 2 mEq/liter. Hyperkalemia has never been observed. Hypokalemia was significantly correlated with hyperleukocytosis ($P < 0.01$), with increased blood lysozyme ($P < 0.02$), and with increased urine lysozyme ($P < 0.01$).

2.3.4 Hemostasis

Sixty-one patients had investigations of hemostasis before treatment: 16 patients had findings consistent with the occurrence of disseminated intravascular coagulation (DIC) (factor V under 50%, fibrinogen under 2 g/liter, and increase of fibrinogen degradation products over 40 µg/ml). The presence of DIC before treatment was significantly correlated with hyperleukocytosis ($P < 0.001$). Only 6 of the 16 patients with DIC had clinical bleeding, and three of them received heparin and platelet transfusion.

2.3.5 Cytogenetic

In five out of six patients with the poorly differentiated type (M5, I) an abnormality of the long arm of chromosome 11 was observed by Dr. Berger. No abnormality was detected in two patients of the well-differentiated type (M5, II).

2.3.6 Induction Treatment and Study of Complete Remission

Of the 88 patients, four died just after admission before receiving chemotherapy. These were all men over 50 years old, three of them with white cell counts above 140 000/mm^3. Thus, 84 patients received induction chemotherapy as described in Sect. 1. Fifty-eight patients (72%) achieved complete remission. Thirteen patients (16%) failed, and 13 (16%) died in aplasia.

Of the 18 patients treated with daunorubicin, 9 (50%) achieved complete remission, while three failed and six died during induction. Of the 55 patients treated with rubidazone, 42 (76%) had complete remission, while eight failed and five died during induction. Seven of the eight patients treated by combinations including rubidazone achieved complete remission. None of the three patients treated without anthracyclines achieved a remission and they all died during induction.

The median duration of complete remission (CR) in the 58 patients was 195 days. Figure 2 shows the actuarial duration of CR for three groups of treated patients: the median CR for nine patients treated by dounorubicin was 65 days; the median CR for the 10 patients treated with rubidazone was 167 days; and for the 32 patients treated with rubidazone who received central nervous system prophylaxis, it was 254 days. The differences in median CR for the three groups were not statistically significant.

The duration of CR was not significantly influenced by the following parameters: age, sex, and tumoral syndrome. It was influenced by WBC, and the patient with an initial WBC count under 20 000/mm^3 had a significantly longer CR.

2.3.7 Study of Relapses

Of the 58 patients who achieved complete remission 35 had relapsed, three had died in complete remission (unknown cause), and 12 were still alive in complete remission. Among the 35 initial relapses, 21 involved the bone marrow, six the CNS, and five the skin. Three patients had a first relapse involving bone marrow and skin; one of these three patients also had testicular infiltration and meningeal leukemia. This last patient had had meningeal involvement at the time of initial diagnosis. None of the six patients who had a meningeal relapse were among the 32 patients who received central nervous system irradiation. After a first ex-

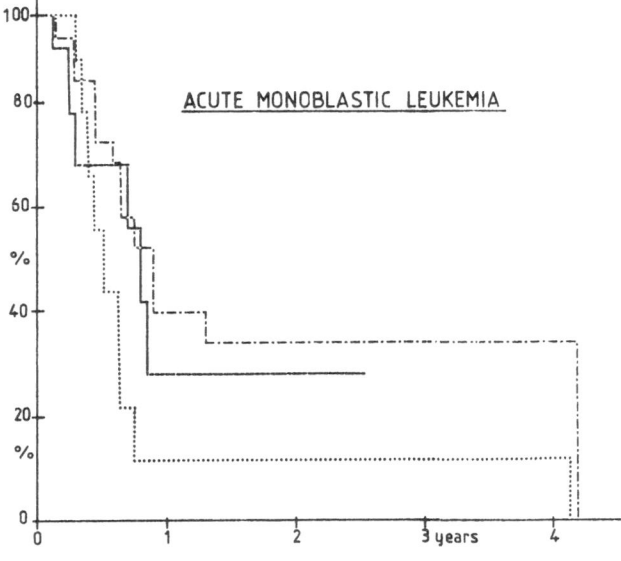

Fig. 2. Duration of complete remission after treatment with daunorubicin ....., with rubidazone —, and with rubidazone and central nervous system prophylaxis

tramedullary relapse (meningeal or skin), bone marrow relapse invariably occurred some weeks later, despite reinforcement of chemotherapy.

3. Discussion

Diagnosis of AMol remains largely based on morphological and cytochemical features. But clinical and biologic characteristics are also consistent with this diagnosis. This study of 88 patients shows that this type of leukemia may occur at any age, without male or female predominance [8, 11]. In this series, two peaks were observed, one of them in childhood and the other after 40 years of age. AMol appears to be a tumoral form of leukemia with a high frequency of skin and gum infiltration, as described before [9, 10, 12]. These extramedullary involvements are probably due to the specific properties of monoblasts, as suggested by Lichtmann and Weed [13]. Hyperleukocytosis was very frequent: the median white cell count for our patients was of 60 000/mm^3. Acute respiratory obstruction has already been described in a case of AMol with a very high white cell count [14] and may have been responsible for leukostasis in three of four patients who died before treatment. Markedly elevated serum and urine lysozyme levels have been reported in AMol [15, 16], and our results confirmed these findings; furthermore, the increased lysozyme activities were significantly correlated with hyperleukocytosis and probably reflected the total leukemic cell mass [16]. The high lysozyme levels before chemotherapy could be related to a spontaneous lysis of blast cells in vivo or to a release from intact leukemic cells [16]. The follow-up of lysozyme levels during remission as a possible early indicator of relapse is now under investigation. Renal tubular dysfunction [17] and hypokalemia have been reported with urine lysozyme, and in this article we showed that they were significantly correlated with increased blood and urine lysozyme. Disseminated intravascular coagulation has been reported in some patients with AMol [18], and in 26% of our patients, coagulation findings indicative of disseminated intravascular coagulation were observed before the start of treatment. Clinical bleeding, however, was less severe than is usually seen in acute promyelocytic leukemia [19]. The cause of disseminated intravascular coagulation could be a spontaneous release of thromboplastic material from monoblastic cells, but this remains to be proved.

The use of daunorubicin alone has produced some progress in induction, since 50% complete remissions have been obtained in our 18 patients treated by this drug. In combination with cytosine arabinoside, daunorubicin has been effective in 47% in a series of 17 treated patients [20]. In a series of ten children who received more than one dose of vinblastine, 60% achieved complete or good partial remission, indicating that vinca alkaloids have a definite effect on the blast cells of AMol [23].

VP 16-213 seems to be an interesting drug, since of eight reported cases, five entered complete remission [18, 22]. By the use of rubidazone, a high rate of complete remissions has been obtained in 55 cases of AMol. The reasons for this might be: 1. a specific antimonoblastic activity of this drug which is still difficult to assess, 2. an easier management with rubidazone, since the necessary doses may bei attained progressively, one ot two additional doses being effective during the regeneration phase without leading to resistance [24], and 3. better supportive therapy of metabolic and hemorrhagic complications. Whatever the reason, 76% of our treated patients entered complete remissions. The complete remission rate did not appear to be influenced by age, sex, organ infiltration, or hyperleukocytosis, but the number of patients may have been too small for such an analysis.

Despite this encouraging remission rate, the duration of complete remission was still short despite monthly reinduction: 195 days for 46 patients who achieved complete remission. For the ten patients of McKenna et al. [18] it was 5 months. In our series, central nervous system irradiation prolonged remission in the group treated by rubidazone.

While six meningeal relapses occurred in the patients not irradiated, none were observed in those who received central nervous system prophylaxis. Irradiation thus seems to be an advance in the treatment of AMol.

While bone marrow remission seems to be relatively easy to achieve, especially in the group treated by rubidazone, the short duration of remission emphasizes the importance of the complete remission. In some cases, for instance, bone marrow remission was obtained while skin or gum infiltration had not completely disappeared [25]. So it may be assumed that persistant blast cell sanctuaries, such as skin and/or gums, were not much affected by initial induction chemotherapy and were responsible for subsequent relapses. The study of the first relapse site was also very instructive, since extramedullary sites were involved in several cases. The question thus arises whether a more intensive induction with several drugs active against monoblasts, such as rubidazone, cytosine arabinoside, and VM 26 or vincristine in combination, might be more efficient and prolong the duration of the complete remission. Such a protocol has now been activated, including cerebral nervous system prophylaxis.

References

1. Bennet JM, Catovsky, Daniel MT, Flandrin G, Galton DAG, Gralnick HR, Sultan C (1976) Proposals for the classification of the acute leukaemias. French American British (FAB) Cooperative Group. Br J Haematol 33:451–458
2. Flandrin G, Daniel MT (1973) Practical value of cytochemical studies for the classification of acute leukemias. In: Mathe G, Pouillard P, Schwarzenberg L (eds) Nomenclature, methodology and results of clinical trials in acute leukemia. Springer, Berlin Heidelberg New York, pp 43–56
3. Bernard J, Boiron M, Lortholary P, Levy JP (1965) The very acute leukemias. Cancer Res 25:1675–1676
4. Daniel MT, Flandrin G, Lejeune G, Liso P, Lortholary P (1971) Specific esterases of monocytes: Their use in the classification of acute leukemias. Nouv Rev Fr Hematol 11:233–240
5. Fischer R, Schmalzl F (1964) Über die Hemmbarkeit der Esterase-Aktivität in Blutmonocyten durch Natriumfluorid. Klin Wochenschr 42:751
6. Peto R, Pike MC, Armitage P, Brewlow NE, Cox DR, Howard SV, Mantel N, McPherson K, Peto J, Smith PG (1976) Design and analysis of randomized clinical trials requiring prolonged observation of each patient. I – Introduction and design. Br J Cancer 34:585–607
7. Peto R, Pike MC, Armitage P, Brewlow NE, Cox DR, Howard SV, Mantel N, McPerson D, Peto J, Smith PG (1977) Design and analysis of randomised clinical trials requiring prolonged observation of each patient. II – Analysis and examples. Br J Cancer 35:1–29
8. Bernard J, Weil M, Flandrin G, Sebaoun G, Daniel MT, Jacquillat C (1975) Clinical study of acute monoblastic leukemia. Proc Am Assoc Cancer Res 16:201

9. Weil M, Sebaoun G, Semama A, Auclerc MF, Jacquillat C (1976) Leucémie aigue monoblastique. Acta Haematol (Basel) 10:156–162
10. Sultan C, Imbert M, Richard MF, Sebaoun G, Marquet M, Brun B, Forgues L (1977) Pure acute monocytic leukemia. A study of 12 cases. Am J Clin Pathol 68:752–757
11. Shaw MT (1978) The distinctive features of acute monocytic leukemia. Am J Hematol 4:97–103
12. Shaw MT, Nordquist RE (1975) Pure monocytic or histomonocytic leukemia: A revised concept. Cancer 35:208–214
13. Lichtmann MA, Weed RI (1972) Peripheral characteristics of leukocytes in monocytic leukemia: possible relationship to clinical manifestation. Blood 40:52–61
14. Pochedly C. Mehta A (1974) Monocytic leukemia as a cause of acute respiratory obstruction. J Pediatr 84:304–307
15. Osserman EF, Lawlor DP (1966) Serum and urinary lysozyme (muramidase) in monocytic and myelomonocytic leukemia. J Exp Med 124:921–930
16. Hansen NE (1974) Lysozyme activity in leukemia. Ser Hematol 7:70–87
17. Muggia FM, Heinemann HO, Farhangi M, Osserman EF (1969) Lysozymuria and renal tubular dysfunction in monocytic and myelomonocytic leukemia. Am J Med 47:351–366
18. McKenna RW, Bloomfield CD, Dick F, Nesbit ME, Brunning RD (1975) Acute monoblastic leukemia: diagnosis and treatment of ten cases. Blood 46:481–494
19. Bernard J, Weil M, Boiron M, Jacquillat C, Flandrin G, Gemon MF (1973) Acute promyelocytic leukemia, results of treatment by daunorubicin. Blood 41:489–496
20. Brun B, Vernant JP, Reyes F, Rochant H, Imbert M, Tulliez M, Sultan C, Dreyfus B (1976) Leucémies aigue monoblastique. Aspects cliniques et thérapeutiques de 20 cas. Ann Med Intern (Paris) 127:807–812
21. Oertc (1970) Essai de recherche d'efficacité de l'association prednisone, vincristine méthyl glyoxal bis et methotrexate sur les leucémies aigues monoblastiques. Eur J Cancer 6:57–59
22. Mathe G, Schwartzenberg L, Puillard P, Weiner R, Oldham R, Jasmin C, Rosenfeld C, Hayat M, Schneider M, Amiel JL, Ledara B, Steresco Musset M, Devassal F (1975) Leucémies aigues et hématosarcomes divers. Essai de traitement par un second dérivé de la podophylotoxine (VP 16-213). Nouv Presse Med 3:385–388
23. Geiser C, Mitus JW (1975) Acute monocytic leukemia in children and its response to vinblastin. Cancer Chemother Rep 59:385–388
24. Jacquillat C, Weil M, Gemon-Auclerc MF, Izrael V, Bussel A, Boiron M, Bernard J (1976) Clinical study of Rubidazone (22 050 RP) a new daunorubicin derived compound in 170 patients with acute leukemias and other malignancies. Cancer 37:653–659
25. Flandrin G, Daniel MT, Blanchet P, Briere J, Bernard J (1971) La leucémie aigue monocytaire. Situation et pronostique actuelle à la lumière des techniques de détermination des estérases spécifiques. Nouv Rev Fr Hematol 11:241–254

Haematology and Blood Transfusion Vol 27
Disorders of the Monocyte Macrophage System
Edited by F. Schmalzl, D. Huhn, H.E. Schaefer
© Springer-Verlag Berlin Heidelberg New York 1981

Histiocytic Medullary Reticulosis – Neoplastic or Atypical Inflammatory Process? A Report of Two Cases with Review of the Literature

R. Budde and H.-E. Schaefer

In 1938 and 1939 Robb-Smith [67] and Scott [73] described for the first time a histiocytic disorder which they distinguished from the large bulk of atypical Hodgkin diseases as an independent entity. Clinically this disease is characterized by a sudden onset, fever, pancytopenia, and by a fatal outcome within the course of a few months. Scott [67] classified this "histiocytic medullary reticulosis" (HMR) in the medullary reticulosis group together with storage diseases. The original description of this entity did not mention at all that it was a malignant tumor and spoke only of a histiocytic proliferation or hyperplasia of unknown etiology.

Since then more than 200 cases of HMR [10, 14, 19, 21, 34, 43, 48, 56, 60, 65, 80, 82, 85, 86] have been described. Due to the unknown pathogenetic mechanisms leading to this disorder and to a certain variability of the clinical course and pathologic–anatomical picture a number of synonymous and related terms [10, 65] were applied which are listed in Table 1.

It is nowadays generally accepted that HMR is a disorder of truely histiocytic origin [19, 20, 27, 48, 61]. Numerous studies using electron microscopical, cytochemical, and immunological tools led to this conclusion and classified the cells of HMR as cells of the mononuclear phagocyte system [12, 20, 27, 61, 83, 84].

On reviewing the relevant literature, these cells were described with the following properties: Electron microscopically the cells show reniform or indented nuclei with marginally condensed heterochromatin, a large cytoplasm with many lysosomes, pinocytotic vesicles, and phagocytosed blood cells, mainly erythrocytes. The cell surface possesses short pseudopodial- or microvilli-like protrusions. There are no Langerhans granula or intercellular junctions [15, 30, 43, 48, 78, 81, 88]. Cytochemically the cells express a positive reaction for non-specific esterase [14, 15], NaF-sensitive naphthol-AS- acetate esterase [14, 30, 43], tartrate-sensitive acid phosphatase [15, 30, 43], and lysozyme [14, 58, 81]. They stain negatively for peroxidase [30] and naphthol-AS-D-chloroacetate esterase [14].

Table 1. Histiocytic medullary reticulosis: Synonymous and related terms

Malignant histiocytosis
Malignant reticulohistiocytosis
Malignant reticulosis
Prohistiocytic medullary reticulosis
Reticulum-celled medullary reticulosis
Aleukemic reticulosis
Reticuloblastomatosis
Reticulose maligne histiocytaire

Table 2. Histiocytic medullary reticulosis: clinical features

General symptoms (fever, chills, wasting, fatigue, night sweats)
Hepatosplenomegaly
Generalized lymph node enlargement
Pancytopenia
Rapidly fatal outcome (in most cases)

Immunological studies revealed surface receptors for Fc fragment of IgG [30], gamma-, kappa-, and lambda-chains of IgG, and the third component of complement. Simultaneously kappa- and lambda-chains were detected in one cell [14], while no J-chains coud be identified [37]. These features demonstrate that the cells do not synthesize immunoglobulins and stress their phagocytotic activity. The main symptoms and signs of the disease are summarized in Table 2.

Risdall et al. [66] recently reported 19 cases of a hemophagocytic histiocytosis. They showed that in 14 of the 19 investigated patients the disease was self-limited and due to viral infection. Their cases strongly resembled those described by Scott and Robb-Smith [73] concerning the clinical course and histological picture. It was therefore our aim to review the publications of HMR with the question in mind as to whether HMR represents a homogenous disease or a rather heterogeneous group comprising at least two different types: a reactive and a neoplastic one. Additionally we would like to include in the study two cases of HMR which we investigated during the last few years.

1. Case Reports

A 50-year-old male suddenly fell ill with general symptoms, slight hepatosplenomegaly, and lymph node enlargement. In the hospital a pancytopenia was found. An iliac crest biopsy was performed and the diagnosis of HMR made. The clinicians began with chemotherapy, including vincristine, cyclophosphamide, and prednisone. Nevertheless, the patient died 4 months later. No autopsy was performed.

A 46-year-old male fell ill with fatigue, fever, and wasting. The family physician observed a massive splenomegaly, which was followed by an enlargement of the liver and lymphnodes. Multiple petechial skin bleedings developed. A massive thrombocytopenia was detected. In the hospital a splenectomy was performed. This organ histologically showed the classical picture of HMR. A combination chemotherapy, similar to the first case, was started, but the patient died within the course of 2 months. An autopsy was performed.

The ante- and post-mortem findings of these cases were nearly identical, so that we restrict ourselves to the description of the main histological changes in the organs involved. The lymph nodes showed a striking hyperplasia of relatively benign-looking histiocytes which have accumulated ingested erythrocytes in the cytoplasm (Fig. 1a). Also atypical forms were found, sometimes binucleated, with a coarse chromatin pattern, prominent nucleoli, and a thick nuclear membrane. A phagocytotic activity of these more malignant looking cells was only rarely observed. A formation of tumor nodules could not be detected; there was no capsule invasion.

As in the lymph node the general architecture of the spleen was not destroyed. The red pulp was crowded with erythrophagocytic histiocytes without any atypia (Fig. 1b). The white pulp resulted to be atrophic. Here, too, there was a lack of tumor nodule formation.

In the liver we observed only few hemophagocytic histiocytes in the sinus and rarely in the portal tracts. As in the other organs, foci of hemorrhagic necrosis were frequently encountered features (Fig. 2).

The bone marrow specimens showed a moderate reduction of fat cells. Here, too, there was a distinct hyperplasia of histiocytes which had phagocytosed cells. The grade of atypia was low (Fig. 3). Only occasionally megakaryocytes were seen, while the granulo- and ery-

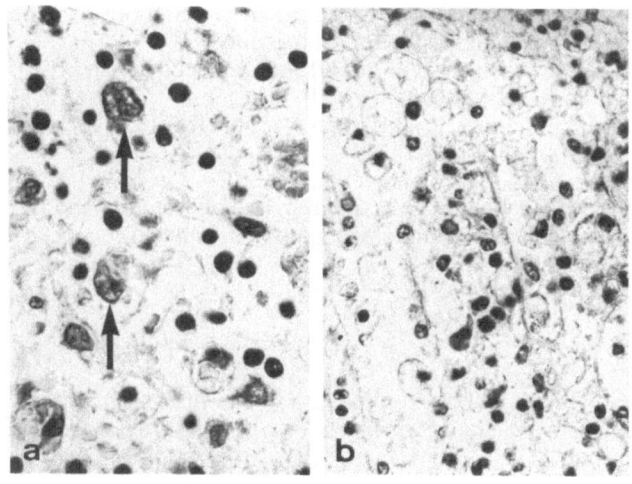

Fig. 1. a Case II: lymphocytic depletion of lymph node tissue with erythrophagocytic histiocytes. Note moderate nuclear atypia of the more immature and less phagocytic histiocytes *(arrows!)*. **b** Case II: surgically removed spleen reveals wide sinus containing many erythrophagocytic histiocytes without frank atypia

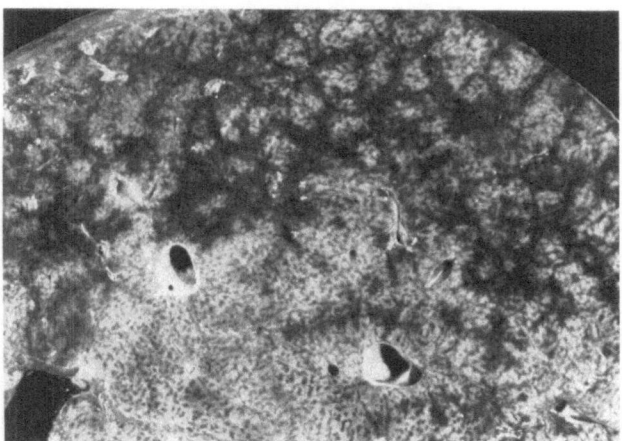

Fig. 2. Case II: disseminated hemorrhagic necrosis in the liver

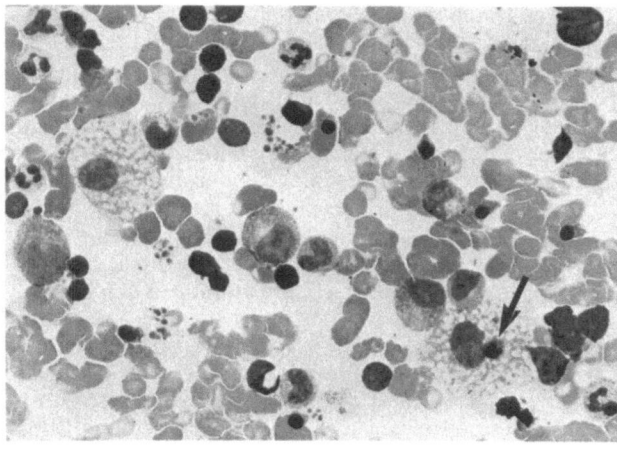

Fig. 3. Case I: two histiocytes with foamy cytoplasm to be found in a bone marrow smear. Note nuclear debris in one histiocyte *(arrow!)*

197

throcytopoiesis seemed to be moderately hypoplastic. Using smear preparations of bone marrow aspiration we performed the following cytochemical reactions:

1. Alpha-naphthylacetate esterase according to Löffler [47] and Davis and Ornstein [22].
2. Acid phosphatase according to Barka and Anderson [1].
3. Acid phosphatase with tartrate inhibition, as described by Schaefer et al. [71].
4. Peroxidase according to Schaefer and Fischer [70].
5. Chloroacetate esterase, as described by Moloney et al. [59].

The histiocytes revealed a positive activity for alpha–naphthylacetate esterase and tartrate-sensitive acid phosphatase. There was no staining with peroxidase and chloroacetate esterase.

2. Discussion

The pathogenesis of this histiocytic proliferation still remains a matter of debate. At least two main alternatives can be discussed: a malignant neoplasia on one hand and an atypical reactive hyperplasia on the other hand.

Our cases and the relatively large number of cases described in the relevant literature express features which speak in favor of both possibilities.

If a disease is defined as malignant because of a fatal course, then HMR represents surely a malignant disorder: most of the patients die within the course of a few months. But reviewing the main publications concerning HMR we only encountered ten cases which were accompanied by the formation of solid tumor nodules composed of atypical hemophagocytic histiocytes with an invasive and destructive growth pattern [17, 29, 36, 49, 50, 52, 62, 86, 87]. Generally these tumors occur in the intestine [17, 29, 36, 49], but they were also described in the spleen [29], skin [62, 86], lung [86], mediastinum [87], nose [50], mesenterium [52], and iliacal crest. In part these cases did not show the classical pathologic-anatomical picture, as described by Scott and Robb-Smith [73], presenting no involvement of lymph nodes and bone marrow [62, 87]. Those cases also show malignant features in which the cytological atypia is striking, while the phagocytotic activity is low [13, 32, 57, 81].

With regard to the development of HMR from acute leukemia, mostly from T-ALL [18, 39, 54, 76, 77, 79] but also from AML [86] and acute monocytic leukemia [15, 43], we do not know whether the histiocytic proliferation is due to malignant cell transformation or has a reactive cause. In this context three single case reports are notworthy: Marshall and Revell [53] described a phagocytotic liver tumor in combination with a reticulum cell sarcoma of the stomach and HMR of lymph nodes, spleen, and bone marrow and presumed three different differentiations of one histiocytic neoplasia. Rabinowitz et al. [68] supposed that HMR represented a well-differentiated reticulum cell sarcoma, and one case of a 14-year-old boy is reported [13] in which HMR appeared after hydrocarbon exposure. These cases, however, differ distinctly from the original description of HMR [73], and it is therefore questionable whether these cases should be designated with this term.

Most published cases of HMR, however, including those investigated by us, possess features which favor a reactive genesis of the disease. First of all, the histiocytes show no or only slight cytological atypia, there is no destruction of the architecture in the organs involved, the organ capsule is not invaded, and except the cases mentioned above, no tumor formation can be detected. The rapid onset and frequently dramatic course is also not characteristic of a malignant neoplasia. Furthermore, the high frequency of HMR in the indigenous population with low social status of Africa (here HMR constitutes 3% of the malignant lymphomas) [35, 72, 74, 75] and the simultaneous occurrence in father and son as described by Boake et al. [8] strongly suggest an environmental factor.

The relevant literature reports 22 cases of HMR in which a sure connection between the disease and an infection with micro-organisms could be shown [38, 45, 66, 72] (see Table 3). Especially Risdall et al. [66] recently reported 19 cases of hemophagocytic histiocytosis, mostly in recipients of renal grafts. Histomorphologically, these cases resembled those

Table 3. Histiocytic medullary reticulosis: reactive type

Demonstration of viruses and other micro-organisms

Cytomegalovirus: Risdall et al. [66]
  Liao et al. [45]
  Kalderon [38]
Herpes virus: Risdall et al. [66]
Epstein-Barr virus: Risdall et al. [66]
  Schumacher and Stass [72]
Adenovirus: Risdall et al. [66]
Toxoplasma gondii: Liao et al. [45]

Special feature
Defective T-cell system: Hirsch et al. [33]

described by Scott and Robb-Smith [73]. In 14 of the 19 cases the disease was self-limited and due to virus infection. In five cases, however, the patients died, even though the viruses could be cultured and no clinical or histological differences were detected when these cases were compared with those in which the patients survived. It is important to stress that most patients described in this paper had received immunosuppressive therapy. It was also the case with the leukemic patients mentioned above, and it is possible that in the latter a viral agent could be responsible for that histiocytic hyperplasia. The same is true for all those cases of HMR after renal grafting [31]. Even though no microorganisms could be detected, two cases of hemophagocytic histiocytosis [16] are noteworthy, because the patients recovered after tuberculostatic therapy.

In this context, it is interesting to note that Hirsch et al. [33] described a case of HMR in combination with a defective T cell system. It seems likely that the coincidental combination of an immunological defect with an otherwise banal infection could lead to this massive proliferation of histiocytes.

Another factor which speaks in favor of a reactive etiology is the similarity of HMR to some of the disorders listed in Table 4. In familial hemophagocytic reticulosis [2, 5–7, 23–26, 41, 51, 55, 64] an immunodeficiency, probably an atypical graft-vs-host reaction or a combined T, B, and M cell defect, is discussed, and lymphadenitis with massive hemophagocytic sinus histiocytosis [44] – which is also called a sinus histiocytosis with massive lymphadenopathy [4, 28, 40, 42, 46, 69] – probably represents an infection with Klebsiella or atypical mycobacteria. In the syndromes of Brambilla et al. [9] and Beck et al. [3] a defective T cell system is described, while in those of Ochs et al. [63] and Carpentieri et al. [11] the B cell system also reveals a deficiency.

In conclusion, we feel that the diseases described under the term HMR do not represent a homogeneous group. Certainly there are at least two different entities: (1) the reactive histiocytic hyperplasia, which in part is self-limited and in part ends fatally, probably due to the

Table 4. Histiocytic medullary reticulosis: related disorders

Familial hemophagocytic reticulosis: Farquhar and Claireaux [23]
Lymphadenitis with massive hemophagocytic sinus histiocytosis: Lennert et al. [44]
Combined immunodeficiency and reticuloendotheliosis with eosinophilia: Ochs et al. [63]
Reversible graft-versus-host reaction as a cause of erythrophagocytic splenomegaly in a child: Beck et al. [3]
Hemophagocytic reticulosis: Carpentieri et al. [11]
Partial albinism and immunodeficiency: Ultrastructural study of hemophagocytosis and bone marrow erythroblasts in one case: Brambilla et al. [9]

coincident combination of immunodeficiency and virus infection, and (2) the malignant tumorous neoplasia of histiocytes.

It is possible, too, that some cases of the first group could represent the prestage of a true neoplastic process. The cases described by Scott and Robb-Smith [73], which undoubtedly correspond to the first type, could thus be regarded as premalignant histiocytoses or the prestage of malignant lymphoma of histiocytic type.

A certain analogy between the reactive and the tumorous variant of HMR on one hand and angioimmunoblastic lymphadenopathy and malignant lymphoma on the other hand seems suggestive. Nevertheless, in our opinion it is important to separate these two hypothetical groups, because chemotherapy appears contraindicated in the case of the "reactive type". It is possible that the fatally ending courses of the first (reactive) variant are due to inadequate therapy.

References

1. Barka T, Anderson PJ (1962) Histochemical methods for acid phosphatase using hexazonium pararosaniline as coupler. J Histochem Cytochem 10:741
2. Barth RF, Vergara GG, Khurana SK, Lowman JT (1972) Rapidly fatal familial histiocytosis associated with eosinophilia and primary immunological deficiency. Lancet 2:503
3. Beck J-D, Weinig JE, Müller-Hermelink HK, Lemmel EM (1977) Reversible graft versus host reaction as a cause of erythrophagocytic splenomegaly in a child? Eur J Pediatr 126:175
4. Becroft DMO, Dix MR, Gillman JC, MacGregor BJL, Shaw RL (1973) Benign sinus histiocytosis with massive lymphadenopathy: transient immunological defects in a child with mediastinal involvement. J Clin Pathol 26:463
5. Bell RJM, Brafield AJE, Barnes ND, France NE (1968) Familial haemophagocytic reticulosis. Arch Dis Child 43:601
6. Bergholz M, Rahlf G, Doering KM (1978) Familial hemophagocytic reticulosis (Farquhar). Pathol Res Pract 163:267
7. Blennow G, Berg B, Brandt L, Messeter L, Loew B, Soenderstroem N (1974) Haemophagocytic reticulosis. A state of chimerism? Arch Dis Child 49:960
8. Boake WC, Card WH, Kimmey JF (1965) Histiocytic medullary reticulosis — Concurrence in in father and son. Arch Intern Med 116:245
9. Brambilla E, Dechelette E, Stroebner P (1980) Partial albinism and immunodeficiency: Ultrastructural study of haemophagocytosis and bone marrow erythroblasts in one case. Pathol Res Pract 167:151
10. Byrne GE Jr, Rappaport H (1973) Malignant histiocytosis. Gan 15:145
11. Carpentieri U, Gustavson LP, Haggard ME, Nichols MM (1979) Haemophagocytic reticulosis. Blut 39:419
12. Carr I, Wright J (1978) The reticuloendothelial and mononuclear phagocyte system and the macrophage. Can Med Assoc J 118:882
13. Case records of the Massachusetts general hospital. Weekly clinicopathological exercises. Case 12-1977 (1977) N Engl J Med 296:673
14. Case records of the Massachusetts general hospital. Weekly clinicopathological exercises. Case 4-1979 (1979) N Engl J Med 300:184
15. Castoldi G, Grusovin GD, Scapoli G, Gualandi M, Spanedda R, Anzanel D (1977) Acute myelomonocytic leukemia terminating in histiocytic medullary reticulosis. Cytochemical, cytogenetic, and electron microscopic studies. Cancer 40:1735
16. Chandra P, Chaudhery SA, Rosner F, Kagen M (1975) Transient histiocytosis with striking phagocytosis of platelets, leukocytes, and erythrocytes. Arch Intern Med 135:989
17. Chawla SK, Lopresti PA, Burdman D, Sileo A, Govoni AF, Smulewicz JJ (1975) Diffuse small bowel involvement in malignant histiocytosis. Am J Gastroenterol 63:129
18. Chen TK, Nesbit ME, McKenna R, Kersey JH (1976) Histiocytic medullary reticulosis in acute lymphocytic leukemia of T cell origin. Am J Dis Child 130:1262
19. Cline MJ, Golde DW (1973) A review and reevaluation of the histiocytic disorders. Am J Med 55:49
20. Cline MJ, Lehrer RI, Territo MC, Golde DW (1978) UCLA Conference — Monocytes and macrophages: Functions and diseases. Ann Intern Med 88:78

21. Crow J, Gumpel JM (1977) Histiocytic medullary reticulosis presenting as rheumatoid arthritis. Proc R Soc Med 70:632
22. Davis BJ, Ornstein L (1959) High resolution enzyme localisation with a new diazo reagent hexazonium pararosaniline. J Histochem Cytochem 7:297
23. Farquhar JW, Claireaux AE (1952) Familial haemophagocytic reticulosis. Arch Dis Child 27:519
24. Farquhar JW, MacGregor AR, Richmond J (1958) Familial haemophagocytic reticulosis. Br Med J 2:1561
25. Fullerton P, Ekert H, Hosking C, Tauro GP (1975) Hemophagocytic reticulosis. A case report with investigations of immune and white cell function. Cancer 36:441
26. Gleichmann E, Gleichmann H (1976) Graft-versus-host reaction: A pathogenetic principle for development of drug allergy, autoimmunity, and malignant lymphoma in non-chimeric individuals. Hypothesis. Z Krebsforsch 85:91
27. Golde DW (1975) Disorders of mononuclear phagocyte proliferation, maturation, and function. Clin Haematol 4:705
28. Haas RJ, Helmig MS, Prechtel K (1978) Sinus histiocytosis with massive lymphadenopathy and paraparesis: Remission with chemotherapy. Cancer 42:77
29. Hardmeier T, Hedinger C, Thalmann H (1969) Zur sogenannten histiocytären medullären Reticulose von Scott und Robb-Smith. Schweiz Med Wochenschr 99:806
30. Harousseau JL, Degos L, Daniel MT, Flandrin G (1979) Leukemic phase of malignant histiocytosis (Arguments in favour of the histiomonocytic origin of the abnormal cells). Med Pediatr Oncol 6:339
31. Hernandez-Nieto L, Bombi JA, Caralps A, Aranalde JM (1977) Histiocytic medullary reticulosis (Robb-Smith's disease) in renal transplant patient. Lancet 1:261
32. Hidvegi DF, Nunez R, Alonso C (1979) Cytomorphologic and cytochemical aspects of malignant histiocytosis with spindle-cell differentiation − A case report. Acta Cytol (Baltimore) 23:402
33. Hirsch J, Ungar B, King WE, Lubbe TR (1964) Histiocytic medullary reticulosis: A case report. Australas Ann Med 13:269
34. Ho FC, Todd D (1978) Malignant histiocytosis. Report of five chinese patients. Cancer 42:2450
35. Hutt MS, Lowenthal MN, Fine J, Jones IG, Nag J, O'Riordan EC (1977) Histiocytic medullary reticulosis (malignant histiocytosis) in Zambia. J Trop Med Hyg 78:239
36. Isaacson P, Wright DH (1978) Malignant histiocytosis in the intestine. Its relationship to malabsorption and ulcerative jejunitis. Hum Pathol 9:661
37. Isaacson P, Wright DH (1979) Anomalous staining pattern in immunohistologic studies of malignant lymphoma. J Histochem Cytochem 8:1197
38. Kalderon AE (1971) Histiocytic medullary reticulosis associated with cytomegalic inclusion disease. Cancer 27:659
39. Karcher DS, Head DR, Mullins JD (1978) Malignant histiocytosis occuring in patients with acute lymphocytic leukemia. Cancer 41:1967
40. Karpas A, Arno J, Cawley J (1973) Sinus histiocytosis with massive lymphadenopathy − Properties of cultured histiocytes. Eur J Cancer 9:729
41. Ladisch S, Poplack DG, Holiman B, Blaese RM (1978) Immunodeficiency in familial erythrophagocytic lymphohistiocytosis. Lancet 1:581
42. Lampert F, Lennert K (1976) Sinus histiocytosis with massive lymphadenopathy: Fifteen new cases. Cancer 37:783
43. Lampert IA, Catovsky D, Bergier N (1978) Malignant histiocytosis: A clinico-pathological study of 12 cases. Br J Haematol 40:65
44. Lennert K, Niedorf HR, Blümcke S, Hardmeier T (1972) Lymphadenitis with massive hemophagocytic sinus histiocytosis. Virchows Arch [Cell Pathol] 10:14
45. Liao KT, Rosai J, Daneshbod K (1972) Malignant histiocytosis with cutaneous involvement and eosinophilia. Am J Clin Pathol 57:438
46. Lober M, Rawlings W, Newell GR, Reed RJ (1973) Sinus histiocytosis with massive lymphadenopathy. Report of a case associated with elevated EBV antibody titers. Cancer 32:421
47. Löffler H (1961) Zytochemischer Nachweis von unspezifischer Esterase in Ausstrichen. Klin Wochenschr 39:1220
48. Lombardi L, Carbone A, Pilotti S, Rilke F (1978) Malignant histiocytosis: a histological and ultrastructural study of lymph nodes in six cases. Histopathology 2:315
49. MacGillivray JB, Duthie JS (1977) Malignant histiocytosis (histiocytic medullary reticulosis) with spindle cell differentiation and tumour formation. J Clin Pathol 30:120

50. Mogi G, Maeda S, Yoshida T (1976) Histiocytic medullary reticulosis with involvement of the nose. Laryngoscope 86:1752
51. Marrian VJ, Sanerkin NG (1963) Familial histiocytic reticulosis (familial haemophagocytic reticulosis). J Clin Pathol 16:65
52. Marshall AHE (1956) Histiocytic medullary reticulosis. J Pathol Bacteriol 71:61
53. Marshall AHE, Revell PA (1977) Malignant histiocytosis with associated "reticulum cell sarcoma". J Pathol 122:9
54. Marvin RF, Hill D (1975) Acute lymphoblastic leukemia and histiocytic medullary reticulosis. JAMA 233:1258
55. McClure PD, Strachan P, Saunders EF (1974) Hypofibrinogenemia and thrombocytopenia in familial hemophagocytic reticulosis. J Pediatr 85:67
56. Medford FE (1965) Histiocytic medullary reticulosis. Report of a case. Arch Intern Med 116:589
57. Meister P, Huhn D, Nathrath W (1980) Malignant histiocytosis. Immunohistochemical characterisation on paraffin embedded tissue. Virchows Arch [Pathol Anat] 385:233
58. Mendelsohn G, Eggleston JC, Mann RB (1980) Relationship of lysozyme (muramidase) to histiocytic differentiation in malignant histiocytosis. An immunohistochemical study. Cancer 45:273
59. Moloney WC, McPherson K, Fliegelman L (1960) Esterase activity in leukocytes demonstrated by the use of naphthol-AS-D-chloro-acetate substrate. J Histochem Cytochem 8:200
60. Natelson EA, Lynch EC, Hettig RA, Alfrey CP Jr (1968) Histiocytic medullary reticulosis. The role of phagocytosis in pancytopenia. Arch Intern Med 122:223
61. Nezelof C, Jaubert F (1968) Histiocytic and/or reticulum cell neoplasias. Recent Results Cancer Res 64:118
62. Nishio K, Koda H, Vrabe H (1975) Über einen Fall von Histiocytic Medullary Reticulosis. Arch Dermatol Forsch 251:259
63. Ochs HD, Davis SD, Mickelson E, Lerner KG, Wedgwood RJ (1974) Combined immunodeficiency and reticuloendotheliosis with eosinophilia. J Pediatr 85:463
64. Perry MC, Harrison EG Jr, Burgert EO Jr, Gilchrist GS (1976) Familial erythrophagocytic lymphohistiocytosis. Report of two cases and clinicopathologic review. Cancer 38:209
65. Rappaport J (1966) Histiocytoses (Reticuloendothelioses). In: Rappaport H (ed) Tumors of the hematopoetic system. Armed Forces Institute of Pathology, Washington (Atlas of tumor pathology, sect III, fasc 8, p 48)
66. Risdall RJ, McKenna RW, Nesbit ME, Krivit W, Balfour HH Jr, Simmons RL, Brunning RD (1979) Virus-associated hemophagocytic syndrome. A benign histiocytic proliferation distinct from malignant histiocytosis. Cancer 44:993
67. Robb-Smith AHT (1938) Reticulosis and reticulosarcoma: A histological classification. J Pathol Bacteriol 47:457
68. Robinowitz BN, Noguchi S, Bergfeld WF (1977) Tumor cell characterisation of histiocytic medullary reticulosis. Arch Dermatol 113:927
69. Rosai J, Dorfman RF (1973) Sinus histiocytosis with massive lymphadenopathy: a pseudolymphomatous benign disorder. Analysis of 34 cases. Cancer 30:1174
70. Schaefer HE, Fischer R (1968) Der Peroxydasenachweis an Ausstrichpräparaten sowie Gewebsschnitten nach Entkalkung und Paraffineinbettung. Klin Wochenschr 16:1228
71. Schaefer HE, Hellriegel KP, Fischer R (1977) Vorkommen von tartrat-resistenter saurer Phosphatase in verschiedenen Zelltypen des lymphoreticulären und hämatopoetischen Zellsystems. Blut 34:393
72. Schumacher HR, Stass SA (1979) Histiocytic medullary reticulosis. Lancet 1:158
73. Scott RB, Robb-Smith ATH (1939) Histiocytic medullary reticulosis. Lancet 2:194
74. Serck-Hanssen A (1973) Histiocytic medullary reticulosis. Recent Results Cancer Res 41:292
75. Serck-Hanssen A, Purchit GP (1968) Histiocytic medullary reticulosis — Report of 14 cases from Uganda. Br J Cancer 22:506
76. Shreiner DP (1975) Acute lymphoblastic leukemia terminating as histiocytic medullary reticulosis. JAMA 231:838
77. Skarin AT, Karb K, Reynolds ES (1972) Acute leukemia terminating in histiocytic medullary reticulosis. Arch Pathol 93:256
78. Skinnider LF, Ghadially FN (1977) Ultrastructure of cell surface abnormalities in neoplastic histiocytes. Br J Cancer 35:657
79. Skoog DP, Feagler JR (1978) T cell acute lymphocytic leukemia terminating as malignant histiocytosis. Am J Med 64:678
80. Symmers WSTC (1978) Malignant histiocytosis and similar conditions. In: Symmers WSTC (ed)

Systemic pathology, vol 2. Livingstone, Edinburgh London New York, p 858

81. Tubbs RR, Sheibani K, Sebek BA, Savage RA (1980) Malignant histiocytosis. Ultrastructural and immunocytochemical characterisation. Arch Pathol Lab Med 104:26

82. Vaithianathan T, Fishkin S, Gruhn JG (1967) Histiocytic medullary reticulosis. Am J Clin Pathol 47:160

83. Van Furth R (1970) Origin and kinetics of monocytes and macrophages. Semin Hematol 7:125

84. Van Furth R, Cohn ZA, Hirsch JG, Humphrey JH, Spector WG, Langevoort HL (1972) The mononuclear phagocyte system: A new classification of macrophages, monocytes, and their precursor cells. Bull WHO 46:845

85. Vardiman JW, Byrne GE Jr, Rappaport H (1975) Malignant histiocytosis with massive splenomegaly in asymptomatic patients. A possible chronic form of the disease. Cancer 36:419

86. Warnke RA, Kim H, Dorfman RF (1975) Malignant histiocytosis (histiocytic medullary reticulosis) I. Clinicopathologic study of 29 cases. Cancer 35:215

87. Watanabe S, Mikata A, Toyama K, Kitamura K, Minato K (1978) Sarcomatous variant of malignant histiocytosis; a case report and review of the literature. Acta Pathol Jpn 28:963

88. Zawadzki ZA, Pena CE, Fisher ER (1969) Histiocytic medullary reticulosis. Case report with electron microscopic study. Acta Haematol (Basel) 42:50

Haematology and Blood Transfusion Vol 27
Disorders of the Monocyte Macrophage System
Edited by F. Schmalzl, D. Huhn, H.E. Schaefer
© Springer-Verlag Berlin Heidelberg New York 1981

Histiocytic Medullary Reticulosis Re-visited

A. Manoharan and D. Catovsky

1. Summary

The macrophage cell proliferation which results in the clinical syndrome of histiocytic medullary reticulosis (HMR) may represent a true malignancy (malignant histiocytosis, MH) or a reactive process to an underlying infection.

We have reviewed the bone marrow cytology of 19 patients with HMR: 15 had de novo and four HMR supervening on a pre-existing chronic lymphocytic leukaemia (CLL). The bone marrow histiocytes were graded morphologically according to their degree of maturation and their phagocytic activity in vivo. This allowed us to define two groups of patients with de novo HMR: (A) ten cases had predominantly immature, poorly phagocytic cells and were considered to have MH, and (B) five cases in which mature, predominantly large, haemophagocytic cells outnumbered the immature ones by a ratio of 2:1 were considered reactive. Three of these patients had evidence of EBV infection and recovered completely; the other two had systemic bacterial infections. The following features were of no value in distinguishing group A and B cases: percentage of bone marrow macrophages (range 5%–40% in both), systemic symptoms, and severity of the cytopenias. On the other hand, moderate to gross hepatosplenomegaly was seen only in group A (MH).

The cell morphology of the macrophages in the four cases with HMR supervening on CLL was similar to group B, thus suggesting a reactive pathogenetic mechanism. However, the disease was acute and rapidly fatal in all of them.

The implications of our findings in the classification of HMR and on the terminology used to define the disease are discussed.

2. Introduction

Histiocytic medullary reticulosis (HMR) was first described by Scott and Robb-Smith [18] as a clinicopathological entity characterised by fever, weight loss, jaundice, lymphadenopathy, hepatosplenomegaly, cytopenias, and widespread histiocytic tissue infiltration. The clinical course in each of the four patients reported by Scott and Robb-Smith was acute and rapidly progressive. Since this original report more than 200 cases have been described in the literature, some of them under the title "malignant histiocytosis" (MH). The latter term, introduced by Rappaport in 1966 [14], is generally considered as a synonym for HMR [10, 12]. In recent years, two further types of clinical disorders with features similar to those in HMR have been recognised. One occurs as a self-limiting disease secondary to viral and other infections [2, 6, 9, 15, 16], and the other, as a terminal complication in patients with a pre-existing haematological malignancy [3–5, 7, 8, 11–13, 19, 20, 22] or carcinoma [17]. Three points of practical significance arise from these reports: 1. the differential diagnosis between the truly malignant MH and the non-malignant and potentially self-limiting disorder; 2. the nature of the HMR-like syndrome in patients with a pre-existing malignancy, i.e. a malignant condition (second neoplasm) or a syndrome secondary to an opportunistic infection and 3. the relevance of the generally accepted synonymity of the terms HMR and MH.

We have recently analysed the clinical and haematological features of 19 patients with the syndrome of HMR seen at the Hammersmith Hospital in the period from 1971 to 1980. This paper reports our findings and the points previously mentioned.

3. Materials and Methods

Four patients were known to have had chronic lymphocytic leukaemia and developed HMR-like syndrome as a pre-terminal complication [13]; in the remaining 15 patients HMR was the only diagnosis. The study was done in two parts. In the first part, bone marrow aspirates of the 15 patients with de novo HMR were examined by one of us (AM) in a blind fashion, i.e. the clinical details, blood counts, etc. were not known to the reviewer. Nine of these cases were previously reported [10] but were re-analysed for this study. Examination of the marrow included an over-all differential count of at least 200 nucleated cells and also a differential count of 50 mononuclear–macrophage cells, classified according to the maturation stage of the cells as follows:

1. A large blast with features similar to those of a monoblast. Auer rods were not seen (Fig. 1A).

2. A primitive mononuclear cell, slightly smaller than the blast, with an eccentric kidney-shaped nucleus, an open chromatin pattern and a moderate amount of cytoplasm (Fig. 1B).

3. A mononuclear cell, slightly smaller than the preceding cell, showing features of intermediate maturity. The nucleus is also kidney-shaped but with more chromatin condensation. The cytoplasm is abundant and has a greyish foamy appearance with occasional azurophilic granules (Fig. 1C).

4. A large cell with a low nuclear-cytoplasmic ratio and a small round nucleus with a condensed chromatin pattern. The cytoplasm appears greyish and foamy and may be vacuolated. Phagocytosis is minimal, and the cytoplasmic border is usually irregular (Fig. 1D).

5. A very large, mature histiocyte with features similar to those described in 4 but with active phagocytosis (Fig. 1E).

It was also recorded whether or not the macrophages showed haemophagocytosis and whether the degree of phagocytic activity was minimal, moderate or marked. The clinical and laboratory features of the patients, the therapy and the clinical course of the disease were related to the bone marrow findings.

In the second part of the study, the bone marrow aspirate findings from the four patients with a HMR-like syndrome complicating chronic lymphocytic leukaemia (CLL) were compared with those in the previous (de novo HMR) group.

3. Results

The mononuclear–macrophage cells comprised 5%–40% of all the nucleated cells in the bone marrow aspirates in the 15 patients with de novo HMR. The morphological features of these cells, however, distinguished two groups: (A) In ten patients the immature mononuclear cells were mainly of stages 1–3, outnumbering the mature macrophages (stages 4 and 5) by a ratio of at least 2:1; haemophagocytosis by the macrophages was either minimal or absent. (B) In the other five patients the mature macrophages (stages 4 and 5) outnumbered the immature mononuclear cells (maturation stages 1–3) by a ratio of at least 2:1; haemophagocytosis by the macrophages was either moderate or marked.

Table 1 summarises the clinical features of the patients in the two groups (A and B). Three of the five patients in group B had laboratory evidence of Epstein-Barr virus (EBV) infection (positive Paul-Bunnel, rising antibody titres against EBV); the fourth patient had staphylococcal septicemia. This patient had previously a diagnosis of systemic lupus erythematosus (SLE) and was receiving prednisone. The fifth patient in this group had miliary tuberculosis. The three patients with EBV infection recovered completely and are now well at 6 months, 4 years and 4.5 years later; two of these patients received only symptomatic treatment, but the third was given six courses of combination chemotherapy CHOP (cyclophos-

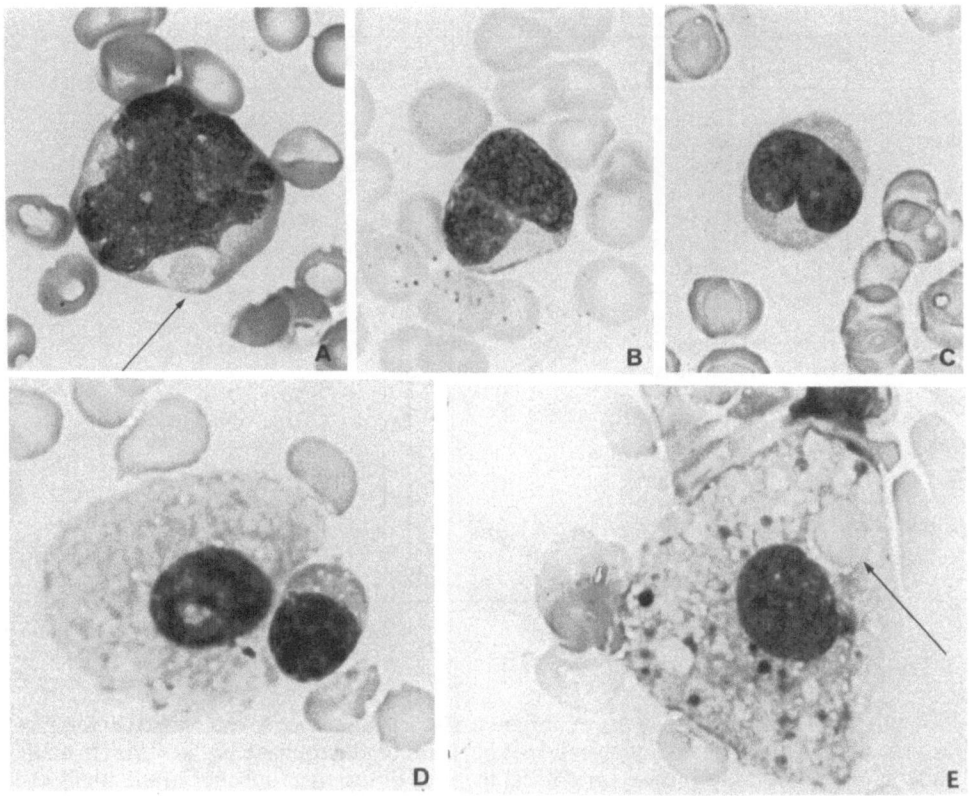

Fig. 1A–E. Cells from bone marrow aspirates of cases with HMR using May-Grünwald-Giemsa stain (× 620). **A** stage 1 mononuclear–macrophage showing features of a monoblast with erythrophagocytosis *(arrow)*; **B** stage 2 mononuclear–macrophage; **C** stage 3 mononuclear–macrophage; **D** stage 4 mononuclear–macrophage; and **E** stage 5 mononuclear–macrophage cell showing a phagocytosed red cell *(arrow)* and cellular debris in the cytoplasm

phamide, doxorubicin, vincristine and prednisone) as the disease was initially classified as MH (case number 12 of the previous series [10]). The patient with SLE and staphylococcal septicemia died from renal failure and disseminated candidiasis. The patient with miliary tuberculosis died 3 days after commencing anti-tuberculous therapy.

There was no documented evidence of any infection in any of the ten group-A patients. Six of them were treated with CHOP, and a clinical response occurred in four. Two of the responders have come off therapy and are alive and well $3\frac{1}{2}$ and 4 years later. One of the other two responders is still on treatment 10 months after diagnosis; the other patient relapsed whilst on treatment and died from peritonitis secondary to multiple intestinal perforations.

The history and clinical features of the four patients with a HMR-like syndrome complicating chronic lymphocytic leukaemia are described in detail elsewhere [13]. The morphological features of the macrophage cells in the bone marrow aspirates of these patients were similar to those seen in the B-group patients described previously; the mature macrophages outnumbered the immature cells by a ratio of at least 2:1, and haemophagocytosis by the mononuclear–macrophage cells was marked in each case.

Table 1. Summary of clinical features in patients with de novo HMR

		Group A	Group B
No. of patients		10	5
Sex:	Male	6	1
	Female	4	4
Age:	< 30 years	0	4
	> 30 years	10	1
Duration of	< 2 months	6	5
history:	> 2 months	4	None
Symptoms:	Fever	8	5
	Malaise	8	5
	Night sweats	7	1
	Weight loss	5	2
	Aches & pains	4	2
	Skin "lesions"	1	None
Signs:	Lymphadenopathy	1	1
	Hepatomegaly	7 (2–10 cm)	2 (1–2 cm)
	Splenomegaly	8 (1–13 cm)	2 (2 cm)
Blood counts:	Anaemia	9	4
	Leucopenia	7	4
	Thrombocytopenia	9	5

4. Discussion

Previous authors have stressed the importance of strict adherence to established cytological criteria of malignancy to distinguish the benign form of macrophage proliferation from the malignant disorder. Distinctive cytological features of the macrophages in the malignant form (MH) include thickening of the nuclear membrane, irregular clumping of the nuclear chromatin, irregular nucleoli and jagged borders [21]. Application of these criteria in an individual patient, however, is often fraught with difficulties because of the morphological heterogeneity of the malignant histiocytes [10] and also because of the varying degree of maturity of tumour cells from case to case [14]. In the present study, we have attempted to overcome these difficulties by classifying the mononuclear–macrophage cells in the bone marrow aspirate according to their maturation stage and assessing the ratio between the immature and mature cells. A predominance of immature cells was seen in patients with the malignant form of disorder, whereas a predominance of mature cells was seen in patients with an underlying infective aetiology. An additional distinguishing feature was the degree of phagocytic activity in the macrophages; in cases with an infection-associated proliferation the macrophages showed more active haemophagocytosis. In the malignant group of patients (MH), on the other hand, the histiocytes showed little haemophagocytosis. We believe that these criteria may be particularly useful in patients in whom diagnostic material from other tissue (e.g. lymph node) can not be obtained.

As shown in Table 1, the incidence of systemic symptoms such as fever, weight loss, etc. was similar in the two groups of patients, i.e. those with the malignant disease (group A) and those with the reactive disorder probably secondary to an infection (group B). Although the incidence of organomegaly was also similar in the two groups, moderate or gross hepato- and/or splenomegaly was noted only in the malignant group. Blood counts, on the other hand, were not useful in the differential diagnosis; severe cytopenias occurred with equal frequency in both groups. The difference in the age of the patients in the two groups (younger in group B) is interesting but may be only of limited value in the differential diagnosis.

In the ten patients with the MH (group A) it was interesting also to note that those with a long history (i.e. more than 2 months) appear to have a good response to therapy and a better prognosis. Alexander and Daniels [1] also documented an improved prognosis in patients who respond to therapy. When we analysed the patients' data from that study [1] we observed that the duration of history was 2 months or less in six of the seven patients who died in less than 1 year. In contrast, the five patients with a history longer than 2 months had survival periods ranging from 8 to 35 months; three of them are still alive, and two show no evidence of disease activity. These findings suggest that the duration of symptoms in patients with MH at presentation may be of prognostic and therapeutic value.

A benign and potentially self-limiting clinical course in patients with an infection-associated macrophage proliferation has been previously reported [2, 15, 16]. The complete recovery seen in our three patients with an underlying EBV infection (group B) reiterates this point. It should be noted, however, that one of these patients was treated with combination chemotherapy CHOP as the disease was initially classified as MH. This patient, who was extremely ill and responded within 48 h after commencing chemotherapy, raises the question of the role of cytotoxic drugs in the management of such patients. The issue can only be solved when more is known about the pathogenesis of this uncommon condition.

The nature of the pre-terminal illness with features similar to those in HMR in patients with a pre-existing malignancy is not clear. Karcher et al. [11] considered this syndrome a true malignancy when seen in patients with acute lymphoblastic leukaemia, whereas Risdall et al. [15] have suggested that these patients may have an opportunistic viral infection related to the immunosuppression by the antileukaemia therapy. Our study showed that the morphology of the bone marrow macrophages from the four patients with the HMR-like syndrome complicating CLL were similar to those seen in patients with the proliferation secondary to infections. Although investigations to rule out a viral infection were inadequate in these patients, our observations suggest an underlying infection probably related to the immunosuppression associated with the disease. It is worth noting that at the time of the development of the HMR-like syndrome two CLL patients also had Richter's syndrome [13]. We believe that the terms HMR and MH could be used as synonyms but only for the truly malignant disease. The non-malignant clinical syndrome secondary to infections is probably best described as a haemophagocytic disorder or an HMR-like syndrome secondary to EBV, H. simplex, etc.

References

1. Alexander M, Daniels JR (1977) Chemotherapy of malignant histiocytosis in adults. Cancer 39:1011
2. Chandra P, Chaudhery SA, Rosner F, Kagen M (1975) Transient histiocytosis with striking phagocytosis of platelets, leukocytes and erythrocytes. Arch Intern Med 135:989
3. Chen TK, Nesbit ME, McKenna R, Kersey JH (1976) Histiocytic medullary reticulosis in acute lymphocytic leukemia of T-cell origin. Am J Dis Child 130:1262
4. Chesney TM (1977) Histiocytic medullary reticulosis (malignant histiocytosis) as a sequel of acute lymphoblastic leukaemia in childhood. Blood [Suppl] 50:186
5. Chesney TM, Florendo NT, Bentley T, Sexton H (1978) Malignant histiocytosis complicating chronic lymphocytic leukemia. Blood [Suppl] 52:243
6. Chih-Fei Y, Chung-Hong TH, Uai-Teh H, Teh-Ts'ung T, Hsi-Hien K, Ch'in-Tong Y, Chieh L (1960) Histiocytic medullary reticulosis. Chin Med J [Engl] 80:466
7. Clark BS, Dawson PJ (1969) Histiocytic medullary reticulosis presenting with a leukemic blood picture. Am J Med 47:314
8. Economopoulos TC, Stathakis N, Stathopoulos E, Alexopoulos C (1979) 'Lennert's lymphoma' terminating as malignant histiocytosis. Scand J Haematol 23:427
9. Fernandes-Costa F, Eintracht I (1979) Histiocytic medullary reticulosis. Lancet 2:204
10. Lampert IA, Catovsky D, Bergier N (1978) Malignant histiocytosis: a clinico-pathological study of 12 cases. Br J Haematol 40:65
11. Karcher DS, Head DR, Mullins JD (1978) Malignant histiocytosis occurring in patients with acute lymphocytic leukemia. Cancer 41:1967
12. Korman LY, Robert-Smith J, Landaw SA, Davey FR (1979) Hodgkin's disease: intramedullary phagocytosis with pancytopenia. Ann Intern Med 91:60
13. Manoharan A, Catovsky D, Lampert IA, Al-Mashadhani Gordon-Smith EC, Galton DAG (1981)

Histiocytic medullary reticulosis complicating chronic lymphocytic leukaemia: malignant or reactive. Scand J Haematol 26:5

14. Rappaport H (ed) (1966) Tumours of the hemopoietic system. Armed Forces Institute of Pathology, Washington (Atlas of tumour pathology, sect III, fasc 8, pp 49, 442)
15. Risdall RJ, McKenna RW, Nesbit ME, Krivit W, Balfour HH, Simmons RL, Brunning RD (1979) Virus-associated hemophagocytic syndrome: a benign histiocytic proliferation distinct from malignant histiocytosis. Cancer 44:993
16. Rosner R (1979) 'Secondary' or 'transient' histiocytic medullary reticulosis. Lancet 1:438
17. Schumacher HR, Stass SA (1979) Histiocytic medullary reticulosis. Lancet 1:158
18. Scott RB, Robb-Smith AHT (1939) Histiocytic medullary reticulosis. Lancet 2:194
19. Shreiner DP (1975) Acute lymphoblastic leukemia terminating as histiocytic medullary reticulosis. JAMA 231:838
20. Skarin AT, Karb K, Reynolds ES (1972) Acute leukemia terminating in histiocytic medullary reticulosis. Arch Pathol 93:256
21. Warnke RA, Kim H, Dorfman RF (1975) Malignant histiocytic (histiocytic medullary reticulosis). I. Clinico-pathological study of 29 cases. Cancer 35:215
22. Wick MR, Li C, Ludwig J, Levitt R, Pierre RV (1980) Malignant histiocytosis as a terminal condition in chronic lymphocytic leukemia. Mayo Clin Proc 55:108

Haematology and Blood Transfusion Vol 27
Disorders of the Monocyte Macrophage System
Edited by F. Schmalzl, D. Huhn, H.E. Schaefer
© Springer-Verlag Berlin Heidelberg New York 1981

Therapy of Malignant Histiocytosis

D. Huhn

Rappaport in 1966 [10] introduced the term "malignant histiocytosis" to describe a disorder of "systemic, progressive, invasive proliferation of atypical histiocytes." An identical disorder was observed by Scott and Robb-Smith in 1939 [11] and termed "histiocytic medullary reticulosis." The most conspicuous clinical features of this disorder are enlargement of lymph nodes, liver, and spleen, and widespread multifocal involvement of the lymphatic and reticulohistiocytic system during the early phase of the disease. In the last 3 years clinical findings and therapeutic results were published [1, 6, 7, 14] and will be presented and discussed in the following.

1. Material and Methods

1.1 Patients

The 16 patients reported on were 16 to 73 years of age (Fig. 1), with diagnosis of malignant histiocytosis confirmed at the Department of Pathology (Prof. Dr. P. Meister) and cared for at the different clinical institutions of the University of Munich between 1974 and 1980. Clinical findings from 14 of these patients were published in 1980 [5], but two patients of this former group were omitted because of loss to follow up.

Histological diagnosis of MH was confirmed by the following supplementary investigations: staining for lysozyme and α_1-antichymotrypsine in paraffine sections by immunoperoxidase method in 13 of our patients [9], cytochemistry in imprints of biopsy material and electron microscopy in eight patients,

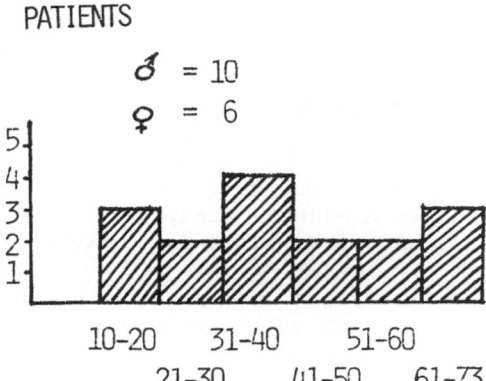

Fig. 1. Malignant histiocytosis: age at diagnosis

Table 1. CHOP-therapy [1]. Repeated after 3 weeks if WBC count > 3500 and platelet count > 100 000 and continued until a total dose of 450 mg/m^2 of adriamycin was reached

Day 1:	Cyclophosphamide	750 mg/m^2 I.V.
	Adriamycin	50 mg/m^2 I.V.
	Vincristine	2 mg I.V.
Days 1–5:	Prednisone	50 mg/m^2 P.O.

in vitro phagocytosis of opsonized moniliasis in two patients, and investigation of the growth of histiocytic cells in diffusion chambers in one case (Dr. B. Lau).

1.2 Methods

Methods comprised immunohistochemical staining for lysozyme and α_1-antichymotrypsine [9], cytochemical staining for acid phosphatase, different esterases, peroxidase, PAS, and electron microscopy [5]; and diffusion chamber tests [8].

For the definition of the clinical stage (Ann Arbor) roentgenographic, scintigraphic, and biopsy investigations were performed as indicated below.

1.3 Therapy

Besides single drug chemotherapy, the following drug combination were used: CHOP (Table 1; [1]); COPP [4, 13]; cyclophosphamide, vincristine, and prednisolone (COP) [2]; and adriamycin, bleomycin, vinblastine, and DTIC (ABVD) [3]. Response to therapy was scored according to the most favorable status achieved during treatment: CR = no evidence of disease; PR = partial regression of at least 50% of all measurable lesions; and PROG = progression.

2. Results

2.1 Morphology

The true histiocytic nature of the malignant cells could be verified by cytochemical tests, and malignant histiocytes marked by various stages of morphological differentiation were observed simultaneously. Cells with a low degree of morphological differentiation were referred to as poorly differentiated, blast-like cells, cells exhibiting a high degree of differentiation, as well-differentiated histiocytic cells. In all patients the various stages of cell differentiation were observed, including all transitional forms, but the prevalence of either poorly or well-differentiated cells was different in individual patients.

Poorly differentiated cells appeared to be blast like; they contained only very few granules. There was a faint but distinct activity of naphthol-AS-acetate esterase and of acid phosphatase.

Well-differentiated cells were marked by polymorphous and often lobulated nuclei; the greyish cytoplasm contained many small azurophilic granules, vesicles, and phagocytosed material. Acid phosphatase and esterase were strongly positive. PAS-stainable material may be present. There was never any activity of peroxidase nor of naphthol-ASD-chloroacetate-esterase in histiocytic cells of either the blast-like or well-differentiated type.

With only few exceptions, malignant histiocytic cells were distinctly positive when stained for lysozyme and α_1-antichymotrypsine.

In two cases phagocytosis of yeast was demonstrated in-vitro. In one patient, malignant histiocytic cells of a predominantly well-differentiated type were introduced into diffusion chambers. After 6, 9, and 13 days of culture a gradual transition to blast-like cells was observed.

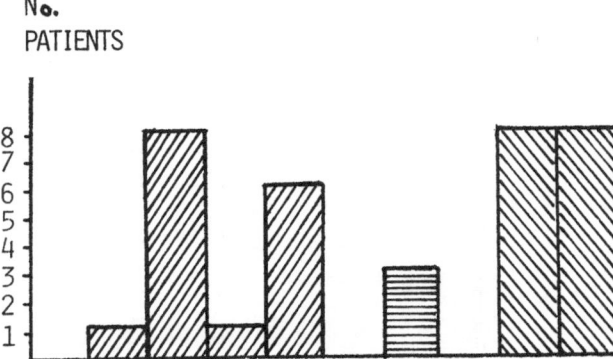

No.
PATIENTS

Fig. 2. Malignant histiocytosis:
stage (Ann Arbor) at diagnosis

2.2 Clinical Data

Eight patients presented with fever, six with loss of weight and two with jaundice. Hemolysis was demonstrated in two patients, anemia in six, leukopenia in two, and thrombopenia in three patients. Immunological parameters were impaired in several patients: lymphocytopenia in six patients and decrease of IgG in three, of IgA in one, and of IgM in three patients.

To define the stage of the disease, the following investigations were performed and revealed involvement by malignant histiocytosis: bone marrow biopsy was positive in 5 of 15 patients investigated; scintigraphy of the skeleton, 2 of 13; of the liver, 3 of 10; of the spleen, 2 of 7; lymphography, 1 of 7; abdominal computerized tomography, 2 of 7; explorative laparotomy, 2 of 2. Stages emerging from these investigations are indicated in Figure 2.

2.3 Therapy

Treatment is summarized in Table 2. *Radiation therapy* was used as initial treatment in five patients (Prof. Dr. v. Lieven, München). This treatment was planned as only "palliative" in two patients (patients No. 4 and 5) and as "curative" in three patients. In the latter three patients local recurrences occurred in spite of doses usually curative in malignant lymphoma, and in two of those patients a long-lasting complete remission (CR) was subsequently achieved by polychemotherapy.

In two patients initial treatment was *splenectomy* (for diagnostic purposes and with preliminary diagnosis of carcinoma of pancreas or non-Hodgkin-lymphoma), resulting in one patient (No. 5) in a remission of 6 months duration (not indicated in Table 2).

In six patients initial treatment was CHOP (Table 1), resulting in a CR in four patients, a PR in two, and in a mean survival of 18 months with four patients still alive. Using polychemotherapy regimens without adriamycin, CR was achieved in one patient (second treatment, patient No. 1) and partial remission (PR) in two patients, while disease was progressive in three patients (including a second treatment in No. 3 and 5).

3. Discussion

Cytochemical and immunohistiochemical investigations may be helpful in diagnosis and differential diagnosis of malignant histiocytosis. Malignant histiocytes are quite similar to the phagocytosing reticular cells of the lymph nodes and bone marrow [5]. In all patients, *mature histiocytic cells* were detected. These cells were characterized by a histiocyte-like

Table 2. Results of therapy in 16 patients[a]

Patient	Stage	1. Treatment	→ result	2. Treatment	→ result	Survival from diagnosis (MO)
1	II A	↯	→ PR	COPP	→ CR	36 +
2	II AE	↯	→ PR	CHOP	→ CR	42 +
3	II B	↯	→ PR	COP	→ PROG	12
4	II BE	↯	→ PROG			3
5	IV B	↯	→ PR	COPP	→ PROG	14
6	I A	CHOP	→ CR			37 +
7	II A	CHOP	→ CR			25 +
8	II A	CHOP	→ CR			18
9	II BE	CHOP	→ PR			3 +
10	III A	CHOP	→ PR			10
11	IV B	CHOP	→ CR			6 +
12	II A	COPP	→ PROG	CHOP	→ PR	4 +
13	IV B	COP	→ PR			13
14	IV B	COP	→ PR			14
15	IV A	ADRIAM., VINCR.	→ PR	ABVD	→ PROG	8
16	IV B	ADRIAM., AraC				1

Mean survival 15 +

[a] C = cyclophosphamide, O = vincristine, P = prednisolone, H = adriamycine, ↯ = irradiation, CR = complete remission, PR = partial remission

morphology in light microscopy, sometimes exhibiting erythrophagocytosis or phagocytosis of opsonized yeast in vitro. These cells were, furthermore, marked by a distinct activity of acid phosphatase and esterase but no peroxidase reactivity. Electron microscopy revealed an abundance of small granules, vesicles, and endoplasmic reticulum. In addition, in all patients investigated less-differentiated, *blast-like cells* could be demonstrated. In these cells esterase was not as prominent as in more mature forms; peroxidase and naphthol-ASD-chloracetate-esterase (as markers for myeloic cells) were completely absent; granules and other cell organelles were rare. So, in malignant histiocytosis transitional forms between histiocytic cells and monocytic–promonocytic cells cannot be demonstrated (as in the development of normal histiocytes).

Taking these findings into consideration, two questions should be discussed. First, is there a justification for regarding different histiocytic cells as "mature" or as "immature" forms? Second, may different morphological subtypes, as seen in our patients, be allocated to one disease entity? The term "mature" or "differentiated" was employed analogously to granulocytes and monocytic precursors; in these cell lines cell maturation is correlated with the development of cell organelles (especially granules) and enzyme activities as observed in malignant histiocytosis. The relationship of different morphological subtypes is underlined by the occurrence of various stages of differentiation simultaneously in one individual patient and the predominance of mature cell types changing gradually from one patient to the other. The changeability of cell morphology, in addition, is demonstrated by the morphological changes seen in diffusion chamber cultures.

Nearly half of our patients presented with *stage III* to *IV*, indicating an early generalization in malignant histiocytosis. Localized organ involvement and B symptoms were rather frequent. All findings are in favor of a natural progression of malignant histiocytosis starting as localized disease, prevalently of the lymph nodes, and spreading early to adjacent or distant organs, especially to bone marrow, liver, and mesenchymal tissues.

Clinical findings reported from other centers indicate a considerable heterogeneity of

Table 3. Results of therapy in malignant histiocytosis[a]

No. Patients	Age (mean)	Stage	Therapy	Results	Author
16 ♂ 13 ♀ 3	39	I-III: 4 IV : 12	Single agent chemotherapy COP CHOP	CR 1/11, PR 2/11 CR 1/5 , PR 1/5 CR 3/7 , PR 2/7 Median survival: 9 mo.	Alexander and Daniels [1]
12 ♂ 5 ♀ 7	37	IV : 12	No therapy CHOP Other agents	3, Survival: 2–6 mo. CR 4/7 PR 1/2 Median survival of treated: 12 mo.	Lampert et al. [7]
22 ♂ 11 ♀ 11	7.5	Bone marrow 5/19	No therapy ↙ CHOP, COH	4, Survival: 1–3 mo. 3, Survival: 1–9 mo. CR 13/15 Median survival after chemotherapy: 24 + mo.	Zucker et al. [14]
16 ♂ 10 ♀ 6	42	I-III: 10 IV : 6	↙ CHOP COP (+ procarbazine) Other agents	PR 1/1 CR 5/7, PR 2/7 CR 1/6, PR 2/6 R 0/2 Median survival: 14 + mo.	Huhn et al. [6]

[a] Abbreviations as in Table 2

patient groups (Table 3). In our patients as in cases reported by Zucker et al. [14] frequency of organ involvement and of dissemination at diagnosis was less than 50%, in patients reported by Alexander [1] about 75%, and in 12 cases published by Lampert et al. [7] 100%. In contrast to this diversity, best *treatment results* in all groups were reported when polychemotherapy with cyclophosphamide, adriamycine, vincristine, and prednisone (CHOP and similar combinations) was tried (Table 3). Therefore, as treatment of choice a combination of irradiation and CHOP-therapy is advocated in stage I and II. In further advanced stages, exclusive chemotherapy will be preferable. The advantage of an additional main bulk irradiation or of a prophylactic CNS treatment cannot yet be evaluated. In patients resistant to CHOP-therapy, high-dose methotrexate in combination with leukovorine rescue was tried with modest success [1]. In some exceptional cases, remissions lasting for some months were achieved with splenectomy [12].

4. Summary

Diagnosis of malignant histiocytosis was confirmed in 16 patients. Stage at diagnosis was I–II in nine and III–IV in seven patients. Poor prognosis and "B" symptoms were correlated to advanced stages. Relapses after radiotherapy were frequent. Polychemotherapy using "CHOP"-combination is recommended for most patients and may in stages I–II be supplemented by primary or secondary involved or extended field irradiation and in more advanced stages by main bulk irradiation. Therapeutic results published from other centers are discussed.
Added after completion of the manuscript:
    When histiocytic cells of our cases are compared with those described by Manoharan and Catovsky (this volume p. 205), the following conclusions may be drawn:
    1. "Poorly differentiated" histiocytic cells as described in our series correspond to stage 1

and/or stage 2 mononuclear macrophages as described by Manoharan. 2. "Well-differentiated" histiocytic cells of our series correspond to stage 2 and/or stage 3 cells. 3. Cases characterized by a substantial histiocytic cell population equivalent to stages 4 and 5 macrophages as defined by Manoharan and marked by small nuclei, condensed chromatin, and abundant greyish and foamy cytoplasm are not included in our series.

References

1. Alexander M, Daniels JR (1977) Chemotherapy of malignant histiocytosis in adults. Cancer 39: 1011–1017
2. Bagley CM Jr, DeVita VT Jr, Berard CW, Canellos GP (1972) Advanced lymphosarcoma: Intensive cyclical combination chemotherapy with cyclophosphamide, vincristine, and prednisone. Ann Intern Med 76:227
3. Bonadonna G, De Lena M, Monfardini S, Rossi A, Brambilla C, Uslenghi C, Zucali R (1975) Combination usage of adriamycin (NSC-123127) in malignant lymphomas. Cancer Chemother Rep 6:381
4. DeVita VT Jr, Serpick AA, Carbone PP (1970) Combination chemotherapy in the treatment of advanced Hodgkin's disease. Ann Intern Med 73:881–895
5. Huhn D, Meister P (1978) Malignant histiocytosis. Morphologic and cytochemical findings. Cancer 42:1341–1349
6. Huhn D, Meister P, Wilmanns W (1980) Malignant histiocytosis. Clinical findings and therapy. Klin Wochenschr 58:31–35
7. Lampert IA, Catovsky D, Bergier N (1978) Malignant histiocytosis: A clinico-pathological study of 12 cases. Br J Haematol 40:65–77
8. Lau B, Jäger G, Thiel E, Rodt H, Huhn D, Pachmann K, Netzel B, Böning L, Thierfelder S, Dörmer P (1979) Growth of the Reh cell line in diffusion chambers. Evidence for differentiation along the T- and B-cell pathway. Scand J Haematol 23:285–292
9. Meister P, Huhn D, Nathrath W (1980) Malignant histiocytosis. Immunohistochemical characterization on paraffin embedded tissue. Virchows Arch [Pathol Anat] 385:233–246
10. Rappaport H (ed) (1966) Tumors of the hematopoietic system. *Armed Forces Institute of Pathology, Washington.* Atlas of tumor pathology, Sect 3, fasc 8, pp 49–63
11. Scott RB, Robb-Smith AHT (1939) Histiocytic medullary reticulosis. Lancet 2:194
12. Vardiman JW, Byrne GE Jr, Rappaport H (1975) Malignant histiocytosis with massive splenomegaly in asymptomatic patients. Cancer 36:419–427
13. Wilmanns W, Wilms K, Koos T (1976) Langzeitremissionen bei fortgeschrittenen Stadien der Lymphogranulomatose nach kombinierter Chemotherapie mit einem modifizierten de Vita-Schema. Lehmann, Munich (Hämatologie und Bluttransfusion, vol 18)
14. Zucker JM, Caillaux JM, Vanel D, Gerard-Marchant R (1980) Malignant histiocytosis in childhood. Clinical study and therapeutic results in 22 cases. Cancer 45:2821–2829

Haematology and Blood Transfusion Vol 27
Disorders of the Monocyte Macrophage System
Edited by F. Schmalzl, D. Huhn, H. E. Schaefer
© Springer-Verlag Berlin Heidelberg New York 1981

Immunohistochemical Marking of Malignant Fibrous Histiocytoma and Malignant Histiocytosis

P. Meister and W. Nathrath

The purpose of our study was to test markers for histiocytic tumor cells which can be applied to formalin-fixed, paraffin-embedded material. Two groups of lesions were studied: [1] malignant fibrous histiocytoma (MFH), which has to be distinguished from other fiber forming sarcomas, and [2] malignant histiocytosis (MH), which has to be recognized among "round cell sarcomas" (including malignant lymphomas) and occasionally even among undifferentiated carcinomas.

1. Material and Methods

In our current material 70 tumors had been classified as MFH based on typical storiform pattern and/or histiocytic activity of tumor cells [19]. An immunohistochemical examination was carried out in 35 MFH. Sixteen cases had been diagnosed as MH because of suggestive features on routine histological preparations, supported by cytochemical findings of acid phosphatase and nonspecific esterase reaction [4, 5, 11]. In 13 cases an immunohistochemical examination on paraffin-embedded material was done.

Lysozyme (muramidase) and alpha$_1$-antichymotrypsin, considered to be markers of histiocytes [6, 7, 13, 16, 17], were studied by the indirect immunoperoxidase technique [1–2, 3, 15, 18, 20]. While lysozyme could be demonstrated without pronase pretreatment, alpha$_1$-antichymotrypsin was unmasked by predigestion with (0.1%) pronase (incubation time ca. 20 min), which also reduces background staining, especially of collagen [14].

Controls were incubated with PBS, without the first and/or without the second antiserum. A positive reaction is expressed by a granular orange-brown marking of tumor cells by 3-amino-9-ethyl-carbazole-AEC [8, 9]. In some cases a predeliction for either cell periphery or perinuclear area was suggested [9].

A semiquantitative evaluation of positive reactions was carried out (+ denotes that few cells are marked and ++ denotes that numerous tumor cells show positive marking [11]).

2. Results

Of 35 MFH, 26 showed positive marking with alpha$_1$-antichymotrypsin. The reaction was graded + in 11 and ++ in 15 cases. With lysozyme the reaction was positive in a total of 16 cases, + in 12 and ++ in 4 [9].

Of 13 previously diagnosed MH, 12 showed positive reactions with alpha$_1$-antichymotrypsin as well, as with lysozyme. In about half of the cases the findings were + or ++, each with either method. The only completely negative lesion possible consisted of immature (monocytic) elements [4, 5].

In comparison cases with histiocytosis X were also studied, showing an almost ubiquitous positive reaction with alpha$_1$-antichymotrypsin and lysozyme [11]. Various granulomas revealed positive or negative findings in epitheloid cells and in multinucleated giant cells [12]. Nonspecific granulation tissue was also examined: here the alpha$_1$-antichymotrypsin reaction was advantageous over the lysozyme reaction, because by the latter not only histiocytes but also polymorphonuclear leukocytes are marked [11].

Whereas positive reactions by both methods can be expected, for instance, with mammary duct and pancreas carcinomas which are compatible with positive reactions in corresponding nonneoplastic cells, a positive reaction with sarcomas was only found in degenerating necrobiotic tumor cells or around foci of necrosis [10]. Here also MFH and MH revealed marking which was stronger than in the remaining tissue [9]. Markings in vital tumor cells were found in some cases with malignant melanoma [9]. Whereas the normal distribution of lysozyme is better known [6, 7, 17], we are now in progress to study that of alpha$_1$-antichymotrypsin.

3. Conclusion

With MFH the marking was more often positive with alpha$_1$-antichymotrypsin than with lysozyme. With MH both markers were simultaneously present with the exception of one possible immature case.

Wrong negative findings could be either due to the immaturity of cells [4] which did not yet show histiocytic differentiation (with MH) or possibly also to the absence of loss of histiocytic differentiation, for instance, with acquisition of fibroblastic qualities (with MFH) [10].

Wrong positive findings can be due to scattered reactive macrophages intermingled with neoplastic cells or with degenerating cells in mesenchymal tumor cells of various histiogenesis. One exception was found with some cases of malignant melanomas, where undamaged tumor cells might also show positive reactions with alpha$_1$-antichymotrypsin as well as with lysozyme (and sometimes even with the PBS control alone) [8-10].

Immunohistochemical marking with alpha$_1$-antichymotrypsin and/or lysozyme on formalin-fixed, paraffin-embedded material appears to be a valuable supportive diagnostic method for the recognition of histiocytic tumors (MFH and especially MH) when evaluated in context with other characteristic histological findings [9].

References

1. Burns J (1975) Background staining and sensitivity of the unlabelled antibody-enzyme (PAP) method. Comparison with the peroxidase labelled antibody Sandwich method using formalin fixed paraffin embedded material. Histochemistry 43:291–294
2. Graham RC, Lundholm U, Karnovsky MJ (1965) Cytochemical demonstration of peroxidase activity with 3-amino-9-ethylcarbazole. J Histochem Cytochem 13:150–152
3. Heyderman E, Neville AM (1977) A shorter immunoperoxidase technique for the demonstration of carcinoembryonic antigen and other cell products. J Clin Pathol 30:138–140
4. Huhn D, Meister P (1978) Malignant histiocytosis. Morphologic and cytochemical findings. Cancer 42:1341–1349
5. Huhn D, Meister P, Thiel E, Bartl R, Theml H (1978) Maligne Histiozytose. Dtsch Med Wochenschr 103:55–61
6. Klockars M, Reitamo S (1975) Tissue distribution of lysozyme in man. J Histochem Cytochem 23:932–940
7. Mason·DY, Taylor CP (1975) The distribution of muramidase (lysozyme) in human tissue. J Clin Pathol 28:124–132
8. Meister P, Nathrath W (1980) Immunhistochemical markers of histiocytic tumours. Hum Pathol 11:300–301
9. Meister P, Nathrath W (1981) Immunohistochemical characterisation of histiocytic tumours. Diagn Histopath 4:79–87

10. Meister P, Konrad EA, Nathrath W, Eder M (1980) Malignant fibrous histiocytoma: histological patterns and cell types. Pathol Res Pract 168:193–212
11. Meister P, Huhn D, Nathrath W (1980) Malignant histiocytosis. Immunhistochemical characterisation on paraffin embedded tissue. Virchows Arch Pathol Anat 385:233–246
12. Meister P, Nathrath W, Konrad E (1980) Immunhistochemische Markierung histiozytärer Tumorzellen. Verh Dtsch Ges Pathol
13. Motoi M, Schwarze EW, Stein H, Lennert K (1978) Lysozymnachweis in histiozytischem Retikulosarkom und maligner Retikulose. Verh Dtsch Ges Pathol 62:514
14. Nathrath W, Meister P (in press) Lysozyme (muramidase) and alpha$_1$-antichymotrypsin as immunohistochemical tumor markers. Acta histochem
15. Neville AM, Grigor KM, Heyderman E (1978) Biological markers and human neoplasia. In: Recent advances in histopathology. Livingstone, Edinburgh London New York, pp 23–44
16. Papadimitriou CS, Stein H, Lennert K (1978) The complexity of immunohistochemical staining pattern of Hodgkin- and Sternberg-Reed cells. – Demonstration of immunoglobulin, albumin, alpha$_1$-antichymotrypsin and lysocyme. Int J Cancer 21:531–541
17. Pinkus GS, Said JW (1977) Profile of intracytoplasmic lysozyme in normal tissues, myeloproliferative disorders, hairy cell leukemia and other pathologic processes. Am J Pathol 89:351–366
18. Schaefer HE, Fischer R (1968) Der Peroxydasenachweis an Ausstrichpräparaten sowie an Gewebsschnitten nach Entkalkung und Paraffineinbettung. Klin Wochenschr 46:1228–1230
19. Weiss S, Enzinger F (1978) Malignant fibrous histiocytoma. Cancer 41:2250–2266
20. Witting C (1977) Immunofluorescence studies on formalin-fixed and paraffin-embedded material. Beitr Pathol 161:288–291

Haematology and Blood Transfusion Vol 27
Disorders of the Monocyte Macrophage System
Edited by F. Schmalzl, D. Huhn, H.E. Schaefer
© Springer-Verlag Berlin Heidelberg New York 1981

Malignant Histiocytic Disorders of the Skin

G. Burg

Histiocytic disorders of the skin can be differentiated according to the following criteria:

1. Grade of malignancy: benign vs. malignant.
2. Number and localization of the lesions: solitary vs. multiple and localized vs. disseminated.
3. Origin: primary skin involvement vs. secondary skin involvement.
4. Natural course: progressive vs. self-healing.

Recent studies have shown that many of the previously described "histiocytic" disorders are in fact large cells (immunoblastic, lymphoblastic, or centroblastic) lymphocytic malignant tumors [26].

In the following diseases histiocytes are the predominant cell type. They often result in a fatal outcome and therefore should be considered as malignant:

Histiocytosis X
Malignant histiocytosis
Multicentric reticulohistiocytosis
Lipogranulomatosis (Farber)[1]
Myelomonocytic leukemia
Reticulohistiocytoma of the Adult's Back (Crosti)
Reticulosarcoma

Recent review articles on histiocytic proliferations in the skin demonstrate that there is a vast spectrum of dermatoses differing in their clinical and histological characteristics [3]. However, the group best characterized and most precisely defined is that of histiocytosis X, whereas others just may be variations of one entity.

1. Histiocytosis X

The term "histiocytosis X" [20] refers to a group of systemic proliferations of histiocytes encompassing Letterer-Siwe disease, Hand-Schüller-Christian disease, and eosinophilic granuloma of the bone, all of which may have cutaneous lesions [12]. On the histological level it may be difficult to differentiate reticulum cells or histiocytes from blasts of the lymphocyte series. In these cases the cytochemical pattern of enzymes in the cells is of great importance.

Skin involvement is frequently seen in Letterer-Siwe disease in young children, usually presenting as eczematous and hemorrhagic lesions. In Hand-Schüller-Christian disease in about one third of the patients erosive, crusty lesions are found in the intertrigines and

---

[1] This rare disease is a storage disorder. However, since the children die during the first months or years of life, it is listed here.

around the orificiae corporis. Eosinophilic granuloma occasionally shows manifestation in the skin presenting as a solitary lesion.

The atypical histiocytes in histiocytosis X contain Langerhans cell granules. However, in contrast to the typical intraepidermal Langerhans cells, the histiocytosis X histiocyte does not protrude long branches but is rather oval or round shaped. The specific granules found in about 50% of the histiocytosis X cells frequently show accumulation at the periphery of the cell [2] and may be longer than those seen in typical Langerhans cells. Electron microscopy has proved to be a valuable tool for confirming the diagnosis of skin lesions as manifestations of histiocytosis X [29].

2. Malignant Histiocytosis (Histiocytic Medullary Reticulosis) [24]

Cutaneous manifestations may occur in about 10% of cases [2] and may be the initial and the predominant manifestation of the disease [4, 13, 21, 22]. They usually consist of papules, 1–10 mm in diameter, and are hemorrhagic in the center. Subcutaneous tumors with normal or crusted and necrotizing overlying skin may also develop.

Histologically, the skin infiltrates are composed of obviously atypical mononuclear round cells with intermingled large histiocytic cells that characteristically show erythrophagocytosis, a rare phenomenon in cutaneous infiltrates [22]. Large amounts of acid phosphatase and nonspecific esterases, but no peroxidase or naphthol-AS-D-chloracetate esterase, are demonstrable. Immunocytologically, the neoplastic cells bear receptors for C_3. They phagocytose latex beads [22].

3. Multicentric Reticulohistiocytosis (Lipoid Dermatoarthritis)

This is a rare and histologically unique disorder, characterized by nodular lesions in the skin, mucosa, subcutaneous tissue, synovia, and, the times, periosteum and bone which result in destructive arthritis [6, 28]. Multicentric reticulohistiocytosis is a systemic disease of unknown cause with a tendency to heal spontaneously. However, it is listed here, since there is a high coincidence with malignant diseases [9]. Pathogenetically, disturbance of the intracellular metabolism within histiocytic cells is hypothesized.

The histology reveals a typical of "histiocytosis gigantocellularis". Granulomatous infiltrates are found in the upper and mid dermis, sometimes in the subcutis which consists of histiocytic cells and multinucleated giant cells with peaked projections of the cytoplasm that stain strongly positive with periodic acid Schiff. Histochemical examination of the material in the giant cells shows lipids (phospholipid and neutral fat) and polyaccharides attached to proteins [15].

4. Lipogranulomatosis (Farber)

This is an extremely rare, inherited, autosomal recessive lipoglycoprotein storage disease (ceramidase deficiency), first described by Farber et al. in 1957 [11]. The major manifestations include nodules, mainly over the wrists and ankles with painful swelling of the joints causing restrictions of movement, mental retardation, and failure to thrive. Further symptoms include hoarseness, noisy breathing, hyperirritability, dyspnea, hepatomegaly, flesh-colored papular eruptions, xanthoma-like lesions on the face and the hands, histiocytic granulomas in pressure areas, and hyporeflexia [2, 23, 25].

In the terminal stage there may be complete paralysis of the extremities. The majority of the patients die by 2 years of age, and therefore this disorder is listed here among malignant diseases. Autopsy findings show involvement of joints, cardiovascular and respiratory tract, central nerve system, and kidney.

Biochemically, great amounts of free ceramid are stored due to an acid ceramidase deficiency [28]. Light microscopy reveals large histiocytes, lymphocytes, fibroblasts, and foam cells. In cryostat sections there is a strong positivity in the periodic acid Schiff reaction and a weak reaction with Sudan stains. Electron microscopy shows "cebra bodies" in endothelial cells, numerous curvilinear bodies in fibroblasts, and in Swann cells unique storage elements resembling bananas or bunches of bananas.

5. Skin Manifestations in Myelomonocytic Leukemia

Cutaneous infiltrates are seen in 20%–25% of cases of acute myelomonocytic leukemia [7, 17]. Chronic myelomonocytic leukemia is a rare condition and thus no good information about cutaneous involvement is available. Many cases formerly designated as "reticulosis" or so-called "reticulosarcomatosis Gottron" in the European dermatological literature are in fact myelomonocytic leukemias [18].

Cutaneous lesions may consist of disseminated brown or red macules or brown-red or blue-red papules. Of considerable diagnostic importance is a diffuse or nodular hyperplasia of the gingiva, which, sometimes occurring with ulceration.

Histologically, the skin lesions show a perivascular, periadnexal or diffuse proliferation, predominantly of mononuclear cells sometimes showing deeply indented and twisted nuclei. The presence of immature eosinophils may serve as a useful diagnostic aid.

Depending on the degree of differentiation, cytochemically the neoplastic cells in skin infiltrates usually contain nonspecific esterases and acid phosphatase; additionally, slight activities of naphthol-AS-D-chloroacetate esterase and peroxidase may be found.

6. Reticulohistiocytoma of the Adult's Back (Crosti)

The exact position of this disease is not completely clear and its distinct entity debatable. It was defined as a peculiar form of well-circumscribed, slowly developing proliferation of neoplastic histiocytic cells in adults [10]. However, this disorder probably comprises a heterogeneous group of diseases including mycosis fungoides *"d'emblée"* and various types of cutaneous B cell lymphomas formerly described as "malignant reticulosis."

7. Reticulosarcoma of the Skin

Most of the lymphoreticular disorders formerly classified as reticulosarcoma have been shown to be proliferations of immunoglobulin-producing immunoblasts [5, 16, 26] or of large cells of follicle centers. On the basis of enzyme cytochemical, electronmicroscopical, and immunological studies, four different types of reticulum cells have been described [19]. There is some evidence that these cell types also may be present in the skin [6]. It remains to be established, however, whether or not each of them may develop a specific "reticulosarcoma".

8. Conclusions

The skin provides a suitable microenvironment for cells of the monocytic-macrophage series to settle, to accumulate, and under certain physiologic conditions to proliferate. The spectrum of skin lesions in malignant histiocytic disorders is vast. They may develop as secondary manifestation of a systemic disease or may be the primary site of manifestation.

Cytochemistry is the most helpful procedure for the differentiation of true histiocytic skin infiltrates from infiltrates that formerly were designated as "malignant reticulosis" and that today are known usually to be cutaneous B cell lymphomas. Especially in the group of large cell ("histiocytic") lymphomas with skin infiltrates the diagnosis should not be made without regard of the enzyme cytochemical pattern of the cells.

References

1. Abele DC, Griffin TB (1972) Histiocytic medullary reticulosis. Arch Dermatol 106:319–329
2. Becker H, Auböck L, Haidvogel M, Bernheimer H (1976) Disseminierte Lipogranulomatose – kasuistischer Bericht des 16. Falles einer Ceramidose. Verh Dtsch Ges Pathol 60:254–258
3. Bonvalet D, Civatte J (1980) Histiocytosis histopathology. G Ital Dermatol Venereol 115:51–58
4. Büchner SA, Rufli T (1980) Maligne Histiozytose mit Hautmanifestationen. Dtsch Med Wochenschr 105:373–377
5. Burg G, Braun-Falco O (to be published) Cutaneous lymphomas, pseudolymphomas, and related disorders. Springer, Berlin Heidelberg New York
6. Burg G, Braun-Falco O, Schmoeckel C, Hoffmann-Fezer G, Fateh-Moghadam A, Herterich J (1978) Preferential microenvironments for B- and T-lymphocytes in the skin. INSERM 80:221–283
7. Burg G, Schmoeckel C, Braun-Falco O, Wolff HH (1978) Monocytic leukemia. Arch Dermatol 114:418–420
8. Caro MR, Senear FE (1952) Reticulohistiocytoma of the skin. Arch Dermatol Syphilol 65:701–713
9. Catterall MD, White JE (1978) Multicentric reticulohistiocytosis and malignant disease. Br J Dermatol 98:221–224
10. Crosti A (1951) Micosi fungoide e reticolo-istiocitomi cutanei maligni. Minerva Dermatol 26:3–11
11. Farber S, Cohen J, Uzman LL (1957) Lipogranulomatosis; a new lipoglycoprotein storage disease. J Mount Sinai Hosp 24:816–837
12. Gebhart W, Knobler R, Niebauer G (1980) Langerhans cells in histiocytosis X. G Ital Dermatol Venereol 115:121–128
13. Hödel S, Auböck L, Kerl H (1978) Maligne Histiozytose. Hautarzt [Suppl 3] 29:93–96
14. Holubar K, Mach K (1966) Histiocytosis giganto-cellularis. Hautarzt 17:440–445
15. Ikezawa Z, Nakajima H (1976) A study of multicentric reticulohistiocytosis (Lipoid dermatoarthritis). J Dermatol 3:289–302
16. Kerl H, Burg G (1979) Immunozytome und immunoblastische Lymphome der Haut. Hautarzt 30:666–672
17. Kerl H, Kresbach H, Hödel S (1978) Klinische und histologische Kriterien zur Diagnose und Klassifikation der Leukämien der Haut. Hautarzt [Suppl 3] 29:97–101
18. Klein UE, Ude P (1975) Monozytenleukämien mit ungewöhnlichem Erkrankungsverlauf. Med Klin 70:613–621
19. Lennert K, Müller-Hermelink HK (1975) Lymphozyten und ihre Funktionsformen-Morphologie, Organisation und immunologische Bedeutung. Verh Anat Ges 69:19–62
20. Lichtenstein L (1953) Histiocytosis X. Integration of eosinophilic granuloma of bone, "Letterer-Siwe-disease" as related manifestations of a single nosologic entity. AMA Arch Pathol 56:84–102
21. Nishio K, Koda H, Urabe H (1975) Über einen Fall von Histiocytic Medullary Reticulosis. Arch Dermatol Res 251:259–269
22. Robinowitz BN, Noguchi S, Bergfeld WF (1977) Tumor cell characterization of histiocytic medullary reticulosis. Arch Dermatol 113:927–929
23. Schmoeckel C, Hohlfeld M (1979) A specific ultrastructural marker for disseminated lipogranulomatosis (Farber). Arch Dermatol Res 266:187–196
24. Scott RB, Robb-Smith AHT (1939) Histiocytic medullary reticulosis. Lancet 2:194–198
25. Sisson TRC (1971) Farber's disease (Lipogranulomatosis). Am J Dis Child 122:513
26. Stein H, Kaiserling E, Lennert K (1974) Evidence for B-cell origin of reticulum cell sarcoma. Virchows Arch [Pathol Anat] 364:51–67
27. Sugita M, Dulaney JT, Moser HW (1972) Ceramidase deficiency in Farber's disease (lipogranulomatosis). Science 178:1100–1102
28. Weber FP, Freudenthal W (1937) Nodular non-diabetic cutaneous xanthomatosis with hypercholesterolaemia and atypical histological features. Proc R Soc Med 30:522–526
29. Wolff HH, Braun-Falco O (1972) Zur Diagnostik und Therapie des Morbus Hand-Schüller-Christian. Hautarzt 23:163–169

Haematology and Blood Transfusion Vol 27
Disorders of the Monocyte Macrophage System
Edited by F. Schmalzl, D. Huhn, H.E. Schaefer
© Springer-Verlag Berlin Heidelberg New York 1981

Childhood Histiocytosis X: Clinical Aspects and Therapeutic Approaches

A.J. Feldges

Childhood histiocytosis X is a disorder of the monocyte–macrophage system with various clinical aspects and immunological abnormalities. The disease may be localized or disseminated, with an acute, subacute, or chronic course which can lead to death or many residual damages. Therapy depends on the extent of the disease and risk factors such as involvement of the lung, the liver, and the hematopoietic system. For localized lesions in the bone no therapy except surgical curettage is necessary. In radiotherapy a dose of 600 rad is usually adequate; the role of radiotherapy as opposed to treatment with surgical measures should be further investigated. For disseminated forms chemotherapy is the treatment of choice. Immunotherapy with thymic extract may be an alternative but must be studied further in the future. The response to specific therapy may be slow; modification or discontinuation depends on host tolerance and toxicity. Prospective controlled trials considering the natural history of the disease, immunological aspects and risk criteria are warranted.

In 1953 Lichtenstein [10] speculated that eosinophilic granuloma of the bone, Hand Schüller Christian syndrome (with the triad: exophthalmus, diabetes insipidus, and membraneous bone defects), and Letterer Siwe syndrome share a common pathogenetic principle and represent a different expression of a single disease of similar pathology and of unknown etiology. He created the term "histiocytosis X" and emphasized that overlapping clinical presentation is quite frequent. A critical analysis by Liebermann et al. in 1969 [11] and an excellent review by Vogel and Vogel in 1972 [17] made clear that in childhood histiocytosis X movement from an isolated eosinophilic granuloma of the bone to a disseminated disorder with multifocal soft tissue – and skeletal – involvement is possible and that the disease may be chronic with episodes of spontaneous remission and exacerbation. In the acute variant, the Letterer Siwe syndrome, however, the systemic disease with various organ involvements leads often to death due to intercurrent infections and bone marrow failure. Unfortunately, other illnesses such as histiocytic medullary reticulosis, the familial erythrophagocytic lymphohistiocytosis, and the malignant histiocytosis were lumped together under the term histiocytosis in the past years. These disorders must be clearly distinguished from childhood histiocytosis X, because they represent pure malignant variants and lack the characteristic cytological features of childhood histiocytosis X (Table 1).

Recent investigation of the morphology and function of mononuclear phagocyte system revealed, however, that the hallmark of childhood histiocytosis X is an infiltrative growth of a well-defined histiocyte, the Langerhans cell, and that the disease is associated with a hyperproliferation of the monocytopoiesis [2]. The Langerhans cell is a macrophage which can be identified with certainty only by electron microscope study, where one recognizes intracytoplasmic X granules. Pathologically the lesions of histiocytosis X appear as reactive infiltrates of histiocytes and lack the cellular atypia and homogeneity of a pure

Table 1. Disorders of the monocyte–macrophage system in children

1. Histiocytosis X, a proliferative disorder of the Langerhans cell[a]
   - Eosinophilic granuloma → unifocal   benign
                              ↘ multifocal
                                 ↑
   - Hand Schüller Christian Syndrome
     (chronic disseminated disease)
   - Letterer-Siwe Syndrome   ↓ "malignant"
     (acute disseminated disease with systemic symptoms)
2. Diseases related to childhood histiocytosis X:
2.1 Reactive benign disorders:
   - Congenital viral infections with immunodeficiencies
   - Graft versus host disease
2.2 Pure malignant disorders:
   - Histiocytic medullary reticulosis
   - Familial lymphohistiocytosis (erythrophagocytic reticulosis)
   - Malignant histiocytosis
   - Monocytic leukemia

[a] Ultrastructurally well-defined histiocyte with intracytoplasmic X-granules

malignant neoplasma, although clinically the disease behaves like a malignancy invading and destroying organs and tissues.

Histiocytosis X must also be distinguished from congenital viral infections like rubella and cytomegaly with immunodeficiency and from graft versus host disease (Table 1). Certain clinical and pathological similarities with these benign reactive lesions led to the suggestion that histiocytosis X may be an immunological disorder. Although routine tests of immunological functions are usually normal, there is evidence for some immunodeficiency, especially in regard to cell-mediated immunity [1]. The presence of autocytotoxic lymphocytes and antibodies against red blood cells in patients with visceral involvement suggested in addition that childhood histiocytosis X might be an autoimmune disorder. Encouraging results with disappearance of all lesions after the treatment with thymic extract have been reported more recently. This form of immunotherapy also corrected the abnormal immunity in these patients [13].

Table 2. Organ dysfunction and high risk criteria in histiocytosis X [7]

1. Liver (1 or more criteria)
    Hypoproteinemia (under 5.5 g/100 ml)
    Hypoalbuminemia (under 2.5 g/100 ml)
    Edema
    Ascites
    Hyperbilirubinemia (total bilirubin over 1.5 mg/100 ml)
2. Lung (in the absence of superimposed infection)
    Tachypnea
    Dyspnea
    Cyanosis
    Pneumothorax
    Pleuraeffusion
3. Hematopoietic system (1 or more)
    Anemia (Hb under 10 g‰/100 ml)
    Leukopenia (leukocytes under 4000/cu mm)
    Thrombocytopenia (thrombocytes under 100 000/cu mm)
4. Age less than 2 years

Table 3. Controlled chemotherapeutic trials in disseminated histiocytosis X[a]

Authors	Cytotoxic drugs	CR/PR[a]	Number of patients
Lahey et al. [9]	VBL	60%	37
Children's	VBL + PRED	60%	35
Cancer Study Group	6MP + PRED	44%	41
	CMB	27%	26
Jones et al. [4]	MTX + PRED	53%	17
CALGB	VCR + PRED	64%	11
Komp et al. [5]	VBL + PRED + CYCLO	88%[b]	17
SWOG			
Starling et al. [16]	VCR	50%	3
SWOG	VBL	55%	11
	CYCLO	63%	14
Smith et al. [15]	CMB	60%	10
Australia	MTX + VBL + PRED + CYCLO	80%	6
Feldges et al. [3]	VBL + PRED	100%[c]	10
SAKK			

[a] CR = complete remission; CYCLO = cyclophosphamide; PR = partial remission; VBL = vinblastine; 6MP = 6-mercaptopurine; VCR = vincristine; MTX = methotrexate; PRED = prednisone; CMB = chlorambucil; CALGB = Cancer Acute Leukemia Group B; SWOG = Southwestern Oncology Group; and SAKK = Schweizerische Arbeitsgruppe für klinische Krebsforschung
[b] For children more than 1 year old
[c] For children with only temporary response

The clinical aspects of childhood histiocytosis X are heterogeneous. The disease can be localized or disseminated. Prognosis and treatment are determined by the extent to which the lesions are found throughout the body. Careful assessment of the extent of the disease (staging) is therefore warranted before any form of treatment is started. In 1962 Lahey [6] reported a scoring system to evaluate the extent of the disease. One score represents the involvement of either one of the following organs or systems: skin, liver, spleen, lung, hypophysis, and skeletal and hematopoietic systems. Prognosis depended on the number of organs and systems which were involved. With a score of 1 to 2 the mortality rate was 5%, with 3 to 4, 14%, with 5 to 6, 75%, and with 7 to 9, 100%. Later the same author emphasized that mainly organ dysfunction, e.g., of the lung, the liver, and the hematopoietic system, determines the adverse prognosis of childhood histiocytosis X [7] (Table 2). The prognostic value of age is still a matter of debate [7, 12, 14]. Despite this controversy about the prognostic value of age, there is agreement that organ dysfunction and age of less than 2 years predict significantly for poor survival [6, 12, 14].

Unifocal eosinophilic granuloma of the bone has a high spontaneous healing rate whatever surgical procedure (biopsy or curettage) is performed. Usually reossification is seen within 2 to 5 months. Radiotherapy is indicated for painful lesions, for example in the vertebral bodies where collapse is warranted. Usually 600–800 rads are considered to be adequate. Chronic disseminated forms with involvement of soft tissue and bone have a high incidence of long-term sequelae and must therefore be treated. CNS disabilities are diabetes insipidus, growth stunting, deafness, blindness, mental retardation, and cerebellar dysfunction. In the lung interstitial fibrosis, bronchiectasis, cor pulmonale, and pneumonia are late sequelae of the disease, and in the liver obstructive hepatopathy, and portal fibrosis are reported. Pathological fracture, osteoporosis, and vertebral collapse are disabilities of eosinophilic granuloma in the bone. Diabetes insipidus requires in general life-long substitution with vasopressin, and growth hormone has been given to patients with impaired linear growth.

The role of chemotherapy has been investigated in the last three decades. Various agents like antibiotics, steroids, and cytotoxic drugs have been shown to be effective in the treatment of histiocytosis X. Chemotherapy should be restricted to patients with disseminated disease. Pulmonary involvement documented only on X-rays without symptoms is considered as a good or intermediate risk; the same holds for less than three simultaneous bone lesions [12]. For patients with multiple bone lesions and soft tissue involvement chemotherapy should be given considering the residual disabilities which occur in this type of disease. As of today only a few controlled trials for chemotherapy of disseminated histiocytosis X have been reported [3-5, 8, 9, 15, 16]. Alkylating agents, vinca alcaloids, and antimetabolites were shown to be effective. These drugs were used either as a monotherapy or as a combination chemotherapy (Table 3).

Comparing vincristine and vinblastine as a single agent, both drugs were equally effective with a response rate of 50%–55%. With cyclophosphamide the complete and partial remission rate was 63% [16]. When vinblastine was compared with vinblastine and prednisone and prednisone and 6-mercaptopurine, no difference could be shown between the three regimens [8]. Combination chemotherapy appeared to be not more effective than monotherapy, except for children older than 2 years without organ dysfunction [5, 15]. In one study it could be documented that children who receive maintenance therapy with methotrexate after achieving remission did better than children without this form of therapy [4]. Therefore, the modern principle of chemotherapy with induction and maintenance therapy should be considered in the disseminated form of childhood histiocytosis X [3, 15].

References

1. Elema JD, Poppema S (1978) Infantile histiocytosis X (Letterer Siwe disease). Investigation with enzyme histochemical and sheep erythrocyte rosetting techniques. Cancer 42:555
2. Feldges A, Meuret G, Gonzenbach P, Fust G (1979) Histiocytosis X in 9 children: clinical aspects and laboratory evaluations including an analysis of monocytopoiesis. Helv Paediatr Acta 34:107
3. Feldges AJ, Imbach P, Plüss HJ, Sartorius J, Wagner HP, Wyss M (1980) Therapie der disseminierten Histiocytosis X. Schweiz Med Wochenschr 110:912
4. Jones B, Kung F, Chevalier L, Forman EN, Rausen A, Koch K, Despoto F, Maurer H, Jacquillat C, Degnan TJ, Pluess HJ, Desorges J, Patterson RB, Glidewell O, Holland J (1974) Chemotherapy of reticuloendotheliosis. Comparison of methotrexate + prednisone. Cancer 34:1011
5. Komp DM, Vietti TJ, Berry DH, Starling KA, Haggard ME, George SL (1977) Combination chemotherapy in histiocytosis X. Med Pediatr Oncol 3:267
6. Lahey ME (1962) Prognosis in reticuloendotheliosis in children. J Pediatr 60:664
7. Lahey ME (1975) Histiocytosis X – an analysis of prognostic factors. J Pediatr 87:184
8. Lahey ME (1975) Histiocytosis X – comparison of three treatment regimens. J Pediatr 87:179
9. Lahey ME, Heyn RM, Newton WA, Shore N, Smitz WB, Leikin S, Hammond D (1979) Histiocytosis X – Clinical trial of chlorambucil: A report from Children's Cancer Study Group. Med Pediatr Oncol 7:197
10. Lichtenstein L (1953) Histiocytosis X. Integration of eosinophilic granuloma of bone, "Letterer Siwe disease" and "Schüller Christian disease" as related manifestations of a single nosologic entity. Arch Pathol 56:84
11. Liebermann PH, Jones CR, Dargeon HWK, Begg CF (1969) A reappraisal of eosinophilic granuloma of bone, Hand Schüller Christian syndrome and Letterer Siwe syndrome. Medicine (Baltimore) 48:375
12. Nezelof C, Frileux-Herbet F, Cronier-Sachot J (1979) Disseminated histiocytosis X. Analysis of prognostic factors based on a retrospective study of 150 cases. Cancer 44:1824
13. Osband M, Lipton J, Vawter G, Levey R, Parkman R (1978) Treatment of histiocytosis X with calf thymus extract. Blood 50:268
14. Sims DG (1977) Histiocytosis X. Follow-up of 43 cases. Arch Dis Child 52:433
15. Smith PJ, Ekert H, Campbell PE (1976) Improved prognosis in disseminated histiocytosis. Med Oncol 2:371

16. Starling KA, Donaldson MH, Haggard ME, Vietti TE, Sutow WW (1972) Therapy of histiocytosis X with vincristine, vinblastine and cyclophosphamide. Am J Dis Child 123:105
17. Vogel J, Vogel P (1972) Idiopathic histiocytosis: a discussion of eosinophilic granuloma, the Hand Schüller Christian syndrome and the Letterer Siwe syndrome. Semin Hematol 9:349

Haematology and Blood Transfusion Vol 27
Disorders of the Monocyte Macrophage System
Edited by F. Schmalzl, D. Huhn, H.E. Schaefer
© Springer-Verlag Berlin Heidelberg New York 1981

Therapy in Pulmonary Histiocytosis X

D. Huhn, G. König, J. Weig and W. Schneller

The term "histiocytosis X" encompasses 3 major entities: Letterer-Siwe disease, Hand-Schüller-Christian disease, and eosinophilic granuloma of the bone. Most investigators now consider these three entities as variable expressions of the same basis abnormality [16]. Histiocytosis X may be defined as a proliferative disorder of histiocytes, of unknown etiology, pathogenetically akin to an inflammatory reaction with multiple clinical manifestations mirroring the widespread distribution of the histiocytic system in the body [6].

There are two clinical forms of *lung involvement* by histiocytosis X: as a generalized disease or as a separate entity (primary pulmonary histiocytosis X). Pulmonary histiocytosis X may be considered a rather rare disorder. In a clinic specialized for care of pulmonary disorders, within the same 6-year period 15 cases of pulmonary histiocytosis were seen while 274 cases of histologically confirmed sarcoidosis were observed [19]. And of 57 patients examined at the Mayo Clinic from 1962 through 1973 who had a diagnosis of histiocytosis X and whose chest roentgenograms were available for review, pulmonary histiocytosis X was detected in only seven patients [5].

Generally, an unfavorable *prognosis* is associated with pulmonary histiocytosis X. Five of seven adult patients were impaired by marked deterioration of pulmonary function [6]. The mortality rate in patients (mostly children) suffering from pulmonary histiocytosis X is in the range of 50% [6, 17, 18]. Treatment, therefore, seems legitimate, and several trials using corticosteroids alone or in combination with different cytostatics were performed with rather controversial results [4, 7, 12–14, 18, 19, 21].

Evaluation of treatment results is rendered more difficult by the varying spontaneous course of histiocytosis X, by the inequality of patients investigated according to age and spread of the disease, and by the lack of unequivocal parameters for judgment of remission of the disease. In six adult patients observed in our institution from 1978 to 1980 and suffering from pulmonary histiocytosis X (four primary pulmonary histiocytosis X and two additional to osseous involvement) extensive clinical investigations were performed to recognize the stage of the disease and the impairment of pulmonary function. In five patients therapy with intermittent cycles of prednisolone in combination with either vinblastine or with alkylating agents was performed. The course of the disease was observed by clinical, laboratory, and roentgenologic methods as well as by serial lung function tests.

1. Patients and Methods

Three patients were male, three female; age ranged from 24 to 44 years (mean 34 years). All patients smoked at least ten cigarettes daily, five had contact with different chemicals (petrol, benzene, trichloroethylene, vapors of dry cleaning, or accidental exposure to high concentrations of vapors of a toilet cleaning agent). All patients suffered from coughing and dyspnea on exertion, and two, from fever. In

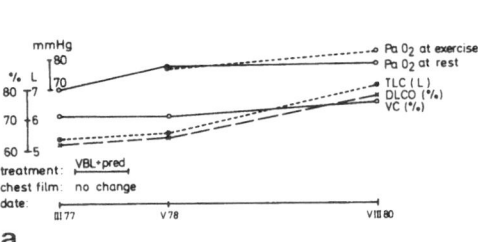

a

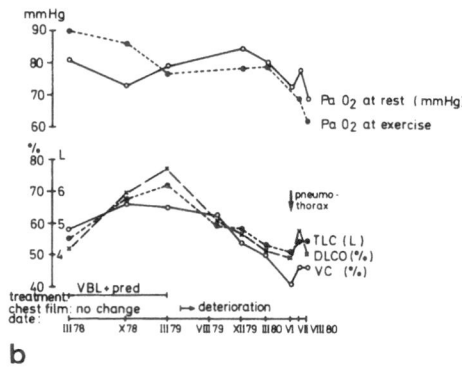

b

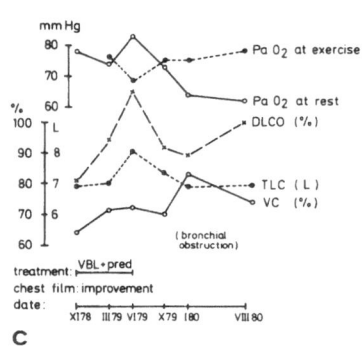

c

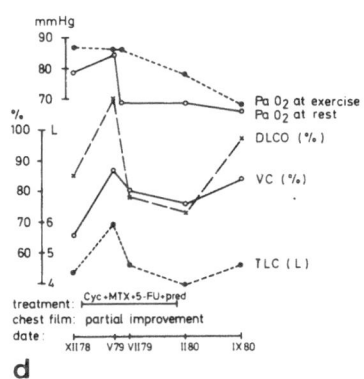

d

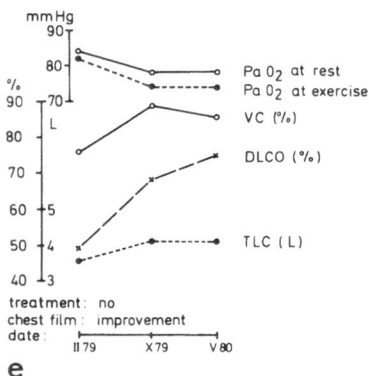

e

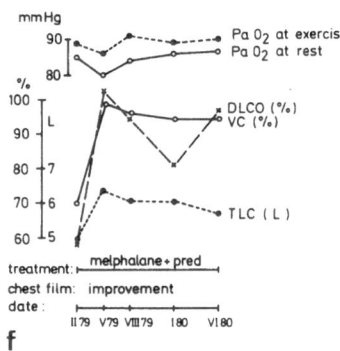

f

five patients *diagnosis* was confirmed by open lung biopsy. In one of these five patients involvement of the skeleton was established in addition by X-ray and biopsy. In one patient diagnosis of histiocytosis X was confirmed by biopsy from a lesion of the skull, and diagnosis of involvement of the lung was based on clinical and roentgenological findings only.

In all patients the severity and the course of the *lung involvement* were controlled at the time of diagnosis and during the following course by repeated radiographs, including conventional or computerized tomography. Studies of pulmonary function included body plethysmography, spirometry, CO transfer factor (DLCO), and blood gas analysis at rest and during bicycle exercise in the sitting position.

For diagnosis of the *stage of the disease*, the following investigations were performed in all patients: radiographs of the skull including maxillar and mandibular bones, skeletal scintigraphs (technetium phosphate) (Prof. U. Büll, Munich), and bone marrow histology in 4 patients by the Burkhardt or Jamshidi technique (Dr. R. Bartl, Munich). In all patients routine laboratory tests included quantitative immunoglobulins, acid and alkaline serum phosphatase, and endocrinological tests for exclusion of pituitary dysfunction (Inst. Klin. Chemie, Prof. M. Knedel).

Treatment was started when symptoms were present and when no spontaneous improvement occurred within an observation period of several weeks. In some patients short trials with cortico-steroids had been tried. In three patients treatment consisted of 10 mg vinblastine on day 1 and 100 mg prednisolone on days 1 to 5 at 2-week intervals (duration of treatment is indicated in Fig. 1). In one patient, in addition, diagnosis of carcinoma of the breast was made and treated by radical mastectomy. Staging investigations revealed stage T4 (because of infiltration of the skin) NoMo. Adjuvant treatment of the carcinoma of the breast *and* treatment of the pulmonary histiocytosis X was started with intermittent cyclophosphamide, 5-fluorouracil, and methotrexate according to Bonadonna. Five days of 50 mg prednisolone daily were added to each 4-week cycle (Fig. 1d). In one patient treatment with melphalan and prednisolone according to Alexanian was started: 4-day cycles of 0.25 mg per kg and day of melphalan and 2.0 mg per kg and day of prednisolone (tapering prednisolone down to 0.25 mg on days 5 to 8) were repeated in 6-week intervals (Fig. 1f).

2. Results

Roentgenographic changes were characterized by diffuse bilateral infiltrates, producing a mainly reticular or reticulonodular pattern. There was no predilection for any portion of the lung. Advanced stages marked by bullae formation or honey combing were not present in our series. Nodular densities, in particular, could be demonstrated convincingly by conventional and by computerized tomography.

Study of *pulmonary function* at diagnosis (Fig. 1) had the following results: the diffusing capacity for CO was reduced to less than 70% of the predicted value in four patients (G.W., H.O., G.S., and B.K.), borderline reduced in one patient (G.S.), and below 70% of the reference value in the remaining five. Total lung capacity should be interpreted according to the trend in each patient. In patient H.O. lung function worsened since III-79. This was demonstrated by a fall of DLCO, VC, TLC, and of PaO_2 at exercise. In the later course there was a deterioration of clinical symptoms and of the chest films, too.

Laboratory tests did not reveal any changes diagnostic of histiocytosis X or instructive for the understanding of its pathogenesis. Alkaline and acid serum phosphatases, calcium, and blood cells were within normal limits in all patients. The mean value for monocytes of all patients was 323 per mm^3; eosinophils were 1% or less in all patients. Serum IgG, IgA, and IgM were low in patient B.K. (6.2 g/liter, 1.0 g/liter, 0.45 g/liter). In the remaining patients all immunoglobulin fractions were within normal limits, and the mean values in all patients were 9.8 g/liter for IgG, 1.9 g/liter for IgA, and 1.5 g/liter for IgM. IgE was distinctly elevated in three patients (G.W. 890 U/ml, H.O. 348 U/ml, and E.S. 690 U/ml) and normal in the remaining patients. Serologic tests for detection of possible pneumonitis-causing agents gave exclusively negative results.

The *response to therapy* was varied. Three patients were treated with intermittent application of vinblastine and prednisolone for periods of 6 to 12 months (Fig. 1a–c). During this period of treatment, no change in chest X-ray was seen in G.W. and H.O. while improvement was noted in G.E. Pulmonary function improved in H.O. and slightly in G.E., and no change was noted in G.W. After treatment was stopped, pulmonary function of

---

Fig. 1a–f. Course of the disease in six patients suffering from pulmonary histiocytosis X. TCL = total lung capacity; DLCO = diffusing capacity for CO; VC = vital capacity; VBL = vinblastine; pred = predniso(lo)ne; Cyc = cyclophosphamide; MTX = methothrexate; 5-FU = fluorouracil

G.W. improved slightly, but new osseous lesions appeared only 4 months after therapy was finished and responded to localized irradiation. In H.O., pulmonary roentgenography and function tests disclosed distinct deterioration. Patient E.S. was treated with cyclophosphamide, methotrexate, fluorouracil, and prednisolone. The chest film improved in part. Compared to XII-78, lung function tests suggested a slight improvement (Fig. 1d). Patient B.K. improved distinctly when intermittent treatment with melphalan and prednisolone was started, and improved values were kept during maintenance therapy (Fig. 1f). The same improvement spontaneously occurred in Patient G.S. (Fig. 1e).

3. Discussion

Etiology and pathogenesis of histiocytosis X remain questionable. Investigations of immunological parameters were performed in a few patients only and revealed normal [11] or low [23] levels of IgG, as confirmed in our patients. IgE was found to be low in two patients [11] but elevated in three. Monocytosis, as described in children suffering from disseminated histiocytosis X [7], was not confirmed in our patients.

It has been discussed whether the purely pulmonary form of histiocytosis X might be just another manner in which the lung reacts to aggression by inhaled particles, this uncommon type of response being due to an unusual pattern of immunologic reactivity on the part of the host [23]. This assumption might be supported by the anamnesis of our patients: all of them had contact with chemicals or were heavy smokers. Serological tests were unproductive.

Langerhans granules are found constantly in normal histiocytes of the epidermis and very frequently in histiocytic cells of histiocytosis X, regardless of the tissue involved [2]. The nature of these organelles remains enigmatic, but their specifity for diagnosis of histiocytosis X is undisputed. When a series of 128 human open lung biopsies were reviewed with the primary goal of identifying Langerhans granules, these particular structures were detected in 18 of 20 patients suffering from histiocytosis X but in only 5 of 23 patients suffering from interstitial pneumonitis or fibrosis and in 1 of 2 patients suffering from bronchioloalveolar carcinoma [2]. Concerning the occurrence of Langerhans granules in some cases of bronchioloalveolar carcinoma, the hypothesis that those cases of bronchioloalveolar carcinoma in which Langerhans granules are present may arise in localized or diffuse pulmonary scarring caused by nonrecognized pulmonary histiocytosis X was discussed [10].

Primary pulmonary involvement in histiocytosis X is a rare event. Differential diagnosis includes sarcoidosis, tuberculosis, pneumocytis carinii infection (which also may complicate pulmonary histiocytosis [8]), carcinomatosis, and the broad spectrum of inflammatory pulmonary disorders resulting in diffuse fibrosis. *Diagnosis*, therefore, can usually only be ensured by open lung biopsy. To avoid this major procedure, an attempt should be made to verify histiocytosis X by the electron microscopic demonstration of the characteristic histiocytes containing Langerhans granules in material obtained by bronchioalveolar lavage. Using this measure, diagnosis was confirmed in four patients [3]. In one of our patients (H.O.), bronchioalveolar lavage was performed after termination of cytostatic treatment, and no characteristic histiocytic cells were detected.

Impaired *pulmonary function* is considered to be an indicator of poor prognosis in pulmonary histiocytosis X [6] and therefore may be taken as a sign to begin treatment. In addition, the further course of the disease and a possible response to treatment may be controlled by repeated examinations of pulmonary function. At the beginning of the disease, impairment of pulmonary function is caused by poorly demarcated infiltrations, situated mainly interstitially and surrounding small airways and their associated vessels. Primarily, therefore, diffusion will be restricted, resulting in a decrease of diffusing capacity for carbon monoxide in 74% of patients tested [4]. In the same group of patients, vital capacity was decreased in 59% of cases, the ratio of residual volume to total lung capacity was pathologic in 61%; normal values were obtained in only 15% of patients tested. In another group of patients [12], vital capacity and total lung capacity were decreased at diagnosis in four patients, diffusing

Table 1. Therapeutic results in pulmonary histiocytosis X[a]

No. of cases	Age group	Pulmonary involvement	Treatment	Results Im-proved	No change	Pro-gress.	Died	Author
4	Adult	Primary	Cort	2	2			24
5	Adult	Primary	Cort	4			1	12
6	Child	Generalized	VBL	3		1	2	17
5	Child		Cort	1		1	3	20
6	Adult	Primary	Cort + Alk	1			2	
			Cort + VBL				2	
			Alk			1		
8	Adult	Primary	4 × no treatment	1	3			9
			4 × Cort	2	2			
12	Adult	Primary	Cort	12				19
1	Child	Generalized	Cort, Alk, MTX, VBL				1	8
6 Child 61 Adult	Child Adult	Primary generalized	Cort, Alk, VBL	9	27	14	17	4
30	Child	Generalized	Cort (+ VBL, + other cytost.)	Alive: 11			19	18
Total: 144				35	34 Alive: 97	17	47	

[a] Cort = corticosteroids, in most cases prednisone; Alk = alkylating cytostatics, in most cases cyclophosphamide; VBL = vinblastine; MTX = methotrexate

capacity of CO in three patients tested; pathological values improved coincidently with successful therapy. None of our patients had entirely normal values at diagnosis. In most patients both tests, i.e. diffusing capacity for CO and vital capacity, gave pathological results. PaO_2 of the arterial blood at rest and especially at exercise reached pathological levels in advanced disease only. During the further course of the disease the pulmonary function was reflected uniformly by all tests that were performed, with the exception of PaO_2 at rest.

Pulmonary involvement in histiocytosis X is rather ominous; clinical symptoms as a result of impaired pulmonary function, in particular, may raise the death rate to more than 80% [18]. Different *treatment* schedules, therefore, were tried in the past: in most cases corticosteroids alone or in combination with alkylating cytostatics, antimetabolites, and vinca alkaloids. In spite of this, the death rate in patients observed long enough to be evaluated is about 33% (Table 1), and definitive improvement is obtained in only about 25% of patients. In our own six cases, long-lasting improvement was obtained in one patient when treated with melphalan and prednisolone, and questionable or short-term improvement occurred in two further patients treated with vinblastine and prednisolone, whilst pathological findings in one patient cleared spontaneously.

When treatment results obtained in children suffering from multiple osseous or disseminated organ involvement, but without pulmonary manifestation, are taken into con-

Table 2. Results of treatment in histiocytosis X. Most patients suffered from multiple osseous lesions and/or disseminated organ involvement, except primary pulmonary histiocytosis X[a]

Number of patients (all children)	Therapy	Course	Author
10	Pred	Improvement 8/10	1
28	Pred + methotrexate	CR 8/17 PR 1/17	13
	Pred + vincristine	CR 2/11 PR 5/11	
65	VBL	CR 3/21 PR 7/21	15
	VLB + pred	CR 7/20 PR 5/20	
	6-MP + pred	CR 5/24 PR 6/24	
14	Chlorambucil	CR 5/10 PR 1/10	21
	VBL and others	CR 4/4	
10	VBL + pred; maintenance 6-MP	CR 8/10 PR 2/10	7

[a] Pred = prednisone; VBL = vinblastine; 6-MP = 6-mercaptopurine; CR = complete remission; PR = partial remission

sideration, the therapeutic value of corticosteroids and cytostatic agents appears somewhat more favorable (Table 2). Taking these experiences into consideration the following *conclusions* may be drawn. In asymptomatic pulmonary histiocytosis X the patient may be observed for several weeks, and a specific therapy withheld. When no spontaneous improvement occurs and when symptoms are present or lung function is impaired, treatment with prednisone should be started and continued with decreasing dosis for several months. Concerning doses of corticosteroids and duration of therapy, the recommendations of the committee on therapy of sarcoidosis may be taken as a basis [22]. When no remission is obtained, and taking into consideration the poor prognosis and the disabling consequences of active pulmonary histiocytosis X and of the resulting increasing pulmonary fibrosis, a therapeutic trial using corticosteroids in combination with cytostatic agents seems justified. According to the literature, vinca alkaloids, alkylating agents, and antimetabolites appear to be promising agents, without a definite priority for either of these being indicated up to now. Combinations of cytostatics will have to be tested in multicentric studies. Special importance will have to be attached to the follow-up of each patient by repeated controls of radiographic findings, in particular of the pulmonary function which proved to be valuable in recognizing the extent and the course of the disease.

4. Summary

Clinical findings and course of the disease are described in six patients suffering from pulmonary histiocytosis X. Diagnosis was suspected when a reticulonodular pattern was detected by conventional X-ray or by computerized tomography of the lungs. Laboratory tests were altered nonspecifically, and lung function was impaired. Signs of the disease improved spontaneously in one patient and when cytostatics were given in two of five patients. To judge the course of the disease, repeated controls of lung function parameters − besides roentgenographic methods − were of particular value.

References

1. Avioli LV, Lasersohn JT, Lopresti JM (1963) Histiocytosis X (Schüller-Christian disease): A clinicopathological survey, review of ten patients and the results of prednisone therapy. Medicine (Baltimore) 42:119–147

2. Basset F, Soler P, Wyllie L, Mazin F, Turiaf J (1976) Langerhans' cells and lung interstitium. Ann NY Acad Sci 278:599–611
3. Basset F, Soler P, Jaurand MC, Bignon J (1977) Ultrastructural examination of broncho-alveolar lavage for diagnosis of pulmonary histiocytosis X: Preliminary report on 4 cases. Thorax 32:303–306
4. Basset F, Corrin B, Spencer H, Lacronique J, Roth C, Soler P, Battesti J-P, Georges R, Chrétien J (1978) Pulmonary histiocytosis X. Am Rev Respir Dis 118:811–820
5. Carlson RA, Hattery RR, O'Connell EJ, Fontana RS (1976) Pulmonary involvement by histiocytosis X in the pediatric age group. Mayo Clin Proc 51:542–547
6. Enriquez P, Dahlin DC, Hayles AB, Henderson ED (1967) Histiocytosis X: A clinical study. Mayo Clin Proc 42:88–99
7. Feldges AJ, Imbach P, Plüss HJ, Sartorius J, Wagner HP, Wyss M (1980) Therapie der disseminierten Histiozytosis X im Kindesalter. Schweiz Med Wochenschr 110:912–915
8. Gold J, L'Heureux P, Dehner LP (1977) Ultrastructure in the differential diagnosis of pulmonary histiocytosis and pneumocystosis. Arch Pathol Lab Med 101:243–247
9. Gribl P (1975) Das eosinophile Granulom der Lunge (Histiocytosis X). 8 Fälle am Zentralkrankenhaus Gauting in der Zeit vom 31. 3. 1971 bis 17. 9. 1974. Medical dissertation, University of Munich
10. Hammar SP, Hallman KO, Winterbauer R, Bockus D, Remington F (1980) Langerhans cells in bronchiolo-alveolar cell carcinoma: Possible association with pulmonary eosinophilic granuloma (Abstr). Am J Clin Pathol 73:302–303
11. Hiwada K, Konishiike J, Sera Y, Yagura T, Asayama S, Yamamura Y (1975) Primary pulmonary histiocytosis X immunological analysis in two cases. Pneumonologie 152:259–266
12. Hoffman L, Cohn JE, Gaensler EA (1962) Respiratory abnormalities in eosinophilic granuloma of the lung. N Engl J Med 267:577–589
13. Jones B, Kung F, Chevalier L., et al. (1974) Chemotherapy of reticuloendotheliosis. Comparison of methotrexate plus prednisone vs. vincristine plus prednisone. Cancer 34:1011–1017
14. Lahey ME (1962) Prognosis in reticuloendotheliosis in children. J Pediatr 60:664
15. Lahey ME (1975) Histiocytosis X – comparsion of three treatment regimens. J Pediatr 87:179–183
16. Lichtenstein L (1953) "Histiocytosis X"; integration of eosinophilic granuloma of bone, "Letterer-Siwe-Disease", and "Schüller-Christian Disease" as related manifestations of single nosologic entity. Arch Pathol 56:84
17. Lucaya A (1971) Histiocytosis. Am J Dis Child 121:289–295
18. Nezelof C, Frileux-Herbet F, Cronier-Sachot J (1979) Disseminated histiocytosis X. Cancer 44:1824–1838
19. Radenbach KL, Brandt H-J, Freise G, Liebig S, Preussler H (1977) Diagnostische and therapeutische Besonderheiten bei zwölf Fällen von pulmonaler Histiocytosis X 1969–1975. Z Erkr Atmungsorgane 147:26–40
20. Smith M, McCormack LJ, van Ordstrand HS, Mercer RD (1974) "Primary" pulmonary histiocytosis X. Chest 65:176–180
21. Smith PJ, Ekert H, Campbell PE (1976) Improved prognosis in disseminated histiocytosis. Med Pediatr Oncol 2:371–377
22. Turiaf J, Johns CJ, Teirstein AS, Tsuji S, Wurm K (1976) The problem of the treatment of sarcoidosis: Report of the subcommittee on therapy. Ann NY Acad Sci 278:743
23. Villar TG, Avila R, Amaral Marques R (1976) Eosinophilic granuloma of the lung and the extrinsic pulmonary granulomatoses. Ann NY Acad Sci 278:612–617
24. Williams AW, Dunnington WG, Berte SJ (1961) Pulmonary eosinophilic granuloma: A clinical and pathologic discussion. Ann Intern Med 54:30–45

Haematology and Blood Transfusion Vol 27
Disorders of the Monocyte Macrophage System
Edited by F. Schmalzl, D. Huhn, H.E. Schaefer
© Springer-Verlag Berlin Heidelberg New York 1981

Sinus Histiocytosis with Massive Lymphadenopathy and Epidural Involvement

R.J. Haas, M.S.E. Helmig and P. Meister

1. Summary

Sinus histiocytosis with massive lymphadenopathy (SHML) was recognized as a new clinical-pathological entity in 1969. Up to the present 134 cases have been described. The disease is characterized by prominent cervical lymph node enlargement. Microscopic features include marked dilatation of sinuses with intrasinusal histiocytes and lymphophagocytosis. About 70% of the patients reported were affected during the first 2 decades of life. The disease is held to be benign on account of spontaneous resolution in some patients. A follow-up survey of 72 patients showed disappearance of the symptoms 10 years after the original diagnosis in 24 patients. In 42 patients the disease still persisted 6 months to 21 years later. Six patients died, but only one of them as a result of the disease. Extranodal involvement was seen in the orbit, eyelid, respiratory tract, skin, bone, salivary glands, and testis. In two cases, one of which will be reported here, paraparesis resulted from infiltration of the epidural space. Treatment with prednisolone was tried in some cases with excellent results. In our case treatment with prednisolone and vinblastine resulted in the disappearance of the neurological symptoms.

2. Introduction

Sinus histiocytosis with massive lymphadenopathy (SHML) was described as a new clinical-pathological entity in 1969 by Rosai and Dorfman [14]. Up to the present 134 cases of the disease have been reported [1–5, 7, 9–16, 18]. The etiology is unknown. The disease is neither hereditary nor contagious. Although there is a high possibility of a specific infectious process [11, 15], no micro-organisms have been detected by bacterial, viral, or fungal isolation techniques. The disease follows a protracted clinical course and is held to be benign since it is usually self-limiting. The disease can be distinguished from "histiocytosis X" [8, 17, 18]. Massive enlargement of cervical nodes and fever are the most consistent clinical manifestations. Histologically there is a marked dilatation of the subcapsular and medullary sinuses in enlarged lymph nodes filled with numerous proliferating histiocytes.

Extranodal deposits at various sites have also been reported [1, 3, 10, 15, 16]. In two cases spinal epidural involvement caused paraplegia [7, 9]. We have been able to study one of these cases since 1976.

3. Findings

Of the total of 134 cases reported, 52% were affected during the 1st decade of live, and 66%, before the age of 20. The oldest patient was 67 years old. The male-female ratio was 2 : 1. There were 72 white, 47 black, and 15 Asian patients.

Table 1. Sign or symptom in 134 patients with SHML [4, 5, 10, 16]

Sign or symptom	Number of patients	Percentage of patients
Cervical lymphadenopathy	118	81
Other lymph node groups	76	57
Extranodal involvement of:		
Upper respiratory tract	17	13
Eyelid and orbit	13	10
Skin	11	8
Bone	5	4
Salivary glands	3	2
Epidural	2	2
Testis	1	1
Fever	68	52
Anemia	69	52
Hypergammaglobulinemia	64	48
Neutrophilia	54	40
Leucocytosis	46	34
Nasal discharge	9	7

4. Microscopic Findings

Characteristically the enlarged lymph nodes show marked capsular and pericapsular fibrosis. In the early stage of the disease there is dilatation of the subcapsular and medullary sinuses. The sinuses are filled with numerous histiocytes. These cells are often gigantic, have a large foamy cytoplasm, and frequently contain phagocytosed lymphocytes, granulocytes, plasma cells, or erythrocytes. There are many plasma cells in the pulp. In later stages of the disease proliferation of a mixed histiocytic cell population may result in effacement of the nodal architecture. Most of these features are reproduced when the disease involves extranodal sites, although fibrosis is common [15, 16]. In a case in which emergency laminectomy was performed [9] epidural infiltration with histiocytes, plasma cells, and lymphocytes was seen. The same infiltrates were seen in the bone included in the biopsy material.

5. Clinical Features

The clinical features of SHLM in these 134 patients are listed in Table 1. As can be noted, painless cervical lymphadenopathy was the prominent sign in most cases. However, enlargement of other nodes was also a common finding. In some cases there were extranodal deposits at various sites, including the soft tissue of the orbit and eyelid, the upper respiratory tract, skin, bone, salivary glands, and testis. It is interesting to note that significant hepatosplenomegaly is absent in SHML. Fever, anemia, leucocytosis, an elevated ESR, and hypergammaglobulinemia are present in most cases. The clinical course of the disease in those patients who were followed up was benign: Sanchez et al. [16] have reported on follow-up results obtained in 65 out of 113 cases, Lampert and Lennert [10] reported on the results of 7 out of 15 cases. Of these 72 patients, 24 were alive and well, lymphadenopathy and symptoms disappearing 6 months to 10 years after the original diagnosis. Forty-two patients were alive with persistence of the disease 6 months to 21 years later. Six patients died, but only one of them as a direct result of SHML. Autopsy revealed bilateral bronchopneumonia. Histologically, the typical features of SHML were present in all lymph nodes examined in the mucosa and submucosa of the respiratory tract, including the nasal cavity, sinuses, and trachea [15].

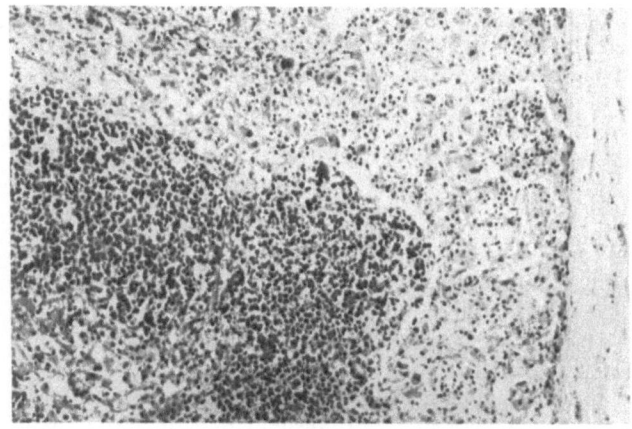

Fig. 1. Widened sinus with predominatingly large, pale histiocytes and lymphocytes. *Left lower corner,* lymph follicle, *right margin,* capsule of Lymph node. H&E, × 83

In connection with the above data it should be stressed that the disorder is not necessarily restricted to lymph nodes and may even be progressive. Evidence of this is provided by two patients with SHML in whom spinal epidural involvement caused paraplegia. We were able to study one of these patients.

6. Case Report

A 3½-year-old girl was examined because of a painless bilateral enlarged submandibular nodular mass in 1968. Pathological laboratory findings included an elevated ESR and leucocytosis with neutrophilia of 13 000 cells/mm^3. In June 1970 lymph node biopsy resulted in the diagnosis of SHML. Cellular and humoral immunity appeared unaltered in a skin test and lymphocyte transformation test in culture. In 1976 the patient suffered from ptosis of the right eyelid, deafness of the right ear, atrophy of the tongue, and left-sided palatoplegia. There was paralysis of the right arm and also loss of pain below the level of the T_2 dermatome; the patient was unable to walk unsupported. Biopsy of a submandibular lymph node resulted again in the diagnosis of SHML (Figs. 1 and 2). X-ray of the lungs showed prominent hilar markings. Computer tomography demonstrated extensive nodular masses in the supraclavicular region and the upper spinal column. Myelography revealed an incomplete extradural block at the C_5–T_2 level and the C_2 level. Treatment with 2 mg/kg pred-

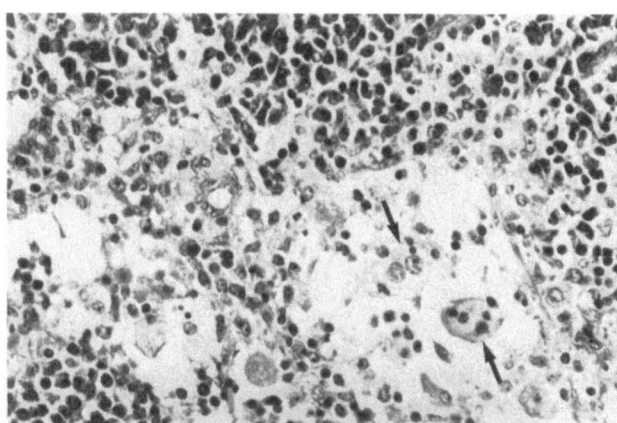

Fig. 2. Sinus with large histiocytes, some showing phagocytosis of erythrocytes and *leucocytes* (→). In addition lymphocytes and plasma cells (*upper margin* of figure). H&E, × 170

nisolone daily and 0.3 mg/kg vinblastine once a week for 8 weeks resulted in total disappearance of the neurological symptoms except deafness in the right ear. The nodular masses decreased drastically. Myelography revealed only irregularity of the spinal wall between C_7 and T_2. Examination of the ocular fundi revealed enlargement of the left optic disc from 1979 onwards. The field of vision was normal. Computer tomography showed a large, high-density mass in the suprasellar and parasellar regions, especially on the right side.

7. Discussion

The differentiation between SHML and other forms of histiocytosis such as "differentiated histiocytosis" or histiocytosis X may pose problems. Despite a resemblance noted by some authors, differentiation is generally facilitated by morphological features [10, 15, 18]. The histiocytes of SHML have large round nuclei and prominent nucleoli, whereas the histiocytes of histiocytosis X have large convoluted, folded nuclei and inconspicuous nucleoli. SHML histiocytes often include massive hemophagocytic sinus histiocytes, and Langerhans' granules have never been identified ultrastructurally. In SHML eosinophiles are virtually absent.

The clinical-pathological entity of SHML may be best defined as "chronic lymphadenitis with massive hemophagocytic sinus histiocytosis" [11]. Although benign in its clinical course the process is not restricted to lymph nodes as shown by cases in which various extranodal sites were involved. The majority of the cases reported are children. About 50% of the patients developed symptoms before the age of 10. Some authors believe that the disease favors Negroes [10, 16] since 47 cases have been reported in blacks, a relatively high percentage considering the diagnostic facilities available in African countries. There is a predominance of males.

The etiology of the disease is unknown. Rosai and Dorfman [15] suggest an abnormal immunologic response to an infectious factor. However, a defect of cellular immune function could be detected in only one case reported by Becroft et al. [2]. In most cases, including the case we have described, the absolute lymphocyte counts in peripheral blood were below $1000/\text{mm}^3$, but this phenomenon may be due to increased consumption of lymphocytes in the lymph nodes by the histiocytes. In 15 of 22 cases reported by Sanchez et al. [16] an elevation of Epstein-Barr virus serum antibody titers was detected. It has also been speculated [10] that rare Klebsiella organisms could cause the disease, since in one case elevated antibody titers to Klebsiella antigen were seen on repeated occasions. However, until now it has not been possible so far to isolate an infectious agent from the biopsy material.

The literature contains no definite indication of an effective treatment for this disorder, although there is a strong suggestion that prednisolone may have excellent results [10]. In our case treatment with prednisolone and vinblastine based on therapeutic trials in cases of histiocytosis X [17] was tried successfully.

Although a dramatic decrease in the nodular masses and the infiltrates in the walls of the spinal cord could be demonstrated, it is reasonable to believe that the suprasellar masses seen on computer tomography are features of SHML.

The only other case of SHML in which spinal epidural involvement has caused paraplegia was reported by Kessler et al. [9]. In this case an epidural tumor extending from C_7 to T_3 was removed by emergency laminectomy. Histologically involvement of vertebral bone tissue was also observed. Surgery resulted in almost complete disappearance of paraplegia.

References

1. Azoury FJ, Reed RJ (1966) Histiocytosis. N Engl J Med 274:928
2. Becroft DMO, Dix MR, Gillman JC, Mac Gregor BJL, Shaw RL (1973) Benign sinus histiocytosis with massive lymphadenopathy – Transient immunological defects in a child with mediastinal involvement. J Clin Pathol 36:463

3. Codling BW, Soni KC, Barry DR, Martin-Walker W (1972) Histiocytosis presenting as swelling of orbit and eyelid. Br J Ophthalmol 56:517
4. Destombes P (1965) Adenites avec surcharge lipidique de l'enfant ou de l'adulte jeune, observées aux Antilles et au Mali (Quatre observations). Bull Soc Pathol Exot 58:1169
5. Destombes P, Destombes M, Martin L (1972) Histiocytose lipidique ganglionaire pseudotumorale. Nouvelle observation chez une jeune Martiniquaise. Bull Soc Pathol Exot 65:488
6. Enriquez P, Dahlin DC, Hayles AB, Henderson ED (1967) Histiocytosis X: A clinical study. Mayo Clin Proc 42:88
7. Haas RJ, Helmig MSE, Prechtl K (1978) Sinus histiocytosis with massive lymphadenopathy and paraparesis. Cancer 42:77
8. Haas RJ, Janka GE, Helmig MSE, Meister P (to be published) Sogenannte Histiozytose-X und maligne Histiozytose. Onkologie
9. Kessler F, Srulijes C, Toledo F, Shalit M (1976) Sinus histiocytosis with massive lymphadenopathy and spinal epidural involvement. Cancer 38:1614
10. Lampert F, Lennert K (1976) Sinus histiocytosis with massive lymphadenopathy. Cancer 37:783
11. Lennert K, Niedorf HR, Blümcke S, Hardmeier T (1972) Lymphadenitis with massive hemophagocytic sinus histiocytosis. Virchows Arch [Cell Pathol] 10:14
12. Lober M, Rawlings W, Newell GR, Reed RJ (1973) Sinus histiocytosis with massive lymphadenopathy. Report of a case associated with elevated EBV antibody titers. Cancer 32:421
13. Marie J, Bernard J, Nezelof G (1966) Adenopathies chroniques avec prolifération réticulohistiocytaire et surcharge lipidique. Ann Pediatr 42:2689
14. Rosai J, Dorfman RF (1969) Sinus histiocytosis with massive lymphadenopathy. Arch Pathol 87:63
15. Rosai J, Dorfman RF (1972) Sinus histiocytosis with massive lymphadenopathy − A pseudolympathous benign disorder. Cancer 30:1174
16. Sanchez R, Rosai J, Dorfman RF (1977) Sinus histiocytosis with massive lymphadenopathy: An analysis of 113 cases with special emphasis on its extranodal manifestations. Lab Invest 36:349
17. Siegel JS, Coltman CA (1966) Histiocytosis-X: Response to vinblastine sulfat. JAMA 197:123
18. Sinclair-Smith GC, Kahn LB, Uys CJ (1974) Sinus histiocytosis with massive lymphadenopathy. Report of two additional cases with ultrastructural observations. S Afr Med J 48:451

Haematology and Blood Transfusion Vol 27
Disorders of the Monocyte Macrophage System
Edited by F. Schmalzl, D. Huhn, H.E. Schaefer
© Springer-Verlag Berlin Heidelberg New York 1981

Familial Lymphohistiocytosis

G.E. Janka, B.H. Belohradsky, S. Däumling, J. Müller-Höcker, P. Meister and R.J. Haas

1. Summary

Familial lymphohistiocytosis is a genetically transmitted disease affecting infants and very young children with usually a fatal outcome. Cardinal symptoms are fever, hepatosplenomegaly, and pancytopenia. Histologic examination shows infiltration of all organs with phagocytosing histiocytes and lymphocytes as well as atrophy of the normal lymphoid tissue. The distinction from other histiocytic disorders, i.e., Letterer-Siwe disease or malignant histiocytosis, may be difficult. However, the familial occurrence and characteristic findings in the coagulation system and lipid pattern make familial lymphohistiocytosis a sufficiently distinct clinical entity. This report reviews 79 cases from the literature and adds four of own observations.

2. Introduction

Familial lymphohistiocytosis (FLH) [14] is a genetically transmitted disease affecting infants and very young children and is usually fatal. In recent years this intriguing disorder of the monocyte–macrophage system has been written up under various descriptions such as familial hemophagocytic reticulosis [12], familial erythrophagocytic lymphohistiocytosis [27], and familial reticuloendotheliosis [40]. This confusion terminology clearly illustrates our ignorance about the etiology and pathogenesis of this disease. In its cardinal features, i.e., fever, hepatosplenomegaly, pancytopenia, and widespread histiocytic infiltration of several organs, especially spleen, liver and kidneys [5], FLH resembles malignant histiocytosis. When first described by Farquhar and Claireaux in 1952 [12], the author considered it to be the infantile form of histiocytic medullary reticulosis, a rapidly fatal disease in adults [43]. However, several features, above all its familial occurrence, distinguish FLH from the latter. We have recently observed four cases of FLH in our department and 79 additional cases have come to our knowledge from the literature. A typical case record will be presented and the clinical data and investigations reported in the literature will be summarized.

3. Case Report

M.V. was the third child of healthy parents. Pregnancy and delivery were uneventful. The baby developed normally until the age of 2 months when spiking temperatures were noted. On admission the baby was pale and irritable. Spleen and liver were palpable 3 cm below the costal margin. No lymphadenopathy or rash was present. The remainder of the physical examination was normal.

3.1 Laboratory Findings

Hemoglobin was 5.5 g/dl; hematocrit, 18%; erythrocytes, $2.02 \times 10^6/\mu l$; reticulocytes, 12%; platelets, $40 \times 10^3/\mu l$; leukocytes, $4.4 \times 10^3/\mu l$ (band forms 3%, segmented neutrophils 8%,

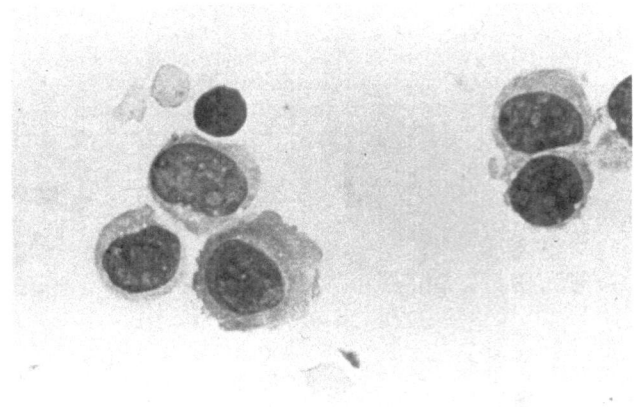

Fig. 1. Histiocytes in pericardial effusion ($\times$ 658)

lymphocytes 87%, monocytes 2%); and sedimentation rate, 4/17. Some of the mononuclear cells were large with a basophilic cytoplasma. Direct and indirect Coombs tests were negative. Bilirubin was 0.76 mg/dl; SGOT 106, IU/l; SGPT, 192 IU/l; haptoglobin, 35 mg/dl; and total protein, 4.3 g/dl. Serum electrophoresis showed albumin 61.4%, alpha-1-globulin 5.9%, alpha-2-globulin 9.7%, beta-globulin 11.2%, gamma-globulin 11.8%, IgG 591 mg/dl, IgA 26 mg/dl, and IgM 53 mg/dl. The distribution of T and B lymphocytes in the peripheral blood was normal. There were normal values for BUN, creatinine, complement C3, C4, C5, erythrocyte enzymes, osmotic resistance of erythrocytes, and hemoglobin electrophoresis. Serum titer for Epstein-Barr virus, rubella, toxoplasmosis, cytomegalovirus, herpes simplex virus, and syphilis were negative. Examination of the cerebrospinal fluid showed seven mononuclear cells per mm^3, some of which were histiocytes. A bone marrow examination revealed increased erythropoiesis without morphologic changes, normal maturation of granulocytes, and slightly increased eosinophils. No increased histiocytes were identified. A chest radiograph showed a normal thymic shadow.

The patient received packed red cells and because of a positive blood culture with Staphylococcus epidermidis she was treated with antibiotics. The fever subsided, and the platelet count rose to normal values, while neutropenia persisted.

Three weeks later the patient was again admitted with fever. Liver and spleen size had increased to 5 and 8 cm respectively. Hemoglobin had dropped to 8.9 g/dl, platelets to 4 $\times$ 10^3/μl, leukocytes were 4.9 $\times$ 10^3/μl with 13% granulocytes. Coagulation studies showed: thromboplastin time 67%, partial thromboplastin time 41.2 s, thrombin time 20.5 s, fibrinogen 105 mg/dl, and normal values for factor II, V, VII, VIII, IX, and X. Lipid values were: cholesterol 123 mg/dl, triglycerides 435 mg/dl. On lipoprotein-electrophoresis the pre-β-lipoproteins were elevated, alpha- and beta-lipoproteins were normal, and chylomicrons were present.

Because an older brother of the patients had died at the age of 10 months with an identical but undiagnosed clinical picture, familial lymphohistiocytosis was suspected and therapy with vinblastine and prednisone was started. Splenectomy was refused. The patients needed frequent blood transfusions which only resulted in a minimal rise in hemoglobin levels but increasing splenic size. The leukopenia became profound with virtually no granulocytes. Liver enzymes rose and icterus developed. The baby showed an opistotonic posture. She died of pseudomonas sepsis at the age of 7 months. Postmortem investigations revealed ascites, pericardial effusion with a protein content of 2.5 g/dl, and 106 mononuclear cells per mm^3, representing histiocytes (Fig. 1). Cells showed positive staining for unspecific esterase and acid phosphatase; phagocytosis of latex particles could be demonstrated. The cerebrospinal fluid had a slightly elevated protein content and 70 histiocytic cells per mm^3. The bone marrow was heavily infiltrated with erythrophagocytosing histiocytes (Fig. 2).

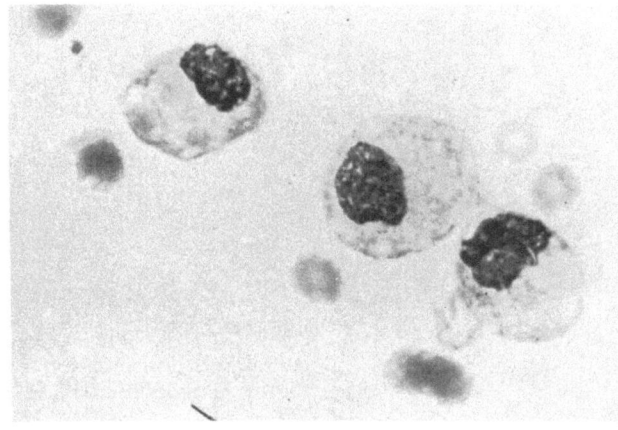

Fig. 2. Erythrophagocytosis of histiocytes in bone marrow aspirate (× 658)

3.2 Autopsy Findings

The examination was restricted to those body regions which were accessible by a median laparotomy section. Only organ pieces could be removed for histological study. No organ weights are therefore available.

The lungs showed an extensive bronchopneumonia with abscess formation. There was a marked hepatosplenomegaly (liver: 18:13:6 cm, spleen: 13:7.5:3.5 cm). The hepatocytes revealed only moderate fatty changes. Within the sinusoids numerous obviously reticulohistiocytic cells were found, often hemosiderin-laden with distinct erythrophagocytosis. The portal tracts were infiltrated by only a few lymphocytes.

The spleen as well as the intra-abdominal lymph nodes exhibited a striking reduction in number of lymphocytes. Especially the thymus showed only scattered lymphocytes in a losely fibrillar and vascular background (Fig. 3). Erythrophagocytosis was mainly seen in the spleen and to a lesser degree also in lymph nodes and in the thymus.

The vertebral bone marrow was extremely hypocellular. However, numerous reticulohistiocytes were present (often hemosiderin laden), with phagocytosis of erythrocytes and leucocytes (Fig. 4). Only in the renal cortex could a small focus with lymphhistiocytic "organ infiltration" be found (Fig. 5).

The second child of this family has proven XO Turner syndrome, and the fourth child who is 2 months old has fallen ill with symptoms characteristic of FLH.

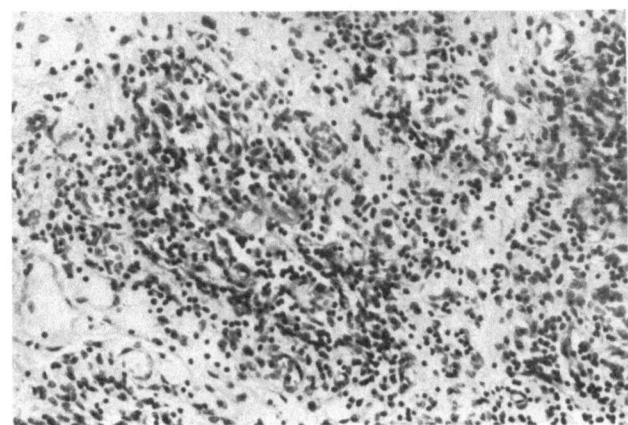

Fig. 3. Hypocellular thymus without dense lymphocytic population (H & H, × 136)

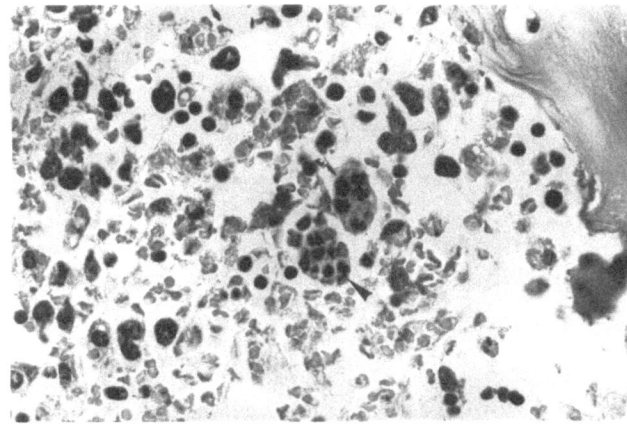

Fig. 4. Hypocellular bone marrow with histiocytes (*arrows*) showing phagocytosis (H & E, × 330)

4. Review of the Literature

Reviews of smaller numbers of patients with FLH have been published in the past [9, 27, 38]. FLH occurs predominantly in siblings, with a slight preponderance of boys. About 20% of the cases are sporadic (Table 1). Up to four affected siblings in one family were reported [13]. In two reports the disease was described in twins [34, 38]. Consanguinity was reported in five cases [6, 26, 28, 31]. In only one case did FLH involve two different generations [46]. In most cases the onset of the disease is within the first $\frac{1}{2}$ year of life, and a period free of symptoms after birth, as in all of our four patients, seems to be typical (Table 2).

The first clinical signs are high temperatures, irritability, and pallor. Upper respiratory infections and gastrointestinal symptoms such as diarrhea and vomiting may be found concomitantly [9]. As in malignant histiocytosis the most prominent features are fever, hepatosplenomegaly, and pancytopenia (Table 3). Whereas hepatosplenomegaly is usually present from the beginning, other symptoms such as icterus, rash, or neurologic abnormalities (e.g., convulsions, meningismus, and opisthotonos) frequently develop as the disease progresses. Lymph node enlargement is not a frequent finding. In most of the patients in whom a lumbar puncture was performed an elevated protein content and pleocytosis were documented showing mononuclear cells and even histiocytes. It should be stressed, however, that meningeal involvement is usually not an early finding.

The most prominent laboratory findings are anemia, neutropenia, and thrombocy-

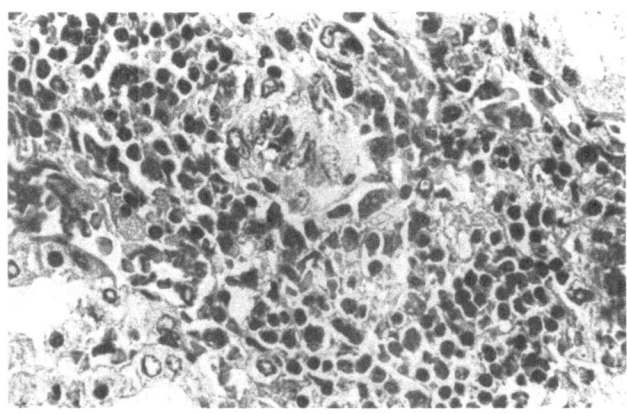

Fig. 5. Renal cortex with small lymphohistiocytic infiltrate (H & E, × 330)

Table 1. Genetic distribution patterns of 83 cases of FLH (including four from our department)

Sporadic	17	
Familial occurrence	66	
siblings		56
relatives		8
Total	No. = 83	

Table 2. Age of onset in FLH in 83 cases (including four from our department)

< 4 weeks	3
4 weeks– 6 months	47
6 weeks–15 months	16
> 15 months	15
not stated	2
	No. = 83

Table 3. Clinical findings in 75 cases of FLH (including four from our department)

Fever	75	(100%)
Hepatomegaly	72	(96%)
Splenomegaly	69	(92%)
Neurologic symptoms	32	(43%)
Icterus	20	(27%)
Lymph node enlargement	19	(25%)
Rash	15	(20%)

Table 4. Hematologic and cytologic findings in FLH (including four cases from our department)

Anemia	69/69	(100%)
Thrombocytopenia	66/66	(100%)
Neutropenia	66/68	(97%)
Leukopenia	55/68	(82%)
CSF pleocytosis	34/43	(79%)

topenia (Table 4). Anemia which is present from the beginning is normochromic and normocytic with moderately elevated reticulocyte counts. With the exception of two patients described by Farquhar [12, 13], the Coombs test is negative. In almost all cases neutropenia and thrombocytopenia is found at diagnosis. Pancytopenia worsens during the course of the disease. In many cases atypical mononuclear cells have been described in the peripheral blood. Bone marrow examination may give a clue to diagnosis if histiocytic infiltration is found [13, 15, 45], but in most cases findings such as increased erythropoiesis or impaired maturation of granulopoiesis are uncharacteristic. In only 14 cases was erythrophagocytosis described [4, 7, 9, 11, 15, 18, 28, 31, 34, 45]. This is in contrast to the striking erythrophagocytosis found in the marrow at autopsy.

Abnormal findings in blood chemistry may include moderately elevated liver enzymes and bilirubin, usually of the conjugated type. Clinical jaundice is more a late complication. An elevated alkaline phosphatase, high lactic dehydrogenase, low haptoglobin, and hyponatremia may be found.

Two abnormalities are sufficiently striking to deserve further comment: The first is hyperlipidemia which was reported in 12 cases [9, 23, 24, 29, 33] and was also present in all three of our cases with appropriate studies. The hyperlipidemia is characterized by a normal cholesterol (exept in one case) [33] and elevated triglycerides up to 1000 mg/dl [23]. In the cases reported by Landrieu and Choulot [24] and deVictor [9] as well in our three patients lipid electrophoresis showed elevated pre-beta-lipoproteins, normal alpha and beta lipoproteins, and a trace of chylomicrons. The cases described by Hagberg et al. [19] and Jaeken et al. [20] as "malignant hyperlipemia" with a clinical picture undistinguishable from FLH should be included.

The second characteristic abnormality concerns the coagulation system. An isolated low fibrinogen level was reported in 19 cases in the literature [3, 7, 9, 23, 26, 31, 36, 38] and was found in three of our four patients. In addition, the cases reported by Donati et al. [10] as Letterer-Siwe disease with hypofibrinogenemia might represent FLH rather than histiocytosis X. The low fibrinogen level is found without evidence of disseminated intravascular coagulation, clotting factors were normal, and fibrin split products, absent.

Familial lymphohistiocytosis is a rapidly fatal disease with a mean survival of usually only a few months following diagnosis. Spontaneous transient improvement has been reported [27] and remission of short duration may be obtained with corticosteroids [4, 28, 33] or splenectomy [26, 29]. Cytotoxic drugs were usually of no convincing benefit, except intrathecal methotrexate in a case with meningeal involvement [25]. Exchange transfusion has also been tried [22]. Noteworthy are two cases with prolonged survival. One reported by Fullerton et al. [15] was treated with cytostatic drugs and survived 19+ months, and one described by Beck et al. [3] was apparently cured 3 years after splenectomy. In all other patients as in our own, the clinical course is rapidly downhill; death is usually due to infection or bleeding.

Pathological investigations in patients with FLH show a lymphohistiocytic infiltration of all organs; atrophy of all lymphoid tissues, especially of the thymus, is a regular finding. The organs which are predominantly infiltrated are spleen, liver, and lymph nodes, but actually no organ is spared, including the brain. Characteristically the infiltrating cells reveal no cytologic evidence of malignancy, they grow in a diffuse pattern, never form tumorous nodules, and the architecture of the organ is preserved [21]. The most striking feature is the phagocytic activity of the histiocytes with special affinity for erythrocytes.

4. Discussion

The systemic histiocytic disorders in childhood encompass several clinical disease pictures which are not always clearly demarcated from each other. Letterer-Siwe disease shares several clinical features with FLH, but typical skin and bone lesions are not found in FLH and histopathology is different. In doubtful cases electron microscopy studies demonstrating the characteristic Langerhans cell granules [2] are helpful. Several cases reported to be familial Letterer-Siwe disease [39, 42] are now considered to represent FLH [21]. The distinction between malignant histiocytosis is less clear [21], especially where nonfamilial cases are concerned [16]. However, in malignant histiocytosis there is a systemic invasive proliferation of morphologically atypical histiocytes, whereas in FLH there is a diffuse infiltration with benign-looking histiocytes and lymphocytes [21, 38].

The characteristic features of FLH are familial occurrence, manifestation during early infancy, the triad of fever, hepatosplenomegaly and pancytopenia, usually a rapidly fatal course, and lymphohistiocytic infiltration of all organs with erythrophagocytosis. Several patients with familial histiocytic disorders have been described who either express only some of the findings of FLH [33, 39] or have additional unusual features [1, 32, 37]. At present a definite classification of these syndromes is not possible; more knowledge about the functional, morphological, and biochemical aspects of the cells involved might provide further insight. Moreover, the macrophage is part of the body's defense against infection and is normally involved in the destruction of hematopoietic cells. Thus reactive histiocytosis have been described in hemolytic anemia, typhoid fever [47, 44], brucellosis [48], leishmaniasis [30], miliary tuberculosis [8], and viral infections [8, 41]. Erythrophagocytosis may be prominent [41], and even "atypical" histiocytes, indistinguishable from malignant histiocytosis, have been described [30].

Some morphological criteria in FLH may serve as arguments against a true neoplastic process [21]. Several theories have been advanced as to the etiology and pathogenesis of FLH. Graft-versus-host disease was proposed by Nezelof and Elichar [35] because of the histiocytic proliferation and atrophy of the lymphoid tissues. In addition the demonstration of fibrin stars in the bone marrow in one patient was interpreted in favor of this theory [26]. However, several findings are not in agreement with the graft-versus-host disease theory: the familial character of the disease and the lack of eosinophilia and skin involvement. In addition, chimerism has never been proven. Gleichmann suggested that alteration of host cells, for example by viruses, causes cells of the host to act against its own altered cells resulting in a GvH-like syndrome [17].

Other authors propose a functional defect of the monocyte, because monocyte-dependent immunologic reactions such as delayed hypersensitivity and lymphocyte proliferation in the presence of antigens and allogenic cells were found to be defective in some patients [22]. Impaired function of mononuclear cells in cytotoxicity assays and depressed monocyte effector function were also described by the same authors. It is possible that the cellular immune deficiency documented in several patients [3, 7, 9, 15, 23] is a consequence of hyperlipidemia since multiple exchange transfusions resulted in complete recovery of monocyte effector function and normal response of lymphocytes to specific antigens [22]. At present the etiology of the hyperlipidemia is poorly understood. Landrieu and Choulot [24] suggested that the gene responsible for the histiocytic disorder either is close to the gene coding for the lipoprotein lipase or that the activity of the enzyme is blocked by the disease process. The isolated low fibrinogen is also unexplained as yet. A hypothesis advanced by DeVictor [9] is the secretion of plasminogen activators by the macrophage and the phagocytosis of fibrinogen degradation products by the same cells.

Although many questions remain open as to the etiology and pathogenesis of FLH, it is in our opinion a sufficiently distinct clinical entity to be considered in any infant with unexplained fever, hepatosplenomegaly, pancytopenia, and a "positive" family history. The diagnostic workup should include a spinal tap with cytocentrifuge examination, coagulation studies, and lipid investigations. The bone marrow and liver biopsies may have to be repeated in case they were nondiagnostic in the early stage of the disease. Detailed immunologic function studies, chromosomes, HLA typing, and studies of monocyte morphology, biochemistry, and function are urgently needed for a better understanding of this disease.

Since FLH seems to be inherited as an autosomal recessive disease genetic counseling of the family is indispensable. The possibility of prenatal diagnosis has so far not been mentioned in the literature.

References

1. Barth RF, Vergara GG, Khurana SK, Lowman JT, Beckwith JB (1972) Rapidly fatal familial histiocytosis with eosinophilia and primary immunologic deficiency. Lancet 2:503
2. Basset F, Turiaf J (1956) Identification par le microscope électronique de particules de nature probablement virale dans les granulomes d'une histiocytose X pulmonaire. C R Acad Sci (Paris) 261:3701
3. Beck JD, Weining JE, Müller-Hermelink HD, Lemmel EM (1977) Reversible graft-versus-host reaction as cause of erythrophagic splenomegaly in a child? Eur J Pediatr 126:175
4. Bell RJM, Brafield AJE, Barnes ND, France NE (1968) Familial haemophagocytic reticulosis. Arch Dis Child 43:601
5. Bergholz M, Rahef G, Doering KM (1978) Familial hemophagocytic reticulosis (Farquhar). Pathol Res Pract 163:267

6. Blennow G (1974) Haemophagocytic reticulosis. Arch Dis Child 49:960
7. Carpentieri U, Gustavson LP, Haggard ME, Nichols MM (1980) Hemophagocytic reticulosis. Blut 39:419
8. Chandra P, Chaudhery A, Rosner F, Kagen M (1975) Transient histiocytosis with striking phagocytosis of platelets, leucocytes, and erythrocytes. Arch Intern Med 135:989
9. De Victor D (1979) La lymphohistiocytose familiale. Medical dissertation, University of Paris
10. Donati MB, Casteels-Van Daele M, Hens G, Vermylen J, Eeckels R (1973) Hypofibrinogenemia in histiocytosis X, type Letterer-Siwe disease. Helv Paediatr Acta 28:603
11. Farber S, Vawter GF (1962) Clinical pathological conference. J Pediatr 61:312
12. Farquhar JW, Claireaux AF (1952) Familial hemophagocytic reticulosis. Arch Dis Child 27:519
13. Farquhar JW, MacGregor A, Richmond J (1958) Familial hemophagocytic reticulosis. Br Med J 2:1561
14. Fauchier C, Benatre A, Regy J-M, Jobard P, Combe P (1970) La lymphohistiocytose familiale. Arch Fr Pediatr 27:51
15. Fullerton P, Ekert H, Hosking C, Tauro CP (1975) Hemophagocytic reticulosis. Cancer 36:441
16. Gilsanz V, Harris GBC (1978) Histiocytic medullary reticulosis in childhood. Radiology 126:463
17. Gleichmann E, Gleichmann H, Wilke W (1976) Autoimmunization and lymphomagenesis in parent F_1 combinations differing at the major histocompatibility complex: Model for spontaneous disease caused by altered self-antigen? Transplant Rev 31:156
18. Goodall HB, Guthrie W, Buist NRM (1965) Familial haemophagocytic reticulosis. Scott Med J 10:425
19. Hagberg B, Hultquist G, Svennerholm L, Voss H (1964) Malignant hyperlipemia in infancy. Am J Dis Child 107:267
20. Jaeken J, Casteels-Van Daele M, Harvengt L, Corbeel L, Broeckaert-Van Orshoven A, van Damme B, Kenis H, Eeckels R (1973) A hyperlipemia syndrome in infancy with rapidly fatal evolution. Helv Paediatr Acta 28:67
21. Koto A, Morecki R, Santorineou M (1976) Congenital hemophagocytic reticulosis. Am J Clin Pathol 65:495
22. Ladisch S (1980) Pathogenesis of cellular immunodeficiency in familial erythrophagocytic lymphohistiocytosis (FEL). 4th Int. Congress of Immunology, Paris abstr. 14. 2. 12
23. Ladisch S, Poplack DG, Holiman B, Blaese RM (1978) Immunodeficiency in familial erythrophagocytic lymphohistiocytosis. Lancet 1:581
24. Landrieu P, Choulot JJ (1976) Réticulose hémophagocytaire avec hypertriglyceridémie. Arch Fr Pediatr 33:497
25. Lilleyman JS (1980) The treatment of familial erythrophagocytic lymphohistiocytosis. Cancer 46: 468
26. Maas B, Goedeke L, Roloff D, Drees N (1974) Familiäre hämophagocytierende Retikulose. Vorstellung eines Krankheitsfalles. Monatsschr Kinderheilkd 122:164
27. MacMahon HE, Bedizel M, Ellis CA (1963) Familial erythrophagocytic lymphohistiocytosis. Pediatrics 32:868
28. Majdalani E, Vassoyan J (1974) A propos d'une observation de lympho-histiocytose familiale "hémophagique". Arch Fr Pediatr 31:297
29. Marrian VJ, Sanerkin NG (1963) Familial histiocytic reticulosis (familial haemophagocytic reticulosis). J Clin Pathol 16:65
30. Matzner Y, Behar A, Beeri E, Gunders AE, Hershko C (1979) Systemic leishmaniasis mimicking malignant histiocytosis. Cancer 43:398
31. McClure PD, Strachan P, Saunders EF (1974) Hypofibrinogenemia and thrombocytopenia in familial hemophagocytic reticulosis. J Pediatr 85:67
32. Miller DR (1966) Familial reticulo-endotheliosis: concurrence of disease in five siblings. Pediatrics 38:986
33. Mozziconacci P, Nezelof C, Attal C, Girard F, Trung P-H, Well J, Desbuqois B, Gadot M (1965) La lymphohistiocytose familiale. Arch Fr Pediatr 22:385
34. Neff JC, Senhauser DA (1967) Familial hemophagocytic reticulosis. Am J Clin Pathol 47:360
35. Nezelof C, Elichar E (1973) La lymphohistiocytose familiale. Révue générale à propos de trois observations. Liens éventuels avec les syndromes secondaires. Nouv Rev Fr Hematol 13:319
36. O'Brien RT, Schwartz AD, Pearson HA, Spencer RP (1972) Reticuloendothelial failure in familial erythrophagocytic lymphohistiocytosis. J Pediatr 81:543
37. Omenn GS (1965) Familial reticuloendotheliosis with eosinophilia. N Engl J Med 273:427
38. Perry MC, Harrison EG, Burgert EO, Gilchrist GS (1976) Familial erythrophagocytic lymphohis-

tiocytosis. Report of two cases and clinicopathologic review. Cancer 38:209
39. Price DL, Woolsey JE, Rosman NP, Richardson EP Jr (1971) Familial lymphohistiocytosis of the nervous system. Arch Neurol 24:270
40. Reese AJM, Levy E (1951) Familial incidence of nonlipid reticuloendotheliosis (Letterer-Siwe disease). Arch Dis Child 26:578
41. Risdall RJ, McKenna RW, Nesbit ME, Krivit W, Balfour HH Jr, Simmons RL, Brunning RD (1979) Virus-associated hemophagocytic syndrome. A benign histiocytic proliferation distinct from malignant histiocytosis. Cancer 44:993
42. Rogers DL, Benson TE (1962) Familial Letterer-Siwe disease. J Pediatr 60:550
43. Scott RB, Robb-Smith AHT (1939) Histiocytic medullary reticulosis. Lancet 2:194
44. Serck-Hanssen A, Purchit GP (1968) Histiocytic medullary reticulosis. Report of 14 cases from Uganda. Br J Cancer 22:506
45. Varadi S, Gordon RR, Abbott D (1964) Haemophagocytic reticulosis diagnosed during life. Acta Haematol (Basel) 31:349
46. Weinberg AG, Rogers LE (1973) Hepatosplenomegaly, pancytopenia, and fever (Clinical-pathological conference). J Pediatr 82:879
47. Wilcox DRC (1952) Hemolytic anemia and reticulosis. Br Med J 1:1322
48. Zuazu JP, Duran JW, Julia AF (1979) Hemophagocytosis in acute brucellosis. N Engl J Med 301:1185

Haematology and Blood Transfusion Vol 27
Disorders of the Monocyte Macrophage System
Edited by F. Schmalzl, D. Huhn, H.E. Schaefer
© Springer-Verlag Berlin Heidelberg New York 1981

Concluding Remarks: The Pathology of the Monocyte–Macrophage System — Present Knowledge and Future Trends

F. Schmalzl, D. Huhn and H.E. Schaefer

The workshop in Innsbruck was organized with the aim to discuss the disorders of the monocyte–macrophage system (MMS) and to consider the general pathophysiological implications of this system as the result from present knowledge. Most of the disorders have been covered for which objective data suggested the involvement of the MMS. For several other diseases, such as sarcoidosis, we can only speculate on the pathogenetic implications of the monocytic cells. If we look at other topics of internal medicine as in cardiopulmonary or renal disorders we recognize that close connections exist between physiology, pathology, clinical features, and therapy. By contrast, up to now only a minor part out of the abundant knowledge on the function of the MMS has gained importance for the clinical disorders of this system. As shown in several papers of this volume we are just beginning to understand the peculiar importance of the MMS and its striking functional capacities. We realize its functional adaptability and selectivity, which allows divergent differentiations of cell individuals involved in chronic inflammatory processes leading to quite different chemical and probably functional properties of these cells, in spite of their close relationship and their identical origin (Dannenberg, this volume). The cellular co-operation leading to humoral and cell-mediated sensitization is partly elucidated, but we are still far from taking real advantage of this information. Few, if any, authors are engaged in studies concerned with the chemical and functional alterations associated with excessive accumulation of neoplastic monocytes — despite the fact that such functionally competent cells may exceed the physiological cell mass by a factor of 100 to 1000 and despite the fact that these "quiescent" cells can be transformed into dangerous effector cells by occasional activation due to severe infections or due to nonspecific stimuli. No serious efforts have been made to investigate the pathology of the monocytes in relation to other phagocytic cells previously defined as reticuloendothelial system (RES) and recently redefined as MPS (mononuclear phagocyte system). This stimulating concept has gained great actuality, but it also stimulated important discussions and controversies in that it indicates a tight connection between all the relevantly phagocytic cells and myelopoiesis (v. Furth, this volume). Clinical investigations contributed very little to the pathophysiology of the monocytes and macrophages, despite the fact that several disorders, such as sarcoidosis or chronic inflammatory disorders and including tuberculosis or rheumatic diseases, may depend upon specific pathogenetic mechanisms in which monocytes play a major role. Little is known about congenital disorders of myelopoiesis, like Kostmann's disease, in which the primary defects of monocyte function may have a determining effect on the clinical course.

Some data exist on the pharmacology of monocytes and macrophages: glucocorticoid effects and alterations on the function of the microtubular system are topics still under discussion. However, this information covers only a few of the multiple functional aspects of these cells, and the unique potent effector mechanisms of monocytes and macrophages are far from being pharmacologically enhanced, suppressed, or modulated in a specific way.

This monograph should contribute to the clinical actualization of the enormous scientific efforts aimed at revealing the pathophysiology of the MMS and at the same time should stimulate further clinical and experimental investigations to get a better understanding of the importance of MMS for human pathology.

Subject Index

259

Haematology and Blood Transfusion

Supplement volumes to the journal „Blut"
Editors: H. Heimpel, D. Huhn,
G. Ruhenstroth-Bauer, W. Stich

A selection

Modern Trends in Human Leukemia IV

Latest Results in Clinical and Biological Research
Including Pediatric Oncology

Editors: R. Neth, R. C. Gallo, T. Graf, K. Mannweiler, K. Winkler

1981. 252 figures, 118 tables. XXVI, 560 pages (Volume 26)
ISBN 3-540-10622-7

This book contains a report on the Wilsede meeting held in Germany in June, 1980. The meeting included the Frederick Stohlman lectures on leukemia, held by G. Klein on the relative role of viral transformation and cytogenetic changes in human and mouse lymphomas, and by H. Kaplan on the biology of Hodgkin's disease. The contributions are divided into clinical, cytogenetic, cell biologic, immunologic, virologic, and molecularbiologic sections. Also included is the Wilsede Joint Meeting on Pediatric Oncology I.
The book will be of value to both clinicians and researchers in many fields of oncology, as well as for investigators and students interested in human leukemia.

Modern Trends in Human Leukemia III

Newest Results in Clinical and Biological Research
9th Scientific Meeting of „Gesellschaft Deutscher Naturforscher und Ärzte"
Together with the "Deutsche Gesellschaft für Hämatologie",
Wilsede, June 19–23, 1978

Editors: R. Neth, R. C. Gallo, P.-H. Hofschneider, K. Mannweiler

1979. 171 figures, 128 tables. XXII, 599 pages (Volume 23)
ISBN 3-540-08999-3

Aplastic Anemia

Pathophysiology and Approaches to Therapy

Editors: H. Heimpel, E. C. Gordon-Smith, W. Heit, B. Kubanek

1979. 81 figures, 71 tables. XIII, 292 pages (Volume 24)
ISBN 3-540-09772-4

Springer-Verlag
Berlin
Heidelberg
New York

Immunobiology of Bone Marrow Transplantation

International Seminar of the Institut für Hämatologie, GSF, Munich, under the auspices of the European Communities, March 8–10, Neuherberg/München

Editors: S. Thierfelder, H. Rodt, H. J. Kolb

1980. 123 figures, 123 tables. XV, 430 pages (Volume 25)
ISBN 3-540-09405-9

Springer Hematology
New and Recent Titles

H. Begemann, J. Rastetter
Atlas of Clinical Hematology

Initiated by L. Heilmeyer, H. Begemann
With contributions on the Ultrastructure of Blood
Cells and their Precursors by D. Huhn and on
Tropical Diseases by W. Mohr
Translated from the German by H. J. Hirsch
3rd, completely revised edition. 1979. 228 figures,
194 in color, 12 tables. XVII, 275 pages
ISBN 3-540-09404-0
Distribution rights for Japan: Maruzen Co. Ltd.,
Tokyo

"This is an exceptionally beautifully produced atlas,
now in its third edition, and reflects continued
credit on its authors and publishers. It has been
completely revised, partly because it has now been
produced by the offset technique. This meant the use
of new photomicrographs, and thus the opportunity
to re-arrange contents... Undoubtedly this book
should be bought by medical libraries, research
institutes and centres of excellence not only in
Europa and the Americas but worldwide. Only a
few individuals will be able to afford a personal copy
but they will surely be fortunate if they can."
(Tropical Diseases Bulletin)

E. Kelemen, W. Calvo, T. M. Fliedner
Atlas of Human Hemopoietic Development

Foreword by M. Bessis
1979. 343 figures, 204 in color, 9 tables.
XIV, 266 pages
ISBN 3-540-08741-9

"This authors, using many morphological methods,
prepared an excellent atlas in which are shown all the
stages of development of hemopoiesis step by step.
The microphotographs collected in this book are of
very high quality and allow a clear understanding
of many morphological nuances.
There are many events in hemopoietic development which are being originally presented by the
authors in this atlas. The author's considerations
and comments devoted to hemopoietic development are in agreement with the progress in this field
and they have gathered a good bibliography..."
(Folia Histochemica et Cytochemica)

Automation in Hematology

What to Measure and Why
Editors: D. W. Ross, G. Brecher, M. Bessis
1981. 106 figures, 45 tables. VIII, 338 pages
ISBN 3-540-10225-6

J. C. Cawley, G. F. Burns, F. G. J. Hayhoe
Hairy Cell Leukemia

1980. 64 figures, 4 tables. IX, 123 pages
(Recent Results in Cancer Research, Volume 72)
ISBN 3-540-09920-4

Clinical Aspects of Blood Viscosity and Cell Deformability

Editors: G. D. O. Lowe, J. C. Barbenel, C. D. Forbes
1981. 79 figures, 20 tables. XV, 262 pages
ISBN 3-540-10299-X

Diffusion Chamber Culture

Hemopoiesis, Cloning of Tumors, Cytogenetic and
Carcinogenic Assays
Editors: E. P. Cronkite, A. L. Carsten
1980. 89 figures, 31 tables. XIV, 277 pages
ISBN 3-540-10064-4

Fundamentals of Immunology

By O. G. Bier, W. Dias Da Silva, D. Götze, I. Mota
1981. 164 figures. VIII, 442 pages
ISBN 3-540-90529-4

Strategies in Clinical Hematology

Editors: R. Gross, K.-P. Hellriegel
1979. 22 figures, 33 tables. X, 140 pages
(Recent Results in Cancer Research, Volume 69)
ISBN 3-540-09578-0

Therapeutic Plasma Exchange

Editors: H. J. Gurland, V. Heinze, H. A. Lee
1981. 35 figures, 52 tables. XII, 237 pages
(78 pages in German)
ISBN 3-540-10590-5

Springer-Verlag
Berlin
Heidelberg
New York